REASONING IN QUANTUM THEORY

TRENDS IN LOGIC
Studia Logica Library

VOLUME 22

SCOPE OF THE SERIES

Trends in Logic is a bookseries covering essentially the same area as the journal *Studia Logica* – that is, contemporary formal logic and its applications and relations to other disciplines. These include artificial intelligence, informatics, cognitive science, philosophy of science, and the philosophy of language. However, this list is not exhaustive, moreover, the range of applications, comparisons and sources of inspiration is open and evolves over time.

Volume Editor
Ryszard Wójcicki

The titles published in this series are listed at the end of this volume.

REASONING IN QUANTUM THEORY

Sharp and Unsharp Quantum Logics

by

M. DALLA CHIARA
University of Florence, Italy

R. GIUNTINI
University of Cagliari, Italy

and

R. GREECHIE
Louisiana Tech University, Ruston, U.S.A.

KLUWER ACADEMIC PUBLISHERS
DORDRECHT / BOSTON / LONDON

A C.I.P. Catalogue record for this book is available from the Library of Congress.

ISBN 978-90-481-6562-9 (PB)
ISBN 1-4020-1979-3 (e-book)

Published by Kluwer Academic Publishers,
P.O. Box 17, 3300 AA Dordrecht, The Netherlands.

Sold and distributed in North, Central and South America
by Kluwer Academic Publishers,
101 Philip Drive, Norwell, MA 02061, U.S.A.

In all other countries, sold and distributed
by Kluwer Academic Publishers,
P.O. Box 322, 3300 AH Dordrecht, The Netherlands.

Printed on acid-free paper

Contents

List of Figures

List of Tables

Preface

The term "quantum logic" has entered our languages as a synonym for something that doesn't make sense to our everyday rationality. Or, somewhat more technically but still in the common literature, it signifies some generic sort of mystification of classical logic understood only by the illuminati. In the technical literature, it is most frequently used to designate the set of projections $\Pi(\mathcal{H})$ on a Hilbert space $\mathcal{H}$ or the set of positive operators $\mathcal{E}(\mathcal{H})$ which fall between the smallest and the largest projections on $\mathcal{H}$ in a suitable ordering - or some algebraic generalization of one of these.

Thus, we have two concrete or standard quantum logics. These are structures closely related to the usual mathematical formalism that underlies the foundation of quantum theory (QT). The set $\Pi(\mathcal{H})$ is the basis for the sharp theory and $\mathcal{E}(\mathcal{H})$ for the unsharp theory, in much the same way that classical logic is based, in its sharp and unsharp manifestations, on (subalgebras of products of) the two-element set $\{0,1\}$ and the real unit interval $[0,1]$, respectively.

There is a still unfolding panorama of structures that generalize these two standard models. The theory of the foundations of quantum mechanics called "quantum logic" studies the standard models and their abstractions.

Confusion has persisted as to just what quantum logic is and how it should be construed as a veritable logic. The purpose of this book is to delineate (what we know of) the quantum logics, to explain of what the panorama of quantum logics consists and to present actual logics whose algebraic or Kripkean semantics are based on the algebraic models that are historically referred to as quantum logics. Our position is that there is not one but that there are many quantum logics. These logics have various models, usually one of the orthomodular structures, which include orthomodular lattices, orthomodular posets, orthoalgebras, and effect algebras.

We present sufficient historical background to give the reader an idea of how the theory developed. However, far more is presented than is needed to simply develop the logic, so that the novice may pick up the motivating aspects of the subject. Readers not interested in the technical logical details of quantum logics may gain an accounting of the mathematical aspects of "quantum logic," the models of the theory and how they relate to one another, by reading only the initial chapters. Readers wanting to learn more about these algebraic structures are referred to (Kalmbach, 1983; Dvurečenskij and Pulmannová, 2000).

Even the purist (non-quantum) logician may have some difficulty reading the technical logic in the second half of the book without the earlier preparation; such readers could, however, begin in the logical sections and refer to the earlier sections as needed. Quantum logicians, however, may proceed directly to the latter chapters.

We are writing for a multidisciplinary audience, and we warn the reader that we at times are too verbose for mathematicians, too pedantic for physicists, too glib for logicians, and too technical for philosophers of science. We assure the reader that we have had the whole readership in mind as we made our compromises and we beg her indulgence.

Here is an outline of the organization of the book. There are two parts. Part I, which consists of the first seven chapters, presents the historical background and the algebraic developments that motivate the syntax and underlie the semantics of the sequel. Part II studies a variety of quantum logics; in this Part, the term "logic" is used in the traditional sense, as a theory for a consequence relation that may hold between well-formed formulas of a given language. Semantical characterizations are introduced, both algebraic semantics and Kripkean semantics.

We set the stage in Chapter 1 by presenting some abstract notions needed later and by sketching the historical underpinnings of the subject. We present some of the basic notions of ordered sets followed by a quick review of Hilbert space and operators thereon. We recall von Neumann's axioms for quantum theory and Birkhoff and von Neumann's inseminal idea that propositions about quantum systems can be viewed as forming a kind of logic more appropriately modeled by projections on a Hilbert space than by a Boolean algebra. Then, we present von Neumann's axiomatic scheme which is followed by a discussion of Mackey's treatment and the abstractions that derived from it.

Chapter 2 presents the abstract axiomatization of sharp quantum theory, as well as the notion of *event*. The events, it is argued, band together to form a σ-complete orthomodular poset. The abstract notion of state and observable are given, and the previously developed structure is rephrased in an equivalent way as an *event-state system*. The interplay between events and states is further developed in *preclusivity spaces* and *similarity spaces*. A preclusivity space is a graph $(X, \perp)$, where the relation $\perp$ is symmetric and irreflexive; the corresponding similarity space is the pair $(X, \mathcal{R})$, where $\mathcal{R} = X \times X - \perp$. Preclusivity spaces invariably induce a closure operator $Y \mapsto Y^{\perp\perp}$, for $Y \subseteq X$, and the family of all closed sets $\{Y : Y = Y^{\perp\perp}\}$ forms a complete ortholattice.

The notion of the **Yes**-set, **Yes**(E), associated with each event E is introduced. **Yes**(E) is the set of all states s with $s(E) = 1$. By regarding states as *possible worlds*, this notion later allows us to introduce a Kripkean semantics.

The main purpose of Chapter 3 is to introduce and characterize the complete ortholattice $\Pi(\mathcal{H})$ of all projections of a Hilbert space. We are led to view $\Pi(\mathcal{H})$ and the states $\mathcal{S}$ thereon as an event-state system, appropriately baptized *the Hilbert event-state system.* The basic lattice theoretic properties of $\Pi(\mathcal{H})$ are given, and we sketch the development of Piron, Keller, Solèr, Holland, and Morash which some regard as the crowning mathematical achievement of the quantum logic approach to the foundations of quantum mechanics, namely the characterization of $\Pi(\mathcal{H})$ by lattice theoretic properties.

In Chapters 4 and 5, we introduce $\mathcal{E}(\mathcal{H})$, the (possibly) unsharp extension of $\Pi(\mathcal{H})$. The underlying set of $\mathcal{E}(\mathcal{H})$ is the set of all the bounded operators trapped between $\mathbb{O}$ and $\mathbb{1}$ in the usual ordering of self-adjoint operators. It forms an *effect algebra*, a type of structure that is then studied in some detail, and it is called *the standard unsharp effect algebra.* Along the way, we introduce *BZ-posets* and *effect BZ-posets*, the Mac Neille completion, and unsharp preclusivity spaces. The BZ-effect algebras admit 2 complements, a Brouwer complement and the fuzzy complement of the effect algebra. Having both allows us to define the necessity and possibility operators.

The Łukasiewicz operations of disjunction and conjunction are introduced and discussed. Axioms for MV algebras are given. These provide an adequate semantic characterization for Łukasiewicz' many valued logics. The quantum version follows, which consists of the axioms for what are called QMV algebras. The QMV algebras that correspond to effect algebras are determined; they are called *quasi-linear QMV algebras.*

Chapter 6 presents the corresponding abstract axiomatics for unsharp QT. Chapter 7 investigates alternative notions for sharpness and the connections between them.

We begin Part II with a discussion of the nature of algebraic semantics and Kripkean semantics. We give the algebraic characterization and the Kripkean characterization for classical logic (**CL**) as well as for intuitionistic logic (**IL**), the former being provided by the Boolean algebras and the latter by Heyting algebras. We review the notions of truth, logical truth, consequence, and logical consequence in both **CL** and **IL**; and we recall that the algebraic semantics and the Kripkean semantics of **CL** characterize the same logic, i.e., the notion of logical truth (resp., logical consequence), for the algebraic semantics is equivalent to that of the Kripkean semantics; the same is true for **IL**.

In Chapter 8, we proceed to sharp quantum logic (**QL**). There are two: **OL** and **OQL**. In **OL** (resp., **OQL**), the algebraic semantics are classified by the class of all algebraic realizations based on ortholattices (resp., orthomodular lattices). The similarity spaces discussed earlier provide the basis for the Kripkean semantics of **OL**. Although the details of the constructions are not the same as for **CL** and **IL**, the algebraic semantics for **OL** (resp., **OQL**) and the Kripkean semantics of **OL** (resp., **OQL**) characterize the same logic. We then specify **OQL** to the Hilbert event-state systems to obtain the algebraic and Kripkean realizations for a quantum system $\mathfrak{S}$ with associated Hilbert space $\mathcal{H}$.

We then turn to a discussion of polynomial implications in **OQL**, arguing that there is a favored one, the so-called Sasaki hook. This conditional may be interpreted as a counterfactual conditional.

In Chapter 9, we introduce the notion of a quasi-model in both the algebraic and the Kripkean semantics of **QL**. This notion allows us to discuss three interesting metalogical properties, called Herbrand-Tarski, Verifiability, and Lindenbaum, whose failure in **QL** represents a significant anomalous feature of quantum logics.

Chapter 10 addresses syntax. In it we provide an axiomatization of **QL** in the natural deduction style, proving soundness and completeness theorems with respect to the Kripkean semantics presented earlier. **OQL**, but not **OL**, admits a material conditional, and we prove a deduction theorem for **OQL**.

In Chapter 11, we prove that orthomodularity is not a first-order property. We also introduce *Hilbert quantum logic* (**HQL**) which is the logic that is semantically based on the class of all Hilbert lattices. We argue that **HQL** is strictly stronger than **OQL**.

In Chapter 12, we extend *sentential quantum logic* presenting an algebraic realization and a Kripkean realization for (first-order) **QL** with its attendant notions of satisfaction, verification, truth, logical truth, consequence in a realization, logical consequence, and rules of inference. We briefly discuss *quantum set theory* in which the usual Boolean-valued models are replaced by complete orthomodular lattices, perhaps the most striking feature of which is the failure of the Leibniz-substitutivity principle. Other approaches to quantum set theory are discussed, where the notion of extensional equality is not characterized by membership. Not only the extensionality axiom but Leibniz's principle of indiscernibles may be violated.

Proceeding from the total logics, those which are syntactically closed under the logical connectives and in which any sentence always has a meaning in both the algebraic and the Kripkean semantics, we present in Chapter 13 *partial classical logic* (**PaCL**) in which sentences need not have a meaning. An interesting feature of **PaCL** is the following: some characteristic classical laws (like distributivity) that are violated in standard quantum logic represent logical truths of **PaCL**.

After presenting an axiomatization of **PaCL** and proving soundness and completeness theorems, we discuss the failure of the Lindenbaum property. We introduce a relativization (to some nonempty class of realizations) of this notion and relate this relativized notion to the relative validity of the logical truths of classical logic as well as to the question of (non-contextual) hidden variable theory.

In Chapter 14, we turn our attention to unsharp quantum logics. In sharp logics both the logical and the semantic version of the noncontradiction principle hold. This means that $\alpha \curlywedge \neg\alpha$ is always false (in **PaCL**, when defined), and it never happens that both α and $\neg\alpha$ are true. In unsharp logics both conditions fail; thus, these logics are the natural logical abstractions of the effect-state systems studied in Part I.

Paraconsistent quantum logic (**PQL**) is presented first. This logic has an algebraic semantics based on bounded involution lattices and a Kripkean semantics in which the accessibility relation is symmetric but not necessarily reflexive, while the propositions behave as in the **OL** case. The other semantic definitions agree with **OL**, *mutatis mutandis*, and the algebraic and Kripkean semantics characterize the same logic. The axiomatization of **PQL** looks like that of the **OL** calculus except that the absurdity rule and the Duns Scotus rule are lacking. As in **OL**, the logic **PQL** satisfies the finite model property and is therefore decidable.

Regular bounded involution lattices give rise to a specialization of **PQL**, naturally called *regular paraconsistent quantum logic* (**RPQL**). The Kripkean realizations of **RPQL** may be represented by ε-preclusivity spaces in which the preclusivity relation is sensitive only to a characteristic approximation degree ε. The realizations are sharp when $0 \leq \varepsilon < \frac{1}{2}$ and unsharp when $\frac{1}{2} \leq \varepsilon \leq 1$.

In Chapter 15, we turn to a stronger unsharp quantum logic, the *Brouwer Zadeh logics*. These are natural abstractions of the class of all BZ-lattices studied in Chapter 4.

In Chapter 16, we study *partial quantum logics* (**PaQL**). There are three types: *unsharp* (**UPaQL**), *weak* (**WPaQL**) and *strong* (**SPaQL**). They admit a semantic characterization corresponding to the class of all effect algebras, orthoalgebras and orthomodular posets, respectively.

The language of **PaQL** consists of a set of atomic sentences and of two primitive connectives, the *negation* $\neg$ and the *exclusive disjunction* $\veebar$ (aut), while the *conjunction* is metalinguistically defined, via the de Morgan law. Whereas all disjunctions and conjunctions are considered "legitimate" from a linguistic point of view, a disjunction $\alpha \veebar \beta$ will have the intended meaning only in case the values of α and β are orthogonal in the corresponding effect algebra, and when this is not the case an arbitrary meaning is assigned.

First, we give the semantics of **UPaQL**, then the axiomatization which, unlike **QL**, admits only inferences from single sentences. Then, we add the Duns Scotus rule to get an axiomatization for **WPaQL**. Finally, we give an axiom which essentially states that the disjunction $\alpha \veebar \beta$ behaves like a supremum whenever α is orthogonal to β, arriving at **SPaQL**. We then discuss soundness and completeness theorems.

We finish the chapter with a discussion of the *Łukasiewicz quantum logic* (**ŁQL**), which generalizes both **OQL** and $\mathbf{L}_{\aleph}$ (Łukasiewicz' infinite many valued logic). The semantics of **ŁQL** is based on QMV algebras.

Finally, in Chapter 17, we discuss *quantum computational logic.* The quantum generalization of a bit is called a quantum bit, or qubit; it is a unit vector in the Hilbert space $\mathbb{C}^2$. Quantum computations, or quantum logical gates, are represented by unitary transformations on qubit systems. Fixing an orthonormal basis, called *the computational basis*, we define an *n-qubit system* to be an element of the n-fold tensor product $\otimes^n\mathbb{C}^2$. Quregisters are qubits or n-qubit systems.

The quantum computational logic is based on certain quantum logical gates defined on quregisters, the Toffoli gate, a reversible conjunction, negation, disjunction, and the square root of negation. The latter is a genuine quantum gate in that it transforms classical registers into quregisters that are superpositions.

The sentential language $Form^{\mathcal{L}}$ is given by negation, conjunction and the square root of negation, with disjunction then defined via the de Morgan law. In quantum computational semantics, the meaning of the linguistic sentences are represented by quregisters.

A quantum computational realization of $Form^{\mathcal{L}}$ is a function Qub from the set of all formulas in $Form^{\mathcal{L}}$ to the union of the Hilbert spaces $\otimes^n\mathbb{C}^2$. The image $Qub(\alpha)$ is called the information value of α. Since $Qub(\alpha)$ is a unit vector of some Hilbert space, it has a probability-value which we use to define a probability-value for α itself. A sentence α is true when its probability-value in a computational realization Qub is 1. The notions of logical truth, consequence, and logical consequence are defined accordingly; the logic characterized by this semantics is called *quantum computational logic* (**QCL**). Distinctions between **QCL** and **QL** are made. It is noted that the flavor of this logic is completely different from that of all the quantum logics that preceded it.

Not addressed herein are the generalized sample spaces of the so called Amherst School, initiated by Foulis and Randall (Foulis, 1999). This theory attempts to generate a logic for any empirical situation, the models for which are precisely the models that we develop for quantum logic. Much of the theory of the algebraic structures that shall concern us was generated by this School.

Nor have we addressed the generalized measure theory that has arisen in which a Boolean algebra of measurable sets is replaced by a more general suitable structure - one of the models that we discuss. While this development may turn out to be of mathematical importance in applications of the theory, it does not directly impact the logic. Nor do we speculate on anticipated applications to the logical underpinnings of other empirical sciences, including biology, economics, political science, and psychology.

What we present is the best that we can do at the moment. We do not present the final definitive way of thinking quantum logically. Although, in most cases, what we do present seems to us natural enough to be lasting. Whatever may come of quantum logics in the future, we hope that a reading of this book will leave no question that quantum logics are (or can be made into) logics in the truest sense of the logical tradition. We hope that we have conveyed the collective enthusiasm of so many of the researchers in this rapidly developing field.

Errors

Errata may be reported by email to `giuntini@unica.it` .
A list of errata will be maintained at the web site:
`http://web.tiscalinet.it/giuntini/rqt/errata.html`

Acknowledgments

Many ideas and results presented in this book are based on long and intense scientific collaborations with other scholars, whose work seems to be deeply "entangled" with our own. We would like to mention *in primis* G. Cattaneo, D. Foulis, S. Gudder, A. Leporati, R. Leporini, P. Mittelstaedt, and S. Pulmannová. All of them are coauthors of many joint articles that have provided momentous inspiration for this book.

Our research has also been greatly enhanced by many stimulating interactions with D. Aerts, E. Beltrametti, P. Busch, J. Butterfield, G. Cassinelli, E. Castellani, A. Dvurečenskij, C. Garola, G. Ghirardi, H. Greuling, P. Lahti, F. Laudisa, D. Mundici, M. Navara, P. Pták, J. Pykacz, M. Rédei, B. Riečan, Z. Riečanova, F. Schroeck, K. Svozil, G. Toraldo di Francia, and with many members of the *International Quantum Structure Association* (IQSA) whose periodical meetings have represented splendid and fruitful occasions for scientific collaboration. Some great figures of the quantum structure community are no longer with us; we take this occasion to remember fondly M.K. Bennett, G. Birkhoff, S. Bugajski, C. Randall, G. Rüttimann.

Finally, we would like to thank warmly F. Paoli, who has carefully read the manuscript and offered precious suggestions and assistance. Our grateful acknowledgement also goes to R. Wójcicki and to the editors of Kluwer who have constantly encouraged our work and patiently waited for the final version of this book.

PART I Mathematical and Physical Background

Introduction

In 1920 Łukasiewicz published a two-page article whose title was "On three-valued logic." The paper proposes a semantic characterization for the logic that has been later called $Ł^3$ (Łukasiewicz' three-valued logic). In spite of the shortness of the paper, all the important points concerning the semantics of $Ł^3$ are already there and can be naturally generalized to the case of a generic number n of truth-values as well as to the case of infinite many values. The conclusion of the article is quite interesting:

> The present author is of the opinion that three-valued logic has above all theoretical importance as an endeavour to construct a system of non-aristotelian logic. Whether the new system of logic has any practical importance will be seen only when the logical phenomena, especially those in the deductive sciences, are thoroughly examined, and when the consequences of the indeterministic philosophy, which is the metaphysical substratum of the new logic, can be compared with empirical data (Łukasiewicz, 1970c).

These days, Łukasiewicz' remark appears to be highly prophetic, at least in two respects. First of all, the practical importance of many-valued logics has gone beyond all reasonable expectations at Łukasiewicz' times. What we call today *fuzzy logics* (natural developments of Łukasiewicz' many-valued logics) gave rise to a number of technological applications. We need only recall that we can buy washing machines and cameras whose suggestive name is just "fuzzy logic."

At the same time, quantum theory (QT) has permitted us to compare the consequences of an *indeterministic philosophy* with *empirical data.* This has been done both at a logico-mathematical level and at an experimental level. The so called *no go theorems*[1] speak to the impossibility of *deterministic completions* of orthodox QT by means of *hidden variable theories.* Interestingly enough, some experiments that have been performed in the Eighties[2] have confirmed the statistical predictions of QT, against the predictions of the most significant hidden variable theories.

[1]See (von Neumann, 1932; Jauch, 1968; Kochen and Specker, 1967; Bell, 1966; Giuntini, 1991a).

[2]See (Aspect, Grangier and Roger, 1981; Aspect and Grangier, 1985).

Łukasiewicz was a contemporary of Heisenberg, Bohr, von Neumann. Strangely enough, however, he very rarely makes explicit references to QT. In spite of this, he seems to be aware of the importance of QT for his indeterministic philosophy. In 1946 he writes a revised version of his paper "On Determinism," an address that he delivered as the rector of the Warsaw University for the inauguration of the academic year 1922/1923. At the very beginning of the article he notices:

> At the time when I gave my address those facts and theories in the field of atomic physics which subsequently led to the undermining of determinism were still unknown. In order not to deviate too much from, and not to interfere with, the original content of the address, I have not amplified my article with arguments drawn from this branch of knowledge (Łukasiewicz, 1970b).

For Łukasiewicz, the basic reason that suggested going beyond classical *bivalent* semantics was a philosophical one.[3] His main argument, developed in the paper "On Determinism" can be sketched as follows:

- **First statement**: *bivalence implies determinism.*
- **Second statement**: *determinism contradicts our basic intuition about necessity and possibility.*
- **Conclusion**: *bivalence has to be refused.*

The argument essentially refers to *temporal sentences* that describe *future contingent events*. A typical example is represented by a sentence like

John will not be at home tomorrow noon.

More formally we can write a temporal sentence as: $\alpha(t)$, to be read as "α is the case at time t." A temporal sentence $\alpha(t)$ may be true or false with respect to a given time t_1, which may either precede or follow t. We will write (according to the usual semantic notation): $\models_{t_1} \alpha(t)$ and $\models_{t_1} \neg\alpha(t)$, respectively.

How can the first statement (*bivalence implies determinism*) be defended? Let us assume the bivalence principle:

any sentence is either true or false (*tertium non datur*).

And let us apply this principle to the case of temporal sentences. Suppose $t_1 < t$. By bivalence we have:

$$\models_{t_1} \alpha(t) \text{ or } \models_{t_1} \neg\alpha(t).$$

In other words, the *event* described by $\alpha(t)$ is already *determined* at time t_1. Suppose: $\models_{t_1} \alpha(t)$. Then, $\alpha(t)$ turns out to describe a *necessary event*. As a consequence, we can conclude that: bivalence implies determinism.

How can we justify the second statement (*determinism contradicts our basic intuition about necessity and possibility*)? Suppose $\alpha(t)$ describes a contingent event (such as the fact asserted by the sentence "John will not

[3]See (Dalla Chiara and Giuntini, 1999).

be at home tomorrow noon"). Now our basic intuition about contingency and necessity seems to require that both sentences $\alpha(t)$ and $\neg\alpha(t)$ must be *possible* at time t_1. As a consequence, future contingent events cannot be determined at any previous time (by definition of contingency).

Strangely enough, from the historical point of view, the abstract researches on fuzzy structures and on quantum structures have undergone quite independent developments for many decades during the 20-th century.

Only after the Eighties, there emerged an interesting convergence between the investigations about fuzzy and quantum structures, in the framework of the so called *unsharp approach to quantum theory*. In this connection a significant conjecture has been proposed: perhaps some apparent *mysteries* of the quantum world should be described as special cases of some more general *fuzzy phenomena*, whose behavior has not yet been fully understood.

The ambiguities of the quantum world can be investigated at different levels. The first level involves the *essential indeterminism* of QT. In order to understand the origin of such indeterminism, from an intuitive point of view, it will be expedient to follow the argument proposed by Birkhoff and von Neumann in their celebrated article "The logic of quantum mechanics" (Birkhoff and von Neumann, 1936). At the very beginning of their paper, Birkhoff and von Neumann observe:

> There is one concept which quantum theory shares alike with classical mechanics and classical electrodynamics. This is the concept of a mathematical "phase-space". According to this concept, any physical system $\mathfrak{S}$ is at each instant hypothetically associated with a "point" in a fixed phase-space Σ; this point is supposed to represent mathematically, the "state" of $\mathfrak{S}$, and the "state" of $\mathfrak{S}$ is supposed to be ascertainable by "maximal" observations.

Maximal pieces of information about physical systems are also called *pure states*. For instance, in classical particle mechanics, a pure state of a single particle can be represented by a sequence of six real numbers $\langle r_1, \ldots, r_6 \rangle$, where the first three numbers correspond to the *position*-coordinates, while the last ones are the *momentum*-components.

As a consequence, the phase-space of a single particle system can be identified with the set $\mathbb{R}^6$, consisting of all sextuples of real numbers. Similarly for the case of compound systems, consisting of a finite number n of particles.

Let us now consider an *experimental proposition* $\mathbf{P}$ about our system, asserting that a given *physical quantity* (also called *observable*) has a certain value (for instance: "the value of position in the x-direction lies in a certain interval"). Such a proposition $\mathbf{P}$ will be naturally associated with a subset X of our phase-space, consisting of all the pure states for which $\mathbf{P}$ holds. In other words, the subsets of Σ seem to represent good mathematical representatives of experimental propositions. These subsets have been called by Birkhoff and von Neumann *physical qualities*. According to alternative terminologies, *physical qualities* are also currently called *physical properties* or *physical propositions* or *physical questions* or *physical events*. At this level of analysis, we will simply say *events*. Of course, the correspondence between the set of all experimental propositions and the set of all events will be many-to-one. When a pure state p belongs to an event X, we can say that the system in state p *verifies* both X and the corresponding experimental proposition.

What about the structure of all events? As is well known, the power set of any set gives rise a *Boolean algebra*. And also the set $\mathcal{F}(\Sigma)$ of all *measurable subsets* of Σ (which is more tractable than the full power set of Σ, from a measure-theoretic point of view) turns out to have a Boolean structure. Hence, we may refer to the following Boolean algebra:

$$\langle \mathcal{F}(\Sigma)\, , \cap\, , \cup\, , {}^{c}\, , \mathbf{0}\, , \mathbf{1} \rangle\, ,$$

where:

1) $\cap\, , \cup\, , {}^{c}$ are, respectively, the operations intersection, union, relative complement;
2) $\mathbf{0}$ is the empty space, while $\mathbf{1}$ is the total space.

According to a standard interpretation, the operations $\cap\, , \cup\, , {}^{c}$ can be naturally regarded as a set-theoretic realization of the classical logical connectives *and*, *or*, *not*. As a consequence, we will obtain a classical semantic behavior:

- a state p verifies a conjunction $X \cap Y$ iff $p \in X \cap Y$ iff p verifies both members;
- p verifies a disjunction $X \cup Y$ iff $p \in X \cup Y$ iff p verifies at least one member;
- p verifies a negation X^c iff $p \notin X$ iff p does not verify X.

In such a framework, classical pure states turn out to satisfy an important condition: they represent *pieces of information* (about the physical system under investigation) that are at the same time *maximal* and *logically complete*. They are *maximal* because they represent a *maximum of information* that cannot be consistently extended to a richer knowledge in the framework of the theory (even a hypothetical *omniscient mind* could not know more about the physical system in question). Furthermore, pure states are *logically complete* in the following sense: they *semantically decide* any event. For any p and X,

$$p \in X \text{ or } p \in X^c .$$

The semantic excluded middle principle is satisfied.

To what extent can such a picture be adequately extended to QT? Birkhoff and von Neumann observe:

> In quantum theory the points of Σ correspond to the so called "wave-functions" and hence Σ is a ... function-space, usually assumed to be Hilbert space.

As opposed to classical mechanics, QT is *essentially probabilistic*. Generally, pure states only assign probability-values to quantum events. This is strongly connected with the *uncertainty relations*, which represent one of the most significant dividing line between the classical and the quantum case. Let ψ represent a pure state (a wave-function) of a quantum system and let **P** be an experimental proposition (for instance "the spin-value in the x-direction is up"). The following cases are possible:

(i) ψ assigns to **P** probability-value 1;
(ii) ψ assigns to **P** probability-value 0;
(iii) ψ assigns to **P** a probability-value different from 1 and from 0.

In the first two cases, we will say that **P** is *true* (*false*) for the system in state ψ; in the third case, **P** will be *semantically indeterminate*. This constitutes the first level of ambiguity or fuzziness.

As a consequence, unlike classical mechanics, in QT pure states turn out to represent pieces of information that are at the same time *maximal* and *logically incomplete*. Such divergence between maximality and logical completeness is the origin of most logical anomalies of the quantum world.

A second level of ambiguity is connected with a possibly fuzzy character of the physical events that are investigated. We can try and illustrate the difference between two "fuzziness-levels" by referring to a nonscientific example. Let us consider the two following sentences, which apparently have no definite truth-value:

I) Hamlet is 1.70 meters tall;
II) Brutus is an honourable man.

The semantic uncertainty involved in the first example seems to depend on the logical incompleteness of the *individual concept* associated to the name "Hamlet." In other words, the property "being 1.70 meters tall" is a *sharp* property. However, our concept of Hamlet is not able to *decide* whether such a property is satisfied or not. Unlike real persons, literary characters have a number of indeterminate properties. On the contrary, the semantic uncertainty involved in the second example, is mainly caused by the ambiguity of the concept "honourable." What does it mean "being honourable?" One need only recall how the ambiguity of the adjective "honourable" plays an important role in the famous Mark Antony's monologue in Shakespeare's "Julius Caesar." Now, orthodox QT generally takes into consideration examples of the first kind (our first level of fuzziness): events are sharp, while all semantic uncertainties are due to the *logical incompleteness* of the individual concepts, that correspond to pure states of quantum objects. A characteristic of *unsharp QT*, instead, is to investigate also examples of the second kind (second level of fuzziness).

In the following chapters (of Part I) we will try to understand how the essential indeterministic and ambiguous features of the quantum world are connected with the deep mathematical and logical structures of QT. We will first present the axiomatic foundations of sharp QT (Chapters 1-3). The unsharp version of the theory will be developed in Chapters 4-7.

CHAPTER 1

The mathematical scenario of quantum theory and von Neumann's axiomatization

In his celebrated book *Mathematische Grundlagen der Quantenmechanik* (von Neumann, 1932),[1] John von Neumann proposed an axiomatic version of sharp QT. This theory is often referred to as *orthodox quantum theory*.

As happens in the case of other physical theories, a number of axioms of QT concern the mathematical interpretation of some basic physical concepts: *physical system*, *state of a physical system*, *observable*, and *event*.

Orthodox QT is developed in the framework of a privileged mathematical "scenario," represented by particular kinds of abstract spaces that are called *Hilbert spaces*. Such spaces have for QT the same role that is played by phase-spaces in the case of classical mechanics. Any physical system $\mathfrak{S}$ is associated to a particular Hilbert space $\mathcal{H}$, which represents the "mathematical environment" of $\mathfrak{S}$. As a consequence, all the basic physical notions concerning $\mathfrak{S}$ have counterparts that "live" in $\mathcal{H}$.

Before presenting the axioms of sharp QT, it will be expedient to recall some fundamental mathematical definitions for the readers who are not familiar with abstract algebra and functional analysis.

1.1. Algebraic structures

Some very important "characters" of the quantum theoretical (and quantum logical) game are represented by particular algebraic structures, that play a relevant role both for sharp and unsharp QT. It is therefore expedient to recall the definitions of some basic algebraic concepts. These will accompany us throughout this book.

DEFINITION 1.1.1. *Poset*
A *partially ordered set* (called also *poset*) is a structure

$$\mathcal{B} = \langle B, \leq \rangle,$$

where: B (the *support* of the structure) is a nonempty set and $\leq$ is a *partial order relation* on B. In other words, $\leq$ satisfies the following conditions for all $a, b, c \in B$:

(i) $a \leq a$ (reflexivity);
(ii) $a \leq b$ and $b \leq a$ implies $a = b$ (antisymmetry);

[1] See also the English version (von Neumann, 1996).

(iii) $a \leq b$ and $b \leq c$ implies $a \leq c$ (transitivity).[2]

DEFINITION 1.1.2. *Bounded poset*
A *bounded poset* is a structure

$$\mathcal{B} = \langle B\,,\, \leq\,,\, \mathbf{0}\,,\mathbf{1}\rangle\,,$$

where:

(i) $\langle B\,,\, \leq\rangle$ is a poset;
(ii) $\mathbf{0}$ and $\mathbf{1}$ are distinct[3] special elements of B: the *minimum* and the *maximum* with respect to $\leq$. In other words, for all $b \in B$:

$$\mathbf{0} \leq b \text{ and } b \leq \mathbf{1}.$$

DEFINITION 1.1.3. *Lattice*
A *lattice* is a poset $\mathcal{B} = \langle B\,,\, \leq\rangle$ in which any pair of elements a, b has a *meet* $a \wedge b$ (also called *infimum*) and a *join* $a \vee b$ (also called *supremum*) such that:

(i) $a \wedge b \leq a, b$, and $\forall c \in B$: $c \leq a, b$ implies $c \leq a \wedge b$;
(ii) $a, b \leq a \vee b$, and $\forall c \in B$: $a, b \leq c$ implies $a \vee b \leq c$.

In any lattice the following condition holds:

$$a \leq b \text{ iff } a \wedge b = a \text{ iff } a \vee b = b.$$

Alternatively, a lattice can be also defined as a structure

$$\mathcal{B} = \langle B\,,\, \wedge\,,\, \vee\rangle\,,$$

where the meet $\wedge$ and the join $\vee$ are binary operations satisfying the following conditions:

(L1) $a \wedge a = a$; $a \vee a = a$ (idempotence);
(L2) $a \wedge b = b \wedge a$; $a \vee b = b \vee a$ (commutativity);
(L3) $a \wedge (b \wedge c) = (a \wedge b) \wedge c$; $a \vee (b \vee c) = (a \vee b) \vee c$ (associativity);
(L4) $a \wedge (a \vee b) = a$; $a \vee (a \wedge b) = a$ (absorption).

(In fact, the idempotence of $\wedge$ and $\vee$, (L1), follows easily from the absorption condition, (L4).)

On a lattice $\mathcal{B} = \langle B\,,\, \wedge\,,\, \vee\rangle$, a partial order relation $\leq$ can be defined in terms of the meet $\wedge$:

$$a \leq b \text{ iff } a \wedge b = a.$$

It follows that:

$$a \leq b \text{ iff } a \vee b = b.$$

Generally, when we know that we are discussing lattices, we will indicate a lattice as a structure

$$\mathcal{B} = \langle B\,,\, \wedge\,,\, \vee\rangle\,.$$

[2]For the sake of simplicity, in the following, we will often indicate a structure by using simply the name of its support.

[3]For our physical and semantic applications, it is expedient to refer to nondegenerate algebraic structures. Hence, we always assume that the minimum and the maximum element of a bounded structure are distinct.

But in some cases, we will prefer the poset notation:

$$\mathcal{B} = \langle B\,, \leq \rangle\,.$$

Suppose the elements of the lattice represent *events* (or *properties* or *propositions*). Then, according to a standard (although sometimes erroneous) logical interpretation, the lattice operations $\wedge$ and $\vee$ will correspond to the connectives *conjunction* (*and*) and *disjunction* (*or*).

Consider a lattice $\mathcal{B} = \langle B\,, \wedge\,, \vee \rangle$ and let X be any set of elements of B. If existing, the *infimum* $\bigwedge X$ and the *supremum* $\bigvee X$ are the elements of B that satisfy the following conditions:

(ia) $\forall a \in X : \bigwedge X \leq a$;
(ib) $\forall c \in B : \forall a \in X[c \leq a]$ implies $c \leq \bigwedge X$;
(iia) $\forall a \in X : a \leq \bigvee X$;
(iib) $\forall c \in B : \forall a \in X[a \leq c]$ implies $\bigvee X \leq c$.

On can show that, when they exist the infimum and the supremum are unique. A lattice is *complete* iff for any set of elements X the infimum $\bigwedge X$ and the supremum $\bigvee X$ exist. A lattice is *σ-complete* iff for any countable set of elements X the infimum $\bigwedge X$ and the supremum $\bigvee X$ exist.

In many interesting situations, a poset (or a lattice) may have a more sophisticated structure: the poset is closed under a unary operation that represents a weak form of logical *negation*. Such a finer structure is represented by a *bounded involution poset*.

DEFINITION 1.1.4. *Bounded involution poset*
A *bounded involution poset* (BIP) is a structure $\mathcal{B} = \langle B\,, \leq\,, '\,, \mathbf{0}\,, \mathbf{1} \rangle$ where:

(i) $\langle B\,, \leq\,, \mathbf{0}\,, \mathbf{1} \rangle$ is a bounded poset;
(ii) $'$ is a unary operation (called *involution* or *generalized complement*) that satisfies the following conditions:
 (a) $a = a''$ (double negation);
 (b) $a \leq b$ implies $b' \leq a'$ (contraposition).

Suppose we have a bounded poset $\mathcal{B} = \langle B\,, \leq\,, '\,, \mathbf{0}\,, \mathbf{1} \rangle$, where $'$ is a unary operation satisfying the double negation principle. Then, the following conditions are equivalent for $\mathcal{B}$:

(1) $a \leq b$ implies $b' \leq a'$;
(2) $a' \wedge b' = (a \vee b)'$ (when one side exists so does the other side);
(3) $a' \vee b' = (a \wedge b)'$ (when one side exists so does the other side).

Conditions (2) and (3) are usually called the *de Morgan laws*.

The presence of a negation-operation permits us to define an *orthogonality relation* $\perp$, that may hold between two elements of a bounded involution poset.

DEFINITION 1.1.5. *Orthogonality*
Let a and b belong to a bounded involution poset. The object a is *orthogonal* to the object b (indicated by $a \perp b$) iff $a \leq b'$.[4]

[4]We find it convenient to refer to the elements of sets as "objects."

A set of elements S is called a *pairwise orthogonal* set iff $\forall a, b \in S$ such that $a \neq b$, $a \perp b$.
A *maximal* set of pairwise orthogonal elements is a set of pairwise orthogonal elements that is not a proper subset of any set of pairwise orthogonal elements.

Let us interpret $\leq$ and $'$ as an *implication-relation* and a *negation-operation*, respectively. Then, the orthogonality relation $\perp$ turns out to describe a *logical incompatibility* relation:

a is orthogonal to b iff a implies the negation of b.

When a is not orthogonal to b we will write:

$$a \not\perp b.$$

The orthogonality relation $\perp$ is sometimes also called *preclusivity*; while its negation $\not\perp$ is also called *accessibility*.

Since, by definition of bounded involution poset, $a \leq b$ implies $b' \leq a'$ (contraposition) and $a = a''$ (double negation), one immediately obtains that $\perp$ is a symmetric relation:

$$a \perp b \text{ implies } b \perp a.$$

Notice that $\mathbf{0} \perp \mathbf{0}$ and that $\perp$ is not necessarily irreflexive. It may happen that an object a (different from the null object $\mathbf{0}$) is orthogonal to itself:

$$a \perp a \text{ (because } a \leq a'\text{)}.$$

In fact, an object may imply its own negation. Objects of this kind are called *self-inconsistent*. Suppose now we have two self-inconsistent objects a and b, and let us ask whether in such a case a is necessarily orthogonal to b. Generally, the answer to this question is negative. There are examples of bounded involution posets (the smallest of which has only 4 elements) such that for some objects a and b:

$$a \perp a \text{ and } b \perp b \text{ and } a \not\perp b.$$

In the particular case where the answer to this question is positive for any pair of elements, we will call our bounded involution poset a *Kleene poset* (or also a *regular* poset).

DEFINITION 1.1.6. *Kleene poset*
A bounded involution poset is a *Kleene poset* (or also a *regular poset*) iff it satisfies the *Kleene condition* for any pair of elements a and b:

$$a \perp a \text{ and } b \perp b \text{ implies } a \perp b.$$

DEFINITION 1.1.7. *Bounded involution lattice*
A *bounded involution lattice* (BIL) is a bounded involution poset that is also a lattice.

By *Kleene lattice* (or *regular lattice*), we will mean a Kleene poset that is also a lattice.

Generally, bounded involution lattices and Kleene lattices may violate both the noncontradiction principle and the excluded middle. In other words, it may happen that:

$$a \wedge a' \neq \mathbf{0} \quad \text{and} \quad a \vee a' \neq \mathbf{1}.$$

According to our logical interpretation, this means that contradictions are not necessarily false, while the disjunction between a proposition and its negation is not necessarily true. As a consequence, Kleene lattices represent an adequate abstract tool that permits us to model unsharp and ambiguous concrete situations. In such situations, the involution operation $'$ is often called *fuzzy complement*. As we will see, Kleene posets and Kleene lattices will play an important role for unsharp QT.

Sharp QT, instead, makes use of stronger algebraic structures, where ambiguous situations (which give rise to violations of the noncontradiction principle) are forbidden. Important examples of *sharp* structures (which satisfy the noncontradiction principle) are represented by *orthoposets* and *ortholattices* (Birkhoff, 1967; Kalmbach, 1983).

DEFINITION 1.1.8. *Orthoposet and ortholattice*
An *orthoposet* (OP) is a bounded involution poset $\mathcal{B} = \langle B\,, \leq\,, '\,, \mathbf{0}\,, \mathbf{1}\rangle$ that satisfies the conditions:

(i) $a \wedge a' = \mathbf{0}$ (noncontradiction principle);
(ii) $a \vee a' = \mathbf{1}$ (excluded middle principle).

An *ortholattice* (OL) is an orthoposet that is also a lattice.

The involution operation $'$ of an orthoposet (ortholattice) is also called *orthocomplementation* (or shortly *orthocomplement*).

A *σ-orthocomplete orthoposet* (*σ-orthocomplete ortholattice*) is an orthoposet (ortholattice) $\mathcal{B}$ such that for any countable set $\{a_i\}_{i\in I}$ of pairwise orthogonal elements the supremum $\bigvee \{a_i\}_{i\in I}$ exists in $\mathcal{B}$.

Another important category of lattices are the lattices that satisfy the *distributivity property*.

DEFINITION 1.1.9. *Distributive lattice*
A lattice $\mathcal{B} = \langle B\,, \wedge\,, \vee\rangle$ is *distributive* iff the meet $\wedge$ is distributed over the join $\vee$ and vice versa. In other words:

(i) $a \wedge (b \vee c) = (a \wedge b) \vee (a \wedge c)$;
(ii) $a \vee (b \wedge c) = (a \vee b) \wedge (a \vee c)$.

Distributive involution lattices are also called *de Morgan lattices*.

In this framework, classical Boolean algebras can be then defined as particular examples of de Morgan lattices.

DEFINITION 1.1.10. *Boolean algebra*
A *Boolean algebra* (BA) is a structure

$$\mathcal{B} = \left\langle B\,, \wedge\,, \vee\,, '\,, \mathbf{0}\,, \mathbf{1}\right\rangle$$

that is at the same time an ortholattice and a de Morgan lattice.

In other words, Boolean algebras are distributive ortholattices.

As we will see, the possibility of violations of the distributivity relations will represent one of the most significant formal properties of the abstract structures arising in the framework of sharp QT. As a consequence, the different forms of sharp quantum logic that we are going to study will turn out to be "strongly nonBoolean." Unsharp QT is also nondistributive in all but special cases; so it too is "strongly nonBoolean."

At the same time, some important quantum structures do satisfy a significant weakening of distributivity, that is represented by the *orthomodularity* property.

Two "leading actors" of the algebraic quantum play are represented by *orthomodular posets* and *orthomodular lattices.*

DEFINITION 1.1.11. *Orthomodular poset and orthomodular lattice*
An *orthomodular poset* (OMP) is an orthoposet

$$\mathcal{B} = \langle B\,,\,\leq\,,\,'\,,\,\mathbf{0}\,,\mathbf{1}\rangle$$

that satisfies the following conditions:

(i) $\forall a, b \in B,\ \ a \perp b$ implies $a \vee b \in B$;
(ii) $\forall a, b \in B,\ \ a \leq b$ implies $b = a \vee (a \vee b')'$.

An *orthomodular lattice* (OML) is an orthomodular poset that is also a lattice.

One can prove that an ortholattice is an orthomodular lattice iff it does not contain the "Benzene ring", whose *Hasse diagram* is given in Figure 1.1.1.[5]

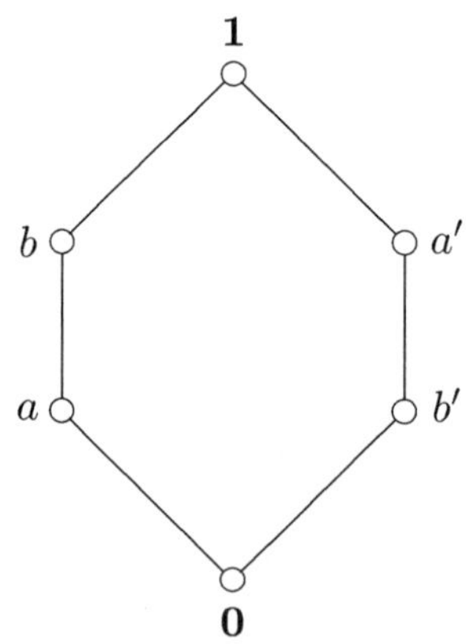

FIGURE 1.1.1. The Benzene ring

[5]Hasse diagrams are the usual way of representing finite posets and lattices. In a Hasse diagram of a poset $\mathcal{B}$ all the elements of B are depicted and an ascending line, say, from a up to b indicates that b is immediately above a in the ordering, i.e., $b > a$ and if $b > c \geq a$ then $c = a$.

Clearly, any distributive ortholattice, i.e., any Boolean algebra, is orthomodular. The smallest orthomodular lattice which is not a Boolean algebra, called $\mathcal{MO}2$, is given in Figure 1.1.2.

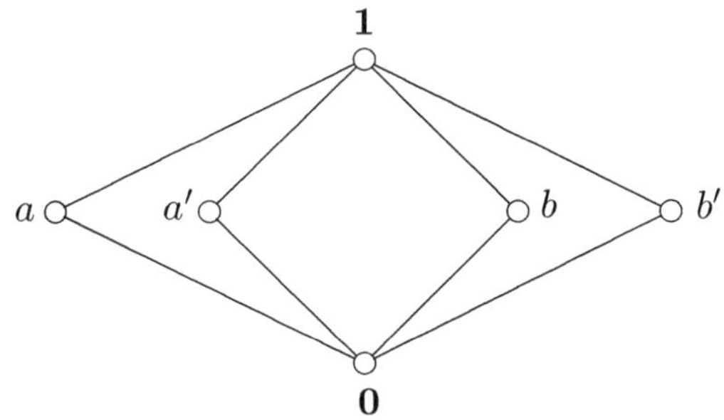

FIGURE 1.1.2. $\mathcal{MO}2$: the smallest OML that is not a BA

Another important property that represents a weakening of distributivity and a strengthening of orthomodularity is represented by *modularity*.

DEFINITION 1.1.12. *Modularity*
A lattice $\mathcal{B}$ is called *modular* iff $\forall a, b \in B$,

$$a \leq b \text{ implies } \forall c \in B[a \vee (c \wedge b) = (a \vee c) \wedge b].$$

Every modular ortholattice is orthomodular, but not the other way around. Furthermore, any distributive lattice is modular.

A bounded poset (lattice) $\mathcal{B}$ may contain some special elements, called *atoms*, that play a significant role.

DEFINITION 1.1.13. *Atom*
An element b of B is called an *atom* of $\mathcal{B}$ iff b *covers* $\mathbf{0}$. In other words, $b \neq \mathbf{0}$ and $\forall c \in B$: $c \leq b$ implies $c = \mathbf{0}$ or $c = b$.

Apparently, atoms are nonzero elements such that no other element lies between them and the lattice-minimum.

DEFINITION 1.1.14. *Atomicity*
A bounded poset $\mathcal{B}$ is *atomic* iff $\forall a \in B - \{\mathbf{0}\}$ there exists an atom b such that $b \leq a$.

Of course, any finite bounded poset is atomic. At the same time, there are examples of infinite bounded posets that are *atomless* (and hence nonatomic), the real interval $[0, 1]$ being the most familiar example.

It turns out that any atomic orthomodular lattice $\mathcal{B}$ is *atomistic* in the sense that any element can be represented as the supremum of a set of atoms, i.e., for any element a there exists a set $\{b_i\}_{i \in I}$ of atoms such that $a = \bigvee \{b_i\}_{i \in I}$.

DEFINITION 1.1.15. *Covering property*
Let a, b be two elements of a lattice $\mathcal{B}$. Recall that b *covers* a iff $a \leq b, a \neq b$, and $\forall c \in B[a \leq c \leq b$ implies $a = c$ or $b = c]$.
A lattice $\mathcal{B}$ satisfies the *covering property* iff $\forall a, b \in B$: if a covers $a \wedge b$, then $a \vee b$ covers b.

It turns out that an atomic lattice $\mathcal{B}$ has the covering property iff for every atom a of $\mathcal{B}$ and for every element $b \in B$ such that $a \wedge b = \mathbf{0}$, the element $a \vee b$ covers b.

One of the most significant quantum relations, *compatibility*, admits a purely algebraic definition.

DEFINITION 1.1.16. *Compatibility*
Let $\mathcal{B}$ be an orthomodular lattice and let a and b be elements of B. The element a is called *compatible* with the element b iff

$$a = (a \wedge b') \vee (a \wedge b).$$

One can show that the compatibility relation is symmetric. Interestingly enough, the proof uses the orthomodular property in an essential way.[6]

Clearly, if $\mathcal{B}$ is a Boolean algebra, then any element is compatible with any other element by distributivity.

One can prove that a, b are compatible in the orthomodular lattice $\mathcal{B}$ iff the subalgebra of $\mathcal{B}$ generated by $\{a, b\}$ is Boolean.[7]

We will see later (Theorem 3.1.2) that this abstract notion of *compatibility* turns out to agree with the concrete notion of *compatibility between observables* in the framework of Hilbert space QT.

DEFINITION 1.1.17. *Irreducibility*
Let $\mathcal{B}$ be an orthomodular lattice. $\mathcal{B}$ is said to be *irreducible* iff

$$\{a \in B : \forall b \in B \, (a \text{ is compatible with } b)\} = \{\mathbf{0}, \mathbf{1}\}.$$

If $\mathcal{B}$ is not irreducible, it is called *reducible*.

Reducible orthomodular lattices can be represented as the cartesian product of "smaller" orthomodular lattices.[8] If $\mathcal{Y}$ and $\mathcal{Z}$ are two orthomodular lattices then their cartesian product may be made into an orthomodular lattice by defining all operations coordinatewise.

Let $\mathcal{B}$ be an orthomodular lattice, let $a \in B$, and let $[0, a] := \{b \in B : \mathbf{0} \leq b \leq a\}$. Each such interval $[0, a]$ can be made into an orthomodular

[6]See, for instance, (Kalmbach, 1983).

[7]See (Kalmbach, 1983). A *subalgebra* of an orthomodular lattice $\mathcal{B} = \langle B, \wedge, \vee, ', \mathbf{0}, \mathbf{1} \rangle$ is a structure $\mathcal{B}^* = \langle B^*, \wedge_*, \vee_*, '^*, \mathbf{0}^*, \mathbf{1}^* \rangle$ where: (i) $B^* \subseteq B$; (ii) $\wedge_*, \vee_*, '^*$ are the restrictions of $\wedge, \vee, '$ to B^*; (iii) $\mathbf{0}^* = \mathbf{0}$ and $\mathbf{1}^* = \mathbf{1}$; (iv) $\mathcal{B}^*$ is an orthomodular lattice. The *subalgebra of $\mathcal{B}$ generated by the elements a, b* is the smallest subalgebra of $\mathcal{B}$ that contains a and b. The notion of *subalgebra* can be naturally generalized to other classes of structures.

[8]Recall that the cartesian product of two sets $Y \times Z$ is the set $\{(y, z) : y \in Y \text{ and } z \in Z\}$ of all ordered pairs with the first element coming from Y and the second from Z.

lattice by imposing on $[0, a]$ the *induced ordering* $\leq_{[0,a]}$ and the relativized orthocomplementation $^{\sharp} : [0, a] \to [0, a]$. These are defined as follows: for $b, c \in [0, a]$, $b \leq_{[0,a]} c$ iff $b \leq c$ in $\mathcal{B}$, and $b^{\sharp} := b' \wedge a$. Doing so makes $[0, a]$ an orthomodular lattice.

Moreover, in a reducible orthomodular lattice $\mathcal{B}$, if an element $a \in B$ is compatible with all the elements of B, then so is a' and $\mathcal{B}$ is isomorphic to[9] the cartesian product of the orthomodular lattices $[0, a] \times [0, a']$, in which case we write $\mathcal{B} \cong [0, a] \times [0, a']$.

DEFINITION 1.1.18. *Separability*
An orthomodular lattice $\mathcal{B}$ is called *separable* iff every set of pairwise orthogonal elements of B is countable.

A useful technique for visualizing orthomodular posets or lattices is based on a notion that has come to be called a *Greechie diagram*.[10] This particular kind of *hypergraph* is essentially based on the fact that a finite Boolean algebra is completely determined by its atoms. The Greechie diagram of, say, a finite orthomodular poset $\mathcal{B}$ consists of a pair $(X, \mathcal{L})$ where X is the set of all atoms of $\mathcal{B}$ and $\mathcal{L}$ is the family of all maximal pairwise orthogonal subsets of X. Thus, $\mathcal{L}$ is in one-to-one correspondence with the set of all maximal Boolean subalgebras, or *blocks*, of $\mathcal{B}$.[11]

For example, the Greechie diagram pictured in Figure 1.1.3 represents

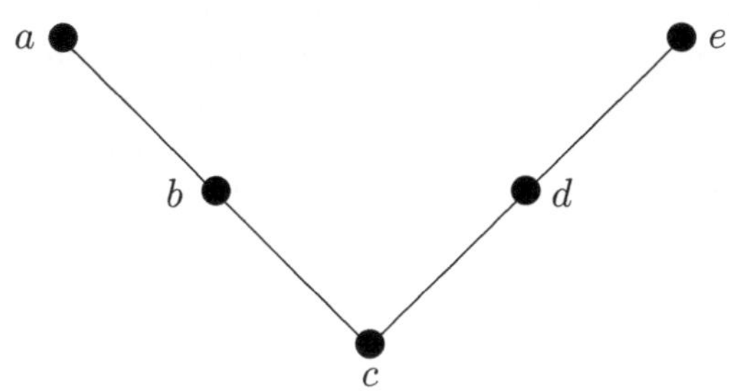

FIGURE 1.1.3. The Greechie diagram of $\mathcal{G}_{12}$

[9]This will be defined shortly; for now read, "essentially the same as."

[10]See (Kalmbach, 1983) and Section 6.5 where a more complete discussion is presented.

[11]A *Boolean subalgebra* of an algebra $\mathcal{B}$ (on which there is a naturally defined ordering and an involution $'$ making it a bounded involution poset with bounds $\mathbf{0}$ and $\mathbf{1}$) is a subalgebra $\mathcal{B}_1$ of $\mathcal{B}$ in which, under the induced natural ordering and involution, $\mathcal{B}_1 = \langle B_1, \wedge, \vee, ', \mathbf{0}, \mathbf{1}\rangle$ is a Boolean algebra. (Our notion of a Boolean subalgebra of an orthomodular poset is rather general because the fundamental operations of the subalgebra may not be precisely the fundamental operations of the enveloping algebra. For example, $\wedge$ is a fundamental operation of a Boolean subalgebra but not of an enveloping orthomodular poset.) A *maximal Boolean subalgebra* of $\mathcal{B}$ is a Boolean subalgebra of $\mathcal{B}$ which is not a proper subalgebra of any Boolean subalgebra of $\mathcal{B}$.

the orthomodular lattice $\mathcal{G}_{12}$, whose Hasse diagram is pictured in Figure 1.1.4.

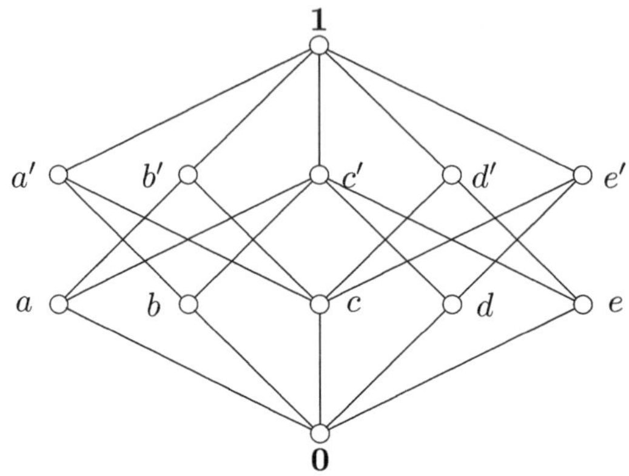

FIGURE 1.1.4. The Hasse diagram of $\mathcal{G}_{12}$

In a Greechie diagram, the slope of the lines means nothing, some "lines" may even be curves. It is instructive to observe that, because c and c' are compatible with all the elements of $\mathcal{G}_{12}$, $\mathcal{G}_{12} \cong \{0,1\} \times \mathcal{MO}2$.

In orthomodular structures the involution $'$ is crucial and must be added, as we did in Figure 1.1.4. The facts that orthomodular posets and lattices are the unions of their maximal Boolean subalgebras and that the involution $'$ is uniquely determined on a Boolean subalgebra underlies the strength of Greechie diagrams. It allows us to depict the n atoms of a finite Boolean subalgebra rather than all 2^n of its elements. Moreover, if an atom a is in many blocks, it is listed only once but is "on" each of the lines corresponding to those blocks. The example given in Figure 2.4.1 later in the text involves blocks of varying sizes, the largest being 2^7 - and there are several of those. It would be difficult to draw meaningful information from the Hasse diagram of that 410-element orthomodular poset.

In many algebraic and logical problems an important role is played by *filters* and *ideals*, that are particular subsets of the support of a lattice.

Let $\mathcal{B} = \langle B, \wedge, \vee, \mathbf{0}, \mathbf{1} \rangle$ be a bounded lattice.

DEFINITION 1.1.19. *Filter and ideal*

(i) A *filter* of $\mathcal{B}$ is a nonempty subset F of B such that
  (a) $\forall a \in F\ \forall b \in B$: $a \vee b \in F$;
  (b) $\forall a \in F\ \forall b \in F$: $a \wedge b \in F$.

(ii) An *ideal* of $\mathcal{B}$ is a nonempty subset I of B such that
  (a) $\forall a \in F\ \forall b \in B$: $a \wedge b \in I$;
  (b) $\forall a \in I\ \forall b \in I$: $a \vee b \in I$.

A *proper* filter (ideal) of $\mathcal{B}$ is a filter (ideal) of $\mathcal{B}$ that does not coincide with the support B.
A *principal* filter (ideal) is a filter (ideal) of $\mathcal{B}$ of the form $[a, \mathbf{1}] = \{x \in B : a \leq x\}$ ($[\mathbf{0}, a] = \{x \in B : x \leq a\}$).

DEFINITION 1.1.20. *Maximal filter and ideal*
A *maximal filter* (*maximal ideal*) of $\mathcal{B}$ is a proper filter (proper ideal) of $\mathcal{B}$ that is not properly included in any proper filter (proper ideal) of $\mathcal{B}$.

DEFINITION 1.1.21. *Complete filter and ideal*
A *complete filter* (*complete ideal*) of a bounded involution lattice $\mathcal{B}$ is a filter (ideal) such that for any $a \in B$, the filter (ideal) contains either a or a'.

Some important connections between structures of the same kind are represented by *homomorphisms*, *embeddings* and *isomorphisms*.
Consider a pair of *similar* structures:

$$\mathcal{B} = \left\langle B\,,\, R_1^{n_1}\,, \ldots, R_i^{n_i}\,, o_1^{m_1}\,, \ldots, o_j^{m_j} \right\rangle \text{ and}$$

$$B^* = \left\langle B^*\,,\, R_1^{*n_1}\,, \ldots, R_i^{*n_i}\,, o_1^{*m_1}\,, \ldots, o_j^{*m_j} \right\rangle.$$

In other words, for any k ($1 \leq k \leq i$), the two relations $R_k^{n_k}$ and $R_k^{*n_k}$ are supposed to have the same *arity* (number of arguments). Similarly, for any k ($1 \leq k \leq j$), the two operations $o_k^{m_k}$ and $o_k^{*m_k}$ have the same *arity*.

DEFINITION 1.1.22. *Homomorphism*
A *homomorphism* from $\mathcal{B}$ into $\mathcal{B}^*$ is a map

$$h : B \mapsto B^*$$

that preserves the relations and the operations. In other words:

(i) for any k ($1 \leq k \leq i$) and any $x_1, \ldots, x_{n_k} \in B$:

$$R_k^{n_k}(x_1, \ldots, x_{n_k}) \text{ implies } R_k^{*n_k}(h(x_1), \ldots, h(x_{n_k}));$$

(ii) for any k ($1 \leq k \leq j$) and any $x_1, \ldots, x_{m_k} \in B$:

$$h(o_k^{m_k}(x_1, \ldots, x_{m_k})) = o_k^{*m_k}(h(x_1), \ldots, h(x_{m_k})).$$

DEFINITION 1.1.23. *Embeddings and Isomorphisms*

(i) An *embedding* of $\mathcal{B}$ into $\mathcal{B}^*$ is a homomorphism h such that h is injective and $h^{-1} \upharpoonright Ran(h)$ (the inverse of h restricted to the range of h) is an injective homomorphism from $Ran(h)$ to $Dom(h)$ (the domain of h).
In other words, for any $x, y \in B$: $x \neq y$ iff $h(x) \neq h(y)$, and both h and h^{-1} preserve existing relations and operations.
(ii) An *isomorphism* from $\mathcal{B}$ to $\mathcal{B}^*$ is an embedding h such that h is *surjective*. In other words, any element of B^* is the h-image of an element of B.

Two similar structures $\mathcal{B}$ and $\mathcal{B}^*$ are called *isomorphic*, written $\mathcal{B} \cong \mathcal{B}^*$, when there is an isomorphism from $\mathcal{B}$ to $\mathcal{B}^*$. An embedding of one orthoposet or ortholattice into another is called an *ortho-embedding.*

As already mentioned, privileged examples of Boolean algebras are represented by set-theoretic structures.

EXAMPLE 1.1.24. Let U be any set. Consider the structure

$$\mathcal{B} = \langle B \,,\, \wedge \,,\, \vee \,,\, ' \,,\, \mathbf{0} \,,\mathbf{1} \rangle \,,$$

where:

- B is the power set $\mathcal{P}(U)$ of U. In other words, $X \in \mathcal{P}(U)$ iff $X \subseteq U$, i.e., any element of X is also an element of U;
- $\wedge$ is the set-theoretic intersection. In other words:

$$X \wedge Y = X \cap Y = \{x \,:\, x \in X \text{ and } x \in Y\}\,;$$

- $\vee$ is the set-theoretic union. In other words:

$$X \vee Y = X \cup Y = \{x \,:\, x \in X \text{ or } x \in Y\}\,;$$

- $'$ is the relative complement with respect to U. In other words:

$$X' = X^c = U - X = \{x \in U \,:\, x \notin X\}\,;$$

- $\mathbf{0}$ is the empty set, while $\mathbf{1}$ is the total set U.

The structure $\mathcal{B}$ turns out to be a Boolean algebra.

In many situations it is useful to consider set-theoretic Boolean algebras whose support B corresponds to a convenient proper subset of the full power set of a given set. In such cases, one usually speaks of *fields of sets.*

EXAMPLE 1.1.25. Let U be any set. Consider the structure

$$\mathcal{B} = \langle B \,,\, \wedge \,,\, \vee \,,\, ' \,,\, \mathbf{0} \,,\mathbf{1} \rangle \,,$$

where:

- B is a *field of sets* $\mathcal{F}(U)$ over U. In other words, $\mathcal{F}(U)$ is a set of subsets of U, such that $\mathcal{F}(U)$ contains $\emptyset$, U, and is closed under the set-theoretic operations intersection $\cap$, union $\cup$ and relative complement ^{c};
- $\wedge$ is the set-theoretic intersection $\cap$;
- $\vee$ is the set-theoretic union $\cup$;
- $'$ is the relative complement ^{c};
- $\mathbf{0}$ is the empty set, while $\mathbf{1}$ is the total set U.

As with the preceding example, the structure $\mathcal{B}$ also turns out to be a Boolean algebra.

From an intuitive point of view, the elements of a set-theoretic Boolean algebra can be interpreted as *extensional* properties or concepts that are *crisp*. These properties are extensional, because they are identified with particular sets. For instance, the numerical property "prime number" is identified with its *extension*: the infinite set $\{2, 3, 5, 7, 11, \ldots\}$ containing precisely all the natural numbers that satisfy the definition of "prime number." What has been traditionally called the *intension* of a given property is here completely neglected. Furthermore, our properties are *crisp*, in the sense that the membership relation is always precisely determined:

$$\forall X \in B \; \forall x \in U: \text{ either } x \in X \text{ or } x \in X'.$$

Tertium non datur! Uncertain or ambiguous membership relations are not admitted.

In many concrete situations, properties and concepts are accompanied by some ambiguous and uncertain borders. Let us think of examples like "beautiful," "honest," "tall," and so on. In such cases, it seems reasonable to assume that the membership relation is only determined for some objects. Generally the question

"does the object x satisfy the property X?"

may have no definite answer.

Situations of this kind have suggested the introduction of the notion of *fuzzy set*, deeply investigated in a rich literature after the pioneering work of Zadeh (1965); this has represented a natural development of Łukasiewicz' approach to many-valued logics. From the intuitive point of view, fuzzy sets can be imagined as collections whose membership relation is not generally precisely determined. In other words, the question

$$x \in X?$$

admits three possible answers: 1) Yes, 2) No; 3) Indeterminate.

Technically, the description of fuzzy sets can be developed by using a convenient generalization of the concept of *characteristic function*.

DEFINITION 1.1.26. *Classical characteristic function*
Let U be any classical set and let X be any subset of U. The *classical characteristic function* of X is the function $\chi_X : U \mapsto \{0, 1\}$ such that

$$\forall x \in U : \chi_X(x) = \begin{cases} 1 & \text{if } x \in X, \\ 0 & \text{otherwise.} \end{cases}$$

From an intuitive point of view, the characteristic function of X can be regarded as a kind of procedure that always gives an answer to the question "does an object x belong to the set X?." The answer will be "1," when the object belongs to X; "0," otherwise. Since there is an obvious one-to-one correspondence between sets and characteristic functions, classical sets are sometimes simply identified with their characteristic functions.

Classical characteristic functions can be naturally generalized to *fuzzy characteristic functions*.

DEFINITION 1.1.27. *Fuzzy characteristic function*
Let U be any classical set: a *fuzzy characteristic function* defined on U is any function

$$f : U \to [0, 1].$$

Note that some fuzzy characteristic functions are classical characteristic functions, namely those with range contained in $\{0, 1\}$. From an intuitive point of view, the number $f(x)$ associated to an object x can be interpreted as a *degree of membership* of x to a "nonclassical subset" X^f of U. As a consequence, such a function $f : U \mapsto [0, 1]$ turns out to describe a set, X^f, with "nebulous borders."

Generally, a fuzzy characteristic function f determines a partition of the original set U into three (classical) subsets:

(1) the set of the objects that definitely belong to X^f:

$$X_1(f) = \{x : f(x) = 1\} ;$$

(2) the set of the objects that definitely do not belong to X^f:

$$X_0(f) = \{x : f(x) = 0\} ;$$

(3) the set of the *uncertain* elements of X^f:

$$X_p(f) = \{x : f(x) \neq 1 \text{ and } f(x) \neq 0\} .$$

The three sets $X_1(f)$, $X_0(f)$ and $X_p(f)$ will be also called the *positive domain*, the *negative domain* and the *possibility domain* of f, respectively.

Of course, $X_p(f)$ will be empty in the particular case where f is a classical characteristic function. From an intuitive point of view, the set $X_p(f)$ can be imagined as a kind of "cloud" that separates the two classical sets $X_1(f)$ and $X_0(f)$.

A fuzzy characteristic function on a set U, will be also called a *fuzzy subset* of U. The *fuzzy power set* of U is the set of all fuzzy subsets of U. By *fuzzy set* we will mean any fuzzy subset of a given set.

Canonical examples of de Morgan lattices that are not Boolean can be obtained by considering structures whose support B is the fuzzy power set of a given set U.

EXAMPLE 1.1.28. Let U be any set. Consider the structure

$$\mathcal{B} = \langle B , \wedge , \vee , ' , \mathbf{0} , \mathbf{1} \rangle ,$$

where:

- B is the set of all fuzzy subsets of U;
- $\forall x \in U : (f \wedge g)(x) = \min(f(x), g(x))$;
- $\forall x \in U : (f \vee g)(x) = \max(f(x), g(x))$;
- $\forall x \in U : f'(x) = 1 - f(x)$;

- **0** is the classical characteristic function of the empty set $\emptyset$, while **1** is the classical characteristic function of the total set U.

The structure $\mathcal{B}$ is a de Morgan lattice that is not Boolean.

If we define $f \leq g$ iff $f \wedge g = f$, one can prove that for any fuzzy sets f and g:

$$f \leq g \text{ iff } f \vee g = g \text{ iff } \forall x \in U[f(x) \leq g(x)].$$

As happens in the Boolean case, in many fuzzy situations it is interesting to consider a structure whose support B is a convenient subset of the set of all possible fuzzy subsets of a given set.

EXAMPLE 1.1.29. Let U be any set. Consider the structure

$$\mathcal{B} = \langle B\,, \wedge\,, \vee\,,{}'\,, \mathbf{0}\,, \mathbf{1} \rangle\,,$$

where:

- B is a set of fuzzy subsets of U that contains the classical characteristic functions of the total set U and of the empty set $\emptyset$. Furthermore, B is closed under the operations $'$, $\wedge$, $\vee$ defined as follows, for any $x \in U$:

$$f'(x) = 1 - f(x);$$

$$(f \wedge g)(x) = \min(f(x), g(x)); \; (f \vee g)(x) = \max(f(x), g(x)).$$

- $\wedge$, $\vee$, $'$ are the operations defined above;
- **0** is the classical characteristic function of the empty set $\emptyset$, while **1** is the classical characteristic function of the total set U.

The structure $\mathcal{B}$ is a bounded de Morgan lattice that is not Boolean, unless $U = \emptyset$.

One immediately sees why fuzzy structures generally give rise to violations of the noncontradiction principle. Suppose that f is the fuzzy set such that for all objects x in U:

$$f(x) = 1/2.$$

This means that the membership relation to the fuzzy set f is completely indeterminate for the object x. As a consequence, we will have:

$$(f \wedge f')(x) = \min(f(x), 1 - f(x)) = 1/2.$$

Since $\mathbf{0}(x) = 0$, we will obtain:

$$f \wedge f' \neq \mathbf{0}.$$

Such a totally ambiguous fuzzy set is often called the *semitransparent fuzzy set* and indicated by $\frac{1}{2}\mathbf{1}$.

Standard fuzzy sets give rise to nonBoolean structures that are distributive. As a consequence, one could say that they represent a kind of "semiclassical" structure. As we will see, important examples of nondistributive structures emerge in the framework of (sharp and unsharp) quantum theory and quantum logic.

1.2. The geometry of quantum theory

Let us finally turn to the most peculiar "characters" of the mathematical scenario of QT: the abstract objects that "live" in the framework of Hilbert spaces. But, before introducing Hilbert spaces, it is expedient to recall the definitions of *group*, *division ring*, *vector space* and *pre-Hilbert space.*

DEFINITION 1.2.1. *Group*
A *group* is a structure $\mathcal{G} = \langle G\,, +\,, -\,, 0\rangle$, where $+$ is a binary operation, $-$ is a unary operation, 0 is a special element. The following conditions hold:

(i) $\langle G\,, +\,, 0\rangle$ is a *monoid*. In other words,
  (a) the operation $+$ is associative:
  $a + (b + c) = (a + b) + c$;
  (b) 0 is the *neutral element*:
  $a + 0 = a$;
(ii) $\forall a \in G$, $-a$ is the *inverse* of a:
  $a + (-a) = 0$.

An *Abelian monoid* (*group*) is a monoid (group) in which the operation $+$ is commutative: $a + b = b + a$. We abbreviate $a + (-b)$ by $a - b$.

DEFINITION 1.2.2. *Partially ordered group*
A *partially ordered group* (or, *po-group*) is a structure $\langle G\,, +\,, 0\,, \leq\rangle$ that satisfies the following conditions:

(i) $\langle G\,, +\,, 0\rangle$ is a group;
(ii) $\langle G\,, \leq\rangle$ is a poset;
(iii) the "group translations" are isotone; in other words, $\forall a, b, c, d \in G$:
  $c \leq d$ implies $a + c \leq a + d$ and $c + b \leq d + b$.

DEFINITION 1.2.3. *Ring*
A *ring* is a structure $\mathbf{D} = \langle D\,, +\,, \cdot\,, -\,, 0\rangle$ that satisfies the following conditions:

(i) $\langle D\,, +\,, 0\rangle$ is an Abelian group;
(ii) the operation $\cdot$ is associative:
  $a \cdot (b \cdot c) = (a \cdot b) \cdot c$;
(iii) the operation $\cdot$ distributes over $+$ on both sides, i.e., $\forall a, b, c \in D$:
  (a) $a \cdot (b + c) = (a \cdot b) + (a \cdot c)$;
  (b) $(a + b) \cdot c = (a \cdot c) + (b \cdot c)$.

If there is an element 1 in D that is neutral for $\cdot$ (i.e., if $\langle D\,, \cdot\,, 1\rangle$ is a monoid), then the ring is called a *ring with unity*.

A ring is *trivial* in case it has only one element, otherwise it is *nontrivial*. It is easy to see that a ring with unity is nontrivial iff $0 \neq 1$.

A *commutative ring* is a ring in which the operation $\cdot$ is commutative.

DEFINITION 1.2.4. *Division ring*
A *division ring* is a nontrivial ring $\mathbf{D}$ with unity such that any nonzero element is *invertible*; in other words, for any $a \in D$ $(a \neq 0)$, there is an element $b \in D$ such that $a \cdot b = b \cdot a = 1$.

DEFINITION 1.2.5. *Field*
A *field* is a commutative division ring.

Both the real numbers ($\mathbb{R}$) and the complex numbers ($\mathbb{C}$) give rise to a field. An example of a genuine division ring (where $\cdot$ is not commutative) is given by the quaternions ($\mathbb{Q}$).

We can now introduce the general notion of *vector space*. Since we are dealing with possibly noncommutative division rings, we shall distinguish between *left vector spaces* and *right vector spaces*.

DEFINITION 1.2.6. *Left vector space*
A *left vector space* over a division ring $\mathbf{D}$ is a structure $\mathcal{V} = \langle V\,, +\,, -\,, \cdot\,, \underline{0}\rangle$ that satisfies the following conditions:

(i) $\langle V\,, +\,, -\,, \underline{0}\rangle$ (the vector structure) is an Abelian group, where $\underline{0}$ (the *null vector*) is the neutral element;
(ii) for any element a of the division ring $\mathbf{D}$ and any vector φ of V, $a\varphi$ (the *scalar product* of a and φ) is a vector in V. The following conditions hold for any $a, b \in D$ and for any $\varphi, \psi \in V$:
  (a) $a(\varphi + \psi) = (a\varphi) + (a\psi)$;
  (b) $(a + b)\varphi = (a\varphi) + (b\varphi)$;
  (c) $a(b\varphi) = (a \cdot b)\varphi$;
  (d) $1\varphi = \varphi$.

The notion of *right vector space* can be defined similarly; it involves expressions of the form φa with the scalars being multiplied on the right. In the following, we will only deal with left vector spaces. Accordingly, we will simply speak of *vector spaces*.

The elements (vectors) of a vector space $\mathcal{V}$ will be indicated by $\varphi, \psi, \chi, \ldots$, while $a, b, c, \ldots$ will represent elements (scalars) of the division ring $\mathbf{D}$. Any finite sum of vectors $\psi_1, \ldots, \psi_n$ will be indicated by $\psi_1 + \ldots + \psi_n$ (or $\sum_{i \in K} \psi_i$, when $K = \{1, \ldots, n\}$.)

On this basis, we can now introduce the notion of *pre-Hilbert space*. Hilbert spaces will be then defined as special cases of pre-Hilbert spaces. For the sake of simplicity, we will first consider pre-Hilbert spaces (and Hilbert spaces) whose division ring is either $\mathbb{R}$ or $\mathbb{C}$.

DEFINITION 1.2.7. *Pre-Hilbert space*
Let $\mathbf{D}$ be the field of the real or the complex numbers. A *pre-Hilbert space* (or *inner space*) over $\mathbf{D}$ is a vector space $\mathcal{V}$ over $\mathbf{D}$, equipped with an inner product

$$\langle\, .|.\, \rangle : V \times V \mapsto D$$

that satisfies the following conditions for any $\varphi, \psi, \chi \in V$ and any $a \in D$:

(i) $\langle\, \varphi|\varphi\, \rangle \geq 0$;
(ii) $\langle\, \varphi|\varphi\, \rangle = 0$ iff $\varphi = \underline{0}$;
(iii) $\langle\, \psi|a\varphi\, \rangle = a\langle\, \psi|\varphi\, \rangle$;
(iv) $\langle\, \varphi|\psi + \chi\, \rangle = \langle\, \varphi|\psi\, \rangle + \langle\, \varphi|\chi\, \rangle$;

(v) $\langle \varphi|\psi \rangle = \langle \psi|\varphi \rangle^*$, where * is the identity if $D = \mathbb{R}$, and the complex conjugation if $D = \mathbb{C}$.

The inner product $\langle .|. \rangle$ permits one to generalize some geometrical notions of ordinary 3-dimensional spaces.

DEFINITION 1.2.8. *Norm of a vector*
The *norm* $\|\varphi\|$ of a vector φ is the number $\langle \varphi|\varphi \rangle^{1/2}$.

Note that the norm of any vector is a real number greater than or equal to 0.

A *unit* (or *normalized*) *vector* is a vector ψ such that $\|\psi\| = 1$.

Two vectors φ, ψ are called *orthogonal* iff $\langle \varphi|\psi \rangle = 0$.

DEFINITION 1.2.9. *Orthonormal set of vectors*
A set $\{\psi_i\}_{i \in I}$ of vectors is called *orthonormal* iff its elements are pairwise *orthogonal* unit vectors. In other words:

(i) $\forall i, j \in I (i \neq j) : \langle \psi_i|\psi_j \rangle = 0$;
(ii) $\forall i \in I : \|\psi_i\| = 1$.

The norm $\|.\|$ induces a *metric* d on the pre-Hilbert space $\mathcal{V}$:

$$d(\psi, \varphi) := \|\psi - \varphi\|.$$

We say that a sequence $\{\psi_i\}_{i \in \mathbb{N}}$ of vectors in V *converges in norm* (or simply *converges*) to a vector φ of V iff $\lim_{i \to \infty} d(\psi_i, \varphi) = 0$. In other words, $\forall \varepsilon > 0 \, \exists n \in \mathbb{N} \, \forall k > n : \; d(\psi_k, \varphi) < \varepsilon$.

A *Cauchy sequence* is a sequence $\{\psi_i\}_{i \in \mathbb{N}}$ of vectors in V such that $\forall \varepsilon > 0 \, \exists n \in \mathbb{N} \, \forall h > n \, \forall k > n : \; d(\psi_h, \psi_k) < \varepsilon$.

It is easy to see that whenever a sequence $\{\psi_i\}_{i \in \mathbb{N}}$ of vectors in V converges to a vector φ of V, then $\{\psi_i\}_{i \in \mathbb{N}}$ is a Cauchy sequence. The crucial question is the converse one: which are the pre-Hilbert spaces in which every Cauchy sequence converges to *an element in the space*?

DEFINITION 1.2.10. *Metrically complete pre-Hilbert space*
A pre-Hilbert space $\mathcal{V}$ with inner product $\langle .|. \rangle$ is *metrically complete* with respect to the metric d induced by $\langle .|. \rangle$ iff every Cauchy sequence of vectors in V converges to a vector of V.

On this basis, we can finally define the notion of Hilbert space.

DEFINITION 1.2.11. *Hilbert space*
A *Hilbert space* is a metrically complete pre-Hilbert space.

A *real* (*complex*) *Hilbert space* is a Hilbert space whose division ring is $\mathbb{R}$ ($\mathbb{C}$). The notion of pre-Hilbert space (Hilbert space) can be generalized to the case where the division ring is represented by $\mathbb{Q}$ (the division ring of all quaternions).

We will now define some basic Hilbert-space notions, which will play an important role in the mathematical formalism of QT.

Consider a Hilbert space $\mathcal{H}$ over a division ring $\mathbf{D}$.

DEFINITION 1.2.12. *(Hilbert) linear combination*
Let $\{\psi_i\}_{i\in I}$ be a set of vectors of $\mathcal{H}$ and let $\{a_i\}_{i\in I} \subseteq D$. A vector ψ is called a *(Hilbert) linear combination* (or *superposition*) of $\{\psi_i\}_{i\in I}$ (with scalars $\{a_i\}_{i\in I}$) iff $\forall\varepsilon \in \mathbb{R}^+$ there is a finite set $J \subseteq I$ such that for any finite subset K of I including J:

$$\|\psi - \sum_{i\in K} a_i\psi_i\| \leq \varepsilon.$$

Apparently, when existing, the linear combination of $\{\varphi_i\}_{i\in I}$ (with scalars $\{a_i\}_{i\in I}$) is unique. We will denote it by $\sum_{i\in I} a_i\psi_i$. When no confusion is possible, the index set I will be omitted.

DEFINITION 1.2.13. *Orthonormal basis*
An *orthonormal basis* of $\mathcal{H}$ is a *maximal orthonormal set* $\{\psi_i\}_{i\in I}$ of $\mathcal{H}$. In other words, $\{\psi_i\}_{i\in I}$ is an orthonormal set such that no orthonormal set includes $\{\psi_i\}_{i\in I}$ as a proper subset.

One can prove that every Hilbert space $\mathcal{H}$ has an orthonormal basis and that all orthonormal bases of $\mathcal{H}$ have the same cardinality. The *dimension* of $\mathcal{H}$ is then defined as the cardinal number of any basis of $\mathcal{H}$.

Let $\{\psi_i\}_{i\in I}$ be any orthonormal basis of $\mathcal{H}$. One can prove that every vector φ of $\mathcal{H}$ can be expressed in the following form:

$$\varphi = \sum_{i\in I}\langle\psi_i|\varphi\rangle\psi_i.$$

Hence, φ is a linear combination of $\{\psi_i\}_{i\in I}$ with scalars $\langle\psi_i|\varphi\rangle$ (the scalars $\langle\psi_i|\varphi\rangle$ are also called *Fourier coefficients.*)
A Hilbert space $\mathcal{H}$ is called *separable* iff $\mathcal{H}$ has a countable orthonormal basis. In the following, we will always refer to separable Hilbert spaces.

DEFINITION 1.2.14. *Closed subspace*
A *closed subspace* of $\mathcal{H}$ is a set X of vectors that satisfies the following conditions:

(i) X is a *subspace* of $\mathcal{H}$. In other words, X is closed under finite linear combinations. Hence,

$$\psi, \varphi \in X \text{ implies } a\psi + b\varphi \in X;$$

(ii) X is closed under limits of Cauchy sequences. In other words: if each element of a Cauchy sequence of vectors belongs to X, then also the limit of the sequence belongs to X.

The set of all closed subspaces of $\mathcal{H}$ will be indicated by $\mathcal{C}(\mathcal{H})$. For any vector ψ, we will indicate by $[\psi]$ the unique 1-dimensional closed subspace that contains ψ.

DEFINITION 1.2.15. *Operator*
An operator of $\mathcal{H}$ is a map

$$A : Dom(\mathcal{H}) \mapsto \mathcal{H},$$

where $Dom(A)$ (the *domain* of A) is a subset of $\mathcal{H}$.

DEFINITION 1.2.16. *Densely defined operator*
A *densely defined operator* of $\mathcal{H}$ is an operator A that satisfies the following condition: $\forall \varepsilon \in \mathbb{R}^+ \ \forall \psi \in \mathcal{H} \ \exists \varphi \in Dom(A) \, [d(\psi, \varphi) < \varepsilon]$, where d represents the metric induced by $\langle . | . \rangle$.

DEFINITION 1.2.17. *Linear operator*
A *linear operator* on $\mathcal{H}$ is an operator A that satisfies the following conditions:

(i) $Dom(A)$ is a closed subspace of $\mathcal{H}$;
(ii) $\forall \psi, \varphi \in Dom(A) \, \forall a, b \in D : A(a\psi + b\varphi) = aA\psi + bA\varphi$.

In other words, a characteristic of linear operators is preserving the linear combinations.

DEFINITION 1.2.18. *Bounded operator*
A linear operator A is called *bounded* iff there exists a positive real number a such that $\forall \psi \in \mathcal{H} : \|A\psi\| \leq a\|\psi\|$.

DEFINITION 1.2.19. *Positive operator*
A bounded operator A is called *positive* iff $\forall \psi \in \mathcal{H} : \langle \psi | A\psi \rangle \geq 0$.

DEFINITION 1.2.20. *The adjoint operator*
Let A be a densely defined linear operator of $\mathcal{H}$. The *adjoint* of A is the unique operator A^* such that

$$\forall \psi \in Dom(A) \forall \varphi \in Dom(A^*) : \langle A\psi | \varphi \rangle = \langle \psi | A^* \varphi \rangle \, .$$

Notice that A^* is not necessarily densely defined.

DEFINITION 1.2.21. *Self-adjoint operator*
A *self-adjoint operator* is a densely defined linear operator A such that $A = A^*$.

If A is self-adjoint, then $\forall \psi, \varphi \in Dom(A) : \langle A\psi | \varphi \rangle = \langle \psi | A\varphi \rangle$.
If A is self-adjoint and everywhere defined (i.e., $Dom(A) = \mathcal{H}$), then A is bounded.

DEFINITION 1.2.22. *Projection operator*
A *projection operator* is an everywhere defined self-adjoint operator P that satisfies the idempotence property: $\forall \psi \in \mathcal{H} : P\psi = PP\psi$.
There are two special projections $\mathbb{O}$ and $\mathbb{I}$ called *the zero* (or *null projection*) and *the identity projection* which are defined as follows: $\forall \psi \in \mathcal{H}$,

$$\mathbb{O}\psi = \underline{0} \text{ and } \mathbb{I}\psi = \psi .$$

Any projection other than $\mathbb{O}$ and $\mathbb{I}$ is called a *nontrivial projection.*

Thus, P is a projection operator if $Dom(P) = \mathcal{H}$ and $P = P^2 = P^*$. The set of all projection operators will be indicated by $\Pi(\mathcal{H})$.

One can prove that the set $\mathcal{C}(\mathcal{H})$ of all closed subspaces and the set $\Pi(\mathcal{H})$ of all projections of $\mathcal{H}$ are in one-to-one correspondence. As we will see, such correspondence will play a very important role in the algebraic and logical structures of QT.

Let X be a closed subspace of $\mathcal{H}$. By the projection theorem[12] every vector $\psi \in \mathcal{H}$ can be uniquely expressed as a linear combination $\psi_1 + \psi_2$, where $\psi_1 \in X$ and ψ_2 is orthogonal to any vector of X. Accordingly, we can define an operator P_X on $\mathcal{H}$ such that

$$\forall\psi \in \mathcal{H} : P_X\psi = \psi_1$$

(in other words, P_X transforms any vector ψ into the "X-component" of ψ)

It turns out that P_X is a projection operator of $\mathcal{H}$.

Conversely, we can associate to any projection P its range,

$$X_P = \{\psi : \exists\varphi(P\varphi = \psi)\},$$

which turns out to be a closed subspace of $\mathcal{H}$.

For any closed subspace X and for any projection P, the following conditions hold:

$$X_{(P_X)} = X; \quad P_{(X_P)} = P.$$

DEFINITION 1.2.23. *Projection-valued measure*
Let $\mathcal{B}(\mathbb{R})$ be the set of all Borel sets of real numbers. A *projection-valued measure* (called also *spectral measure*) is a map

$$M : \mathcal{B}(\mathbb{R}) \mapsto \Pi(\mathcal{H}),$$

that satisfies the following conditions:

- $M(\emptyset) = \mathbb{O}$;
- $M(\mathbb{R}) = \mathbb{1}$;
- for any countable set $\{\Delta_i\}_{i \in I}$ of pairwise disjoint Borel-sets:

$$M(\bigcup\{\Delta_i\}_{i\in I}) = \sum_i M(\Delta_i),$$

  where (in the infinite case) the series converges in the weak operator topology[13] of the set of all bounded operators of $\mathcal{H}$.

One can prove that there exists a one-to-one correspondence between the set of all projection-valued measures and the set of all self-adjoint operators of $\mathcal{H}$. This is the content of the *spectral theorem*.[14] We will indicate by A^M the self-adjoint operator associated with the projection-valued measure M; while M^A will represent the projection-valued measure associated with the self-adjoint operator A.

---

[12]See, for instance, (Reed and Simon, 1972).

[13]For the notion of weak operator topology, see (Reed and Simon, 1972).

[14]See, for instance, (Beltrametti and Cassinelli, 1981).

DEFINITION 1.2.24. *The trace functional*
Let $\{\psi_i\}_{i\in I}$ be any orthonormal basis for $\mathcal{H}$ and let A be a positive operator. The *trace* of A (indicated by $\text{Tr}(A)$) is defined as follows:

$$\text{Tr}(A) := \sum_i \langle \psi_i | A\psi_i \rangle .$$

One can prove that the definition of Tr is independent of the choice of the basis.

For any positive operator A, there exists a unique positive operator B such that: $B^2 = A$. If A is a (not necessarily positive) bounded operator, then A^*A is positive. Let $|A|$ be the unique positive operator such that $|A|^2 = A^*A$. A bounded operator A is called a *trace-class operator* iff $\text{Tr}(|A|) < \infty$.

DEFINITION 1.2.25. *Density operator*
A density operator is a positive, self-adjoint, trace-class operator ρ such that $\text{Tr}(\rho) = 1$.

It is easy to see that, for any vector ψ, the projection $P_{[\psi]}$ onto the 1-dimensional closed subspace $[\psi]$ is a density operator.

DEFINITION 1.2.26. *Unitary operator*
A *unitary operator* is a linear operator U such that:

- $Dom(U) = \mathcal{H}$;
- $UU^* = U^*U = \mathbb{1}$.

One can show that the unitary operators U are precisely the operators that preserve the inner product. In other words, for any $\psi, \varphi \in \mathcal{H}$:

$$\langle \psi | \varphi \rangle = \langle U\psi | U\varphi \rangle .$$

Any pair of Hilbert spaces $\mathcal{H}_1 , \mathcal{H}_2$ gives rise to a new Hilbert space $\mathcal{H}_1 \otimes \mathcal{H}_2$ that represents the *tensor product* of $\mathcal{H}_1$ and $\mathcal{H}_2$. As we will see, tensor products play an important role for the mathematical representation of compound quantum systems. They are also systematically used in the mathematical formalism of quantum computation (see Chapter 17).

DEFINITION 1.2.27. *Tensor product Hilbert space*
Let $\mathcal{H}_1$ and $\mathcal{H}_2$ be two Hilbert spaces over the same field $\mathbf{D}$ (the real or the complex numbers). A Hilbert space $\mathcal{H}$ is the *tensor product* of $\mathcal{H}_1$ and $\mathcal{H}_2$ iff the following conditions are satisfied:

(i) there exists a map $\otimes$ (called *tensor product*) from the cartesian product $\mathcal{H}_1 \times \mathcal{H}_2$ into $\mathcal{H}$ that satisfies the following conditions:
  (a) the tensor product $\otimes$ is linear in each component; in other words, $\forall \psi , \varphi \in \mathcal{H}_1 \, \forall \chi, \delta \in \mathcal{H}_2 \, \forall a, b \in D$:
    (a1) $(a\psi + b\varphi) \otimes \chi = (a\psi) \otimes \chi + (b\varphi) \otimes \chi$;
    (a2) $\psi \otimes (a\chi + b\delta) = \psi \otimes (a\chi) + \psi \otimes (b\delta)$;

(b) the external product with a scalar carries across the tensor product; in other words, $\forall \psi \in \mathcal{H}_1 \, \forall \varphi \in \mathcal{H}_2 \, \forall a \in D$: $a(\psi \otimes \varphi) = (a\psi) \otimes \varphi = \psi \otimes (a\varphi)$;

(ii) every vector of $\mathcal{H}$ can be expressed as a (Hilbert) linear combination of vectors of the set $\{\varphi \otimes \psi \, : \, \varphi \in \mathcal{H}_1 \, , \psi \in \mathcal{H}_2\}$.

One can show that the tensor product is unique up to isomorphism.

As required by condition (ii), every vector of $\mathcal{H}_1 \otimes \mathcal{H}_2$ can be expressed as a linear combination of vectors of the form $\psi \otimes \varphi$ (where $\psi \in \mathcal{H}_1$, $\varphi \in \mathcal{H}_2$). At the same time, there are vectors of $\mathcal{H}_1 \otimes \mathcal{H}_2$ that cannot be written as a single product $\psi \otimes \varphi$ for any $\psi \in \mathcal{H}_1$, $\varphi \in \mathcal{H}_2$. These vectors are called *non-factorized*.

If $\{\psi_i\}_{i \in I}$ and $\{\varphi_j\}_{j \in J}$ are orthonormal basis for $\mathcal{H}_1$ and $\mathcal{H}_2$, respectively, then the set $\{\varphi_i \otimes \psi_j \, : \, i \in I \, , j \in J\}$ is an orthonormal basis of the tensor product Hilbert space $\mathcal{H}_1 \otimes \mathcal{H}_2$. In particular, if $\{\psi_i, \ldots, \psi_n\}$ and $\{\varphi_1, \ldots \varphi_m\}$ are orthonormal basis of the finite dimensional Hilbert spaces $\mathcal{H}_1$, $\mathcal{H}_2$, then every vector $\psi \in \mathcal{H}_1 \otimes \mathcal{H}_2$ can be written as

$$\psi = \sum_{i=1}^{n} \sum_{j=1}^{m} a_{ij} \psi_i \otimes \varphi_j .$$

It follows that the dimension of the tensor product Hilbert space $\mathcal{H}_1 \otimes \mathcal{H}_2$ is the product of the dimensions of $\mathcal{H}_1$ and $\mathcal{H}_2$.

1.3. The axiomatization of orthodox QT

After having briefly sketched the basic "mathematical ingredients" of QT, we can now present the main axioms of the theory (following von Neumann's proposal).

Axiom vN1. *Physical systems*
The mathematical interpretation of any physical system $\mathfrak{S}$ (either a single particle or a compound system of many particles) is a complex separable Hilbert space $\mathcal{H}$.

Axiom vN2. *Pure and mixed states*
Any pure state of the system $\mathfrak{S}$ is mathematically represented by a unit vector ψ of the space $\mathcal{H}$.[15] States that do not necessarily correspond to maximal information are called *mixed states* (or *mixtures*). They are mathematically represented by density operators ρ of $\mathcal{H}$. Apparently, pure states turn out to be limit-cases of mixed states: any unit vector ψ determines a projection $P_{[\psi]}$ (the projection associated to the 1-dimensional closed subspace $[\psi]$ corresponding to ψ). Accordingly, one can also say that any pure

[15] Following a successful notation introduced by Dirac, pure states are often represented by the so called "ket-notation." Accordingly, a pure state is indicated by $|\psi\rangle$ (instead of ψ). We will use Dirac's notation only in Chapter 17, when we will deal with quantum computation.

state is represented by a projection $P_{[\psi]}$. Density operators that cannot be represented in the form $P_{[\psi]}$ are also called *proper mixtures*.

Axiom vN3. *Events*
Any *event* that may hold for $\mathfrak{S}$ is mathematically represented by a projection operator P of $\mathcal{H}$. By the one-to-one correspondence between the set $\Pi(\mathcal{H})$ of all projection operators and the set $\mathcal{C}(\mathcal{H})$ of all closed subspaces X of $\mathcal{H}$, events turn out to be equivalently represented by closed subspaces of $\mathcal{H}$.

Axiom vN4. *Observables*
Any *observable* (or *physical quantity*) that can be measured on $\mathfrak{S}$ is mathematically represented by a *projection-valued measure* M of $\mathcal{H}$ (or, equivalently, by the corresponding self-adjoint operator A^M). For any projection-valued measure M and any Borel set Δ the intended physical interpretation of the projection $M(\Delta)$ will be the event: "the observable represented by M has a value that lies in Δ."[16]

Axiom vN5. *The Born probability*
The probability $p(\rho, P)$ that the system $\mathfrak{S}$ in state ρ satisfies an event P is determined by the following rule (called the *Born-rule*):

$$p(\rho, P) := \mathrm{Tr}(\rho P),$$

where Tr represents the *trace functional*.

One can show that for any state ρ and for any projection P:

$$p(\rho, P) \in [0, 1].$$

Furthermore, for any pure state ψ and any projection P_X (associated to the closed subspace X):

$$p(P_{[\psi]}, P_X) = 1 \text{ iff } \psi \in X \text{ iff } P_X\psi = \psi.$$

Hence, a pure state ψ *certainly verifies* an event corresponding to the closed subspace X (i.e., ψ assigns to X probability-value 1) iff ψ is an element of X.
An interesting case arises when a pure state ψ can be represented as superposition of other pure states ψ_i, where each ψ_i is a *unit-eigenvector* of a given observable (self-adjoint operator) A. In other words:

- each ψ_i is a unit vector;
- each ψ_i is an *eigenvector* of A with corresponding *eigenvalue* a_i ($A\psi_i = a_i\psi_i$);
- $\psi = \sum_i c_i\psi_i$, where $\sum_i |c_i|^2 = 1$.

[16]By abuse of language, we will often say "observable" instead of "mathematical representation of an observable." Similarly in the case of states and events.

In such a case, we will have:

$$p\left(P_{[\psi]}, M^A(\{a_i\})\right) = \mathrm{Tr}\left(P_{[\psi]} M^A(\{a_i\})\right) = |c_i|^2 .$$

In other words, the number $|c_i|^2$ represents the probability that the system in state ψ has for the observable A to have the precise value a_i. As a consequence, ψ *certainly has* the value a_i for the observable A iff ψ is a unit-eigenvector of A with eigenvalue a_i.

In this framework, one can define the notions of *expected value* (or *mean value*) and *variance* of a given observable. Consider an observable M and a state ρ.

DEFINITION 1.3.1. *Expected value*
The expected value of M with respect to the state ρ, $Exp(M, \rho)$, is defined as follows:

$$Exp(M, \rho) := \int_{-\infty}^{+\infty} x \mathrm{Tr}(\rho M(dx)),$$

provided the integral exists.

DEFINITION 1.3.2. *Variance*
The variance of M with respect to the state ρ, $Var(M, \rho)$, is defined as follows:

$$Var(M, \rho) := \int_{-\infty}^{+\infty} [x - Exp(M, \rho)]^2 \mathrm{Tr}(\rho M(dx)),$$

provided $Exp(M, \rho)$ and the integral exist and are finite.

From an intuitive point of view, $Var(M, \rho)$ describes the *dispersion* of the values of the observable M around the mean value.

Specific quantum observables (**position**, **momentum**, **spin**....) in a given direction are then defined as particular self-adjoint operators.

A crucial quantum relation between observables is represented by *compatibility*. From an intuitive point of view, two observables M and N are *compatible* when they are *simultaneously measurable*.

DEFINITION 1.3.3. *Compatible observables*
Two observables M and N are *compatible* iff for any pair of Borel sets Δ and Γ, the two projections $M(\Delta)$ and $M(\Gamma)$ commute:

$$M(\Delta)M(\Gamma) = M(\Gamma)M(\Delta).$$

It turns out that two observables M and N are compatible iff, for any pair of Borel sets Δ and Γ, the operator $M(\Delta)N(\Gamma)$ is a projection.

The existence of pairs of *incompatible* observables can be proved. Canonical examples are represented by the pairs:

- **position** in a given direction d and **momentum** in the same direction.
- **spin** in a given direction d and **spin** in a different direction e.

Pairs of incompatible observables M and N satisfy the general form of Heisenberg *uncertainty principle*:

$\exists \varepsilon > 0$ such that for any pure state ψ, $Var(M, \psi)Var(N, \psi) > \varepsilon$.

As a consequence: *no pure state can assign an exact value to both elements of a pair of incompatible observables.*

Axiom vN6. *The Schrödinger equation*
The spontaneous time-evolution of the states of a quantum system $\mathfrak{S}$ is determined by the Schrödinger equation. According to this equation, for any time interval $[t_0, t_1]$ (where either $t_0 \leq t_1$ or $t_1 \leq t_0$), there exists a unitary operator $U_{[t_0,t_1]}$ that maps the pure states of $\mathfrak{S}$ into pure states of $\mathfrak{S}$. For any pure state ψ, $U_{[t_0,t_1]}\psi$ represents the state of the system at time t_1, provided the system is in state ψ at time t_0. Such a map can be also extended to the case of mixed states.

Axiom vN7. *The reduction-postulate* (von Neumann-Lüders)[17]
Suppose the observer measures an observable represented by the spectral measure M on the system $\mathfrak{S}$ during the time interval $[t_0, t_1]$, by means of a *non-destructive* measurement procedure. Let ρ represent the state of $\mathfrak{S}$ at the initial time t_0. Suppose the result of the measurement is the Borel set Δ. Then, soon after the measurement, at time t_1, the observer will associate to the system the following state:

$$\rho' = \frac{M(\Delta)\rho M(\Delta)}{\mathrm{Tr}(\rho M(\Delta))}.$$

One can show that ρ' assigns probability 1 to the event $M(\Delta)$ ("the observable M has a value lying in Δ"). In other words, the performance of a measurement, induces a state-transformation that takes into account the information obtained by the measuring procedure.

Axiom vN8. *Compound systems*
Suppose we have a compound quantum system $\mathfrak{S}$ consisting of two subsystems $\mathfrak{S}_1$ and $\mathfrak{S}_2$ (for instance, let $\mathfrak{S}_1$ and $\mathfrak{S}_2$ correspond to the two electrons of a helium atom). Let $\mathcal{H}_1$ be the Hilbert space representing the mathematical interpretation of $\mathfrak{S}_1$, while $\mathcal{H}_2$ represent the mathematical interpretation of $\mathfrak{S}_2$. Then, the tensor product $\mathcal{H}_1 \otimes \mathcal{H}_2$ will represent the mathematical interpretation of the compound system $\mathfrak{S}$.

1.4. The "logic" of the quantum events

As we have seen, according to von Neumann's axiomatization, quantum events are mathematically realized by projection operators of a Hilbert space $\mathcal{H}$ (which represents the "mathematical scenario" associated to the quantum system under investigation). Hence, any experimental proposition concerning a quantum object will correspond to a projection in a convenient

[17]This axiom is often called *collapse of the wave-function*.

Hilbert space $\mathcal{H}$. Owing to the one-to-one correspondence between the set of all projections and the set of all closed subspaces, in this framework, experimental propositions turn out to be equivalently represented by closed subspaces of the Hilbert space.

A natural question arises: why are the mere subsets of $\mathcal{H}$ not adequate mathematical representatives for quantum events (as in the phase-space case)? The reason depends on the *superposition principle*, which represents one of the basic dividing lines between the quantum and the classical case.[18]

As opposed to classical mechanics, in quantum mechanics, any unit vector, that is a linear combination of pure states, gives rise to a new pure state. Suppose two pure states ψ_1, ψ_2 are orthogonal and suppose that a pure state ψ is a linear combination of ψ_1, ψ_2. In other words:

$$\psi = c_1\psi_1 + c_2\psi_2 \text{ (where } |c_1|^2 + |c_2|^2 = 1).$$

According to the Born rule (Axiom vN5), this means that a quantum system in state ψ might verify with probability $|c_1|^2$ those events (and those experimental propositions) that are certain for state ψ_1 (and are not certain for ψ) and might verify with probability $|c_2|^2$ those events (and those experimental propositions) that are certain for state ψ_2 (and are not certain for ψ). Suppose now there is given an orthonormal set of pure states $\{\psi_i\}_{i \in I}$, where each ψ_i assigns probability 1 to a given experimental proposition **P**. Consider the linear combination

$$\psi = \sum_i c_i\psi_i \quad (c_i \neq 0 \text{ and } \sum_i |c_i|^2 = 1),$$

which turns out to be a pure state. Then, also ψ will assign probability 1 to the proposition **P**. As a consequence, the mathematical representatives of experimental propositions should be closed under finite and infinite linear combinations. Since, according to Axiom vN2, every pure state ψ can be represented by a projection $P_{[\psi]}$ onto the 1-dimensional subspace $[\psi]$, it follows that the mathematical representative of the experimental proposition **P**, that contains the vector ψ, must also contain any vector $a\psi$ (with $a \in \mathbb{C}$). The closed subspaces of $\mathcal{H}$ are just the mathematical objects that can realize such a role.

What will be the meaning of *negation*, *conjunction* and *disjunction* in the realm of quantum events?

Let us first consider negation, by referring again to Birkhoff and von Neumann. They observe:

[18] As noticed by Beltrametti and Cassinelli (1981): The appearance of these superpositions of pure states is a distinguishing feature of quantum mechanics: it has no analogue in classical mechanics. True, a notion of superposition is also present in classical mechanics (e.g., in classical wave phenomena), but what is superposed in classical examples is quite different from what is superposed in quantum mechanics.

> The mathematical representative of the *negative* of any experimental proposition is the *orthogonal complement* of the mathematical representative of the proposition itself.

For the sake of simplicity, let us refer here to the structure of all closed subspaces. The orthogonal complement X' of a closed subspace X is defined as the set of all vectors that are orthogonal to all elements of X. In other words, $\psi \in X'$ iff $\psi \perp X$ iff, for any $\varphi \in X$, $\langle\psi|\varphi\rangle = 0$. From the point of view of the physical interpretation, the orthogonal complement (called also *orthocomplement*) is particularly interesting, since it satisfies the following property: for any event X and any pure state ψ,

$$p(P_{[\psi]}, X) = 1 \text{ iff } p(P_{[\psi]}, X') = 0;$$

$$p(P_{[\psi]}, X) = 0 \text{ iff } p(P_{[\psi]}, X') = 1.$$

In other words, ψ assigns to an event X probability 1 (0, respectively) iff ψ assigns to the orthocomplement of X probability 0 (1, respectively). As a consequence, one is dealing with an operation that *inverts* the two extreme probability-values, which naturally correspond to the truth-values *truth* and *falsity* (as in the classical truth-table of negation).

As for conjunction, Birkhoff and von Neumann point out that this can be still represented by the set-theoretic intersection (as in the classical case). For, the intersection $X \cap Y$ of two closed subspaces is again a closed subspace. Hence, we will obtain the usual truth-table for the connective *and*:

$$\psi \text{ verifies } X \cap Y \text{ iff } \psi \text{ verifies both members.}$$

Disjunction, however, cannot be represented here as a set-theoretic union. For, generally, the union $X \cup Y$ of two closed subspaces is not a closed subspace, except in special circumstances. In spite of this, we have at our disposal another good representative for the connective *or*: the *supremum* (in the lattice of all closed subspaces) $X \vee Y$ of two closed subspaces, that is the smallest closed subspace including both X and Y. Of course, $X \vee Y$ will include $X \cup Y$.

As a consequence, we obtain the following structure:

$$\langle \mathcal{C}(\mathcal{H})\,, \wedge\,, \vee\,, {}'\,, \mathbf{0}\,, \mathbf{1} \rangle\,,$$

where $\wedge$ is the set-theoretic intersection; $\vee\,,\,'$ are defined as above; while $\mathbf{0}$ and $\mathbf{1}$ represent, respectively, the null subspace (the singleton consisting of the null vector, which is the smallest possible subspace) and the total space $\mathcal{H}$.

The structure $\mathcal{C}(\mathcal{H})$ turns out to simulate a "quasi-Boolean behavior"; however, it is not a Boolean algebra. Something very essential is missing: conjunction and disjunction are not distributive. Generally,

$$X \wedge (Y \vee Z) \neq (X \wedge Y) \vee (X \wedge Z).$$

One can prove that $\mathcal{C}(\mathcal{H})$ belongs to the class of all *orthomodular lattices* (which are not necessarily distributive [see Definition 1.1.11]). If the dimension of $\mathcal{H}$ is finite, then (and only then) $\mathcal{C}(\mathcal{H})$ is also modular (see Definition 1.1.12).

The failure of distributivity is connected with a characteristic property of disjunction in QT. As opposed to classical (bivalent) semantics, a quantum disjunction $X \vee Y$ may be true even if neither member is true. In fact, it may happen that a pure state ψ belongs to a closed subspace $X \vee Y$, even if ψ belongs neither to X nor to Y. As an example consider the situation represented by Figure 1.4.1 (which refers to the Hilbert space $\mathbb{R}^2$): the vector ψ does not belong either to the (1-dimensional) closed subspace X or to the (1-dimensional) closed subspace Y. At the same time ψ does belong to the supremum $X \vee Y$ (which coincides with the total space $\mathbb{R}^2$).

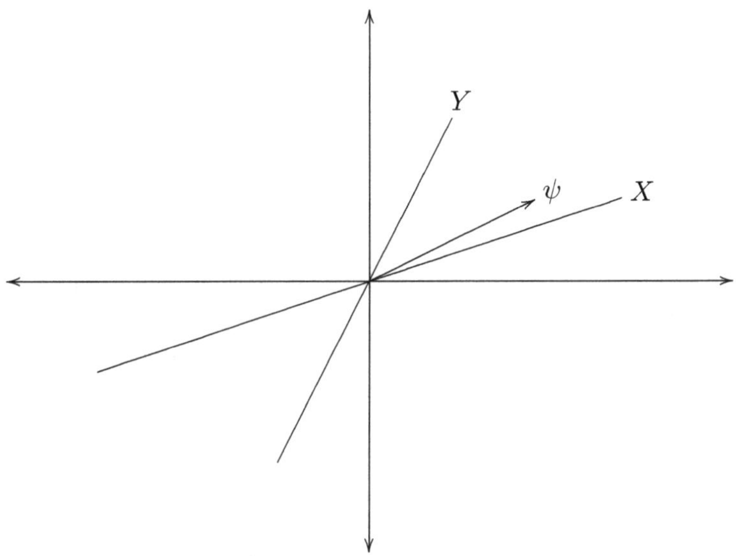

FIGURE 1.4.1. Failure of bivalence in QT

Such a semantic behavior, which may appear *prima facie* somewhat strange, seems to reflect pretty well a number of concrete quantum situations. In QT one is often dealing with alternatives that are semantically determined and true, while both members are, in principle, indeterminate. For instance, suppose we are referring to a spin one-half particle (say an electron) whose spin in a certain direction may assume only two possible values: either *up* or *down*. Now, according to one of the *uncertainty principles*, the spin in the x direction ($spin_x$) and the spin in the y direction ($spin_y$) represent two *incompatible* quantities that cannot be simultaneously measured. Suppose an electron in state ψ verifies the proposition "$spin_x$ is up". As a consequence of the uncertainty principle both propositions "$spin_y$

is up" and "$spin_y$ is down" shall be indeterminate. However the disjunction "either $spin_y$ is up or $spin_y$ is down" must be true.

Interestingly enough, this characteristic feature of quantum logic had been already considered by Aristotle. One of Łukasiewicz' celebrated contributions to the history of logic was the discovery that Aristotle was *the first many-valued logician*. Following this line of thought, one could reasonably add that Aristotle was, in a sense, even *the first quantum logician*. Let us refer to Łukasiewicz' analysis of the 9-th chapter of Aristotle's *De Interpretatione*. We are dealing with the famous example concerning the *sea-battle question*. According to Łukasiewicz' interpretation, Aristotle seems to assert that both the sentence

Tomorrow there will be a sea-battle,

and its negation

Tomorrow there will not be a sea-battle

have today no definite truth-value. At the same time the disjunction

Either tomorrow there will be a sea-battle
or
tomorrow there will not be a sea-battle

is today (and always) true.

In other words, Aristotle seems to be aware of the necessity of distinguishing the *logical law of the excluded middle* from the *semantic bivalence principle*. As a consequence, we obtain the possibility of a typical quantum logical situation:

the truth of a disjunction does not generally imply
the truth of at least one member.

1.5. The logico-algebraic approach to QT

Birkhoff and von Neumann's proposal did not arouse any immediate interest, neither in the logical nor in the physical community. Probably, the quantum logical approach appeared too abstract for the foundational debate about QT, which in the Thirties was generally formulated in a more traditional philosophical language. As an example, we need only recall the famous discussion between Einstein and Bohr. At the same time, the work of logicians was still mainly devoted to classical logic.

Strangely enough, von Neumann himself was not completely satisfied with the basic idea of the quantum logical approach, according to which the lattice of all closed subspaces of a Hilbert space should faithfully represent the structure of quantum propositions. The basic reason was the failure of modularity in the infinite dimensional case. In fact, von Neumann considered this property as an essential condition for a frequency-interpretation of quantum probabilities.[19]

---

[19] An interesting analysis of von Neumann's doubts and critiques can be found in (Rédei, 1996).

Only twenty years later, after the appearance of George Mackey's book *Mathematical Foundations of Quantum Theory* (Mackey, 1957), one has witnessed a "renaissance period" for the logico-algebraic approach to QT. This has been mainly stimulated by the researches of Jauch, Piron, Varadarajan, Suppes, Finkelstein, Foulis, Randall, Greechie, Gudder, Beltrametti, Cassinelli, Mittelstaedt and many others. The new proposals are characterized by a more general approach, based on a kind of abstraction from the Hilbert space structures. The starting point of the new trends can be summarized as follows. Generally, any physical theory **T** determines a class of *event-state* systems $(\mathcal{E}, \mathcal{S})$, where $\mathcal{E}$ contains the events that may occur to a given system, while $\mathcal{S}$ contains the states that such a physical system described by the theory may assume. The question arises: what are the abstract conditions that one should postulate for any pair $(\mathcal{E}, \mathcal{S})$? In the case of QT, having in mind the Hilbert space model, one is naturally led to the following requirement:

- the set $\mathcal{E}$ of events should be a good abstraction of the structure of all closed subspaces in a Hilbert space. As a consequence, $\mathcal{E}$ should be at least a σ-orthocomplete orthomodular lattice (generally nondistributive).
- The set $\mathcal{S}$ of states should be a good abstraction from the *density operators* in a Hilbert space, that represent possible states of physical systems. As a consequence, any state shall determine a *probability measure*, that assigns to any event in $\mathcal{E}$ a value in the interval $[0, 1]$. Both in the concrete and in the abstract case, states may be either *pure* (maximal pieces of information that cannot be consistently extended to a richer knowledge) or *mixtures* (non-maximal pieces of information).

In such a framework two basic problems arise:

I) Is it possible to capture, by means of some abstract conditions that are required for any event-state pair $(\mathcal{E}, \mathcal{S})$, the behavior of the concrete Hilbert space pairs?

II) To what extent should the Hilbert space model be absolutely binding?

The first problem gave rise to a number of attempts to prove a kind of *representation theorem*. More precisely, the main question was: what are the necessary and sufficient conditions for a generic event-state pair $(\mathcal{E}, \mathcal{S})$ that make $\mathcal{E}$ isomorphic to the lattice of all closed subspaces of a Hilbert space? The representation problem has been successfully solved only in 1995 by an important theorem proved by M.P. Solèr (see Chapter 3).

The second problem stimulated the investigation about more and more general quantum structures. Of course, looking for more general structures seems to imply a kind of discontent towards the standard quantum logical approach, based on Hilbert space lattices. The fundamental criticisms that have been raised concern the following items:

1) The standard structures seem to determine a kind of *extensional* collapse. In fact, the closed subspaces of a Hilbert space represent at the same time *physical properties* in an *intensional sense* and the *extensions* thereof (sets of states that certainly verify the properties in question). As happens in classical set-theoretic semantics, there is no mathematical representative for physical properties in an intensional sense. Foulis and Randall (1983) have called such an extensional collapse "the metaphysical disaster" of the standard quantum logical approach.
2) The lattice structure of the closed subspaces automatically renders the quantum event system closed under logical conjunction. This seems to imply some counterintuitive consequences from the physical point of view. Suppose two experimental propositions that concern two incompatible quantities, like "the spin in the x direction is up", "the spin in the y direction is down". In such a situation, the intuition of the quantum physicist seems to suggest the following semantic requirement: the conjunction of such propositions has no definite meaning; for, they cannot be experimentally tested at the same time. As a consequence, a lattice structure for the event system seems to be too strong.

An interesting weakening can be obtained by giving up the lattice condition: generally the *supremum* is assumed to exist only for countable sets of events that are pairwise orthogonal. In other words, the event-structure is supposed to be a σ-orthocomplete orthomodular poset (which is not necessarily a lattice). In the recent quantum logical literature such a structure has been often simply called a *quantum logic*. At the same time, by *standard quantum logic* one usually means the complete orthomodular lattice based on the closed subspaces in a Hilbert space. Needless to say, such a terminology that identifies a *logic* with a particular example of an algebraic structure turns out to be somewhat misleading from the strict logical point of view. As we will see in Part II, different forms of quantum logic, which represent "genuine logics" according to the standard way of thinking of the logical tradition, can be characterized by convenient abstraction from the physical models.

CHAPTER 2

Abstract axiomatic foundations of sharp QT

We will now develop an abstract analysis of the basic conceptual structures of QT in a framework that is relatively independent of the Hilbert space formalism. On this basis, orthodox QT will be reconstructed as a particular model of such a general formal approach.

Let us first discuss, from a general point of view, the basic conceptual "ingredients" that occur in any physical theory, where statistical notions play an important role. We will call such theories *statistical physical theories*.[1]

Any statistical physical theory **T** involves some fundamental physical notions that can be assumed as *primitive* (*undefined*).

- The notion of *state* of the physical system under investigation. From an intuitive point of view, any state represents an abstract object that sums up the observer's information. Generally a state does not refer to a *single* system, but rather to a *class* of preparations of individual samples of a certain physical entity under well defined and repeatable conditions. As a limit case, the class may correspond to a *singleton*.
- The notion of *observable* or *physical quantity* that can be measured on the physical systems under consideration.
- The notion of *probability*.

We will use the following notation.

(i) $\mathcal{S}$ will represent a (nonempty) set of states; while $u, v, w, ...$ (possibly with indices) will range over the elements of $\mathcal{S}$.

(ii) $\mathcal{O}$ will represent a (nonempty) set of observables; while $A, B, C, ...$ will range over the elements of $\mathcal{O}$.

(iii) p will represent a *generalized probability function* that assigns to any triple consisting of a state, of an observable and of a Borel set of the real line, a probability-value in $[0, 1]$. In other words, p will be a map such that:

$$p : \mathcal{S} \times \mathcal{O} \times \mathcal{B}(\mathbb{R}) \mapsto [0, 1]$$

(where $\mathcal{B}(\mathbb{R})$ is the set of all Borel subsets of the real line $\mathbb{R}$). For any triple $(w, A, \Delta) \in \mathcal{S} \times \mathcal{O} \times \mathcal{B}(\mathbb{R})$, the real number

$$p(w, A, \Delta) \in [0, 1]$$

[1]See (Cattaneo and Laudisa, 1994).

> will be interpreted as "the *probability* that a measurement of the physical quantity A for the physical system prepared in state w gives a value contained in the set Δ."

Any triple $(\mathcal{S}, \mathcal{O}, p)$, associated to this canonical physical interpretation, will be called a *state-observable probability system.* Any statistical physical theory $\mathbf{T}$ can be then characterized by a convenient class $\mathbf{K}$ of state-observable probability systems, where each element $(\mathcal{S}, \mathcal{O}, p)$ is supposed to satisfy some well determined conditions.

Suppose that we have any state-observable probability system $(\mathcal{S}, \mathcal{O}, p)$, in the framework of a given statistical theory $\mathbf{T}$, and let us ask: what might be the intuitive meaning of a pair (A, Δ), where A is an observable and Δ is a Borel set? A natural answer to this question is provided by Beltrametti and Cassinelli (1981):

> From the physical point of view, we may picture the pair (A, Δ) as the *experimental device* that is obtained from the instrument used to measure the physical quantity A by attaching to the reading scale a window that isolates the numerical subset Δ; according as the pointer of the instrument does or does not appear within the window, one may answer 'yes' or 'no' to the *question*
>
> Is the measured result of A contained in Δ?

What Beltrametti and Cassinelli call an *experimental device* clearly corresponds to what Birkhoff and von Neumann had called *experimental proposition*. One is dealing with a *question* that can be checked in the case of a physical system in a given state. After the test, the answer will be either "yes" or "no". Accordingly, we will adopt the following terminology: we will say that any pair (A, Δ) represents an *experimental device* (or a *yes-no device* or an *experimental question* or simply a *question*). We will use the letters $q, q_1, q_2, ...$ as ranging over the set $\mathcal{Q}$ of all possible questions.

Thus, any state w will determine a map p_w that assigns to any question (A, Δ) a probability-value in $[0, 1]$. The natural definition of p_w is the following: for any question (A, Δ) in $\mathcal{Q}$,

$$p_w((A, \Delta)) := p(w, A, \Delta).$$

2.1. Mackey's minimal axiomatization of QT

As we have mentioned in the Introduction, Mackey's book *Mathematical Foundations of Quantum Theory* has marked the most significant turning point for the abstract investigations about QT, since the Sixties. It will be expedient to sum up the basic ideas of Mackey's approach.

In this framework, abstract QT can be described as a particular example of a statistical physical theory. As a consequence, one has to define a class $\mathbf{K}$ of state-observable probability systems $(\mathcal{S}, \mathcal{O}, p)$ that are governed by a convenient set of axioms. We will first formally present the axioms; then we will comment on them from an intuitive point of view.

DEFINITION 2.1.1. *Mackey state-observable probability system*
A *Mackey state-observable probability system* is a triple $(\mathcal{S}, \mathcal{O}, p)$ where $\mathcal{S}$ represents a nonempty set of states, $\mathcal{O}$ represents a nonempty set of observables, p represents a probability-function such that:

$$p(w, A, \Delta) \in [0, 1], \quad \forall w \in \mathcal{S}\ \forall A \in \mathcal{O}\ \forall \Delta \in \mathcal{B}(\mathbb{R}),$$

and the following axioms (M1-M7) hold.

Axiom M1.

(i) $\forall w \in \mathcal{S}\ \forall A \in \mathcal{O}:$

$$p(w, A, \emptyset) = 0, \quad p(w, A, \mathbb{R}) = 1.$$

(ii) Let $\{\Delta_n\}_{n \in I}$ be a countable set of pairwise disjoint Borel subsets of the real line (i.e., $\Delta_i \cap \Delta_j = \emptyset$, if $i \neq j$). Then, $\forall w \in \mathcal{S}\ \forall A \in \mathcal{O}:$

$$p(w, A, \bigcup \{\Delta_n\}_{n \in I}) = \sum_n p(w, A, \Delta_n).$$

Axiom M2.

(i) If $\forall w \in \mathcal{S}\ \forall \Delta \in \mathcal{B}(\mathbb{R})\ [p(w, A, \Delta) = p(w, B, \Delta)]$, then $A = B$.
(ii) If $\forall A \in \mathcal{O}\ \forall \Delta \in \mathcal{B}(\mathbb{R})\ [p(w, A, \Delta) = p(v, A, \Delta)]$, then $w = v$.

Axiom M3.

Let $A \in \mathcal{O}$ and let $f : \mathbb{R} \mapsto \mathbb{R}$ be any Borel function. Then, there exists an observable $B \in \mathcal{O}$ such that $\forall w \in \mathcal{S}\, \forall \Delta \in \mathcal{B}(\mathbb{R})$:

$$p(w, B, \Delta) = p(w, A, f^{-1}(\Delta)).$$

The observable B, which is uniquely determined by A (owing to Axiom M2(i)), will be denoted by $f(A)$.

Axiom M4.

Let $\{w_n\}_{n \in I}$ be a countable set of states in $\mathcal{S}$ and let $\{\lambda_n\}_{n \in I}$ be a countable set of nonnegative real numbers such that $\sum_n \lambda_n = 1$. Then, there exists a state $w \in \mathcal{S}$ such that $\forall A \in \mathcal{O}\, \forall \Delta \in \mathcal{B}(\mathbb{R})$:

$$p(w, A, \Delta) = \sum_n \lambda_n p(w_n, A, \Delta).$$

The state w is called a *convex combination* (or *mixture*) of the states w_n, where each λ_n represents the *weight* of the corresponding state w_n.

Axiom M5.

Let $\{A_n\}_{n \in I}$ be a countable set of observables and let $\{\Delta_n\}_{n \in I}$ be a countable set of Borel sets. Suppose the following condition holds:

$$\forall w \in \mathcal{S}\,[p(w, A_i, \Delta_i) + p(w, A_j, \Delta_j) \leq 1]\,, \quad \text{whenever } i \neq j.$$

Then, there exists an observable $A \in \mathcal{O}$ and there exists a Borel set Δ such that

$$p(w, A, \Delta) = \sum_n p(w, A_n, \Delta_n).$$

Based on the above, the notion of *event* can be defined as a special kind of observable.

DEFINITION 2.1.2. *Event*
An observable $A \in \mathcal{O}$ is called an *event* iff $\forall w \in \mathcal{S}$: $p(w, A, \{0, 1\}) = 1$.

In other words, for any state, the probability that an event has either value 1 or value 0 is 1.

We will indicate the set of all events by $\mathcal{E}_v$, and we will use the letters $E, F, ...$ as ranging over $\mathcal{E}_v$.

One immediately realizes that any state w determines a map

$$m_w : \mathcal{E}_v \mapsto [0, 1], \text{ such that } \forall E \in \mathcal{E}_v : m_w(E) = p(w, E, \{1\}).$$

In other words, $m_w(E)$ represents the probability-value that the event E occurs for the physical system in state w.

The set $\mathcal{E}_v$ turns out to be partially ordered by the following relation:

DEFINITION 2.1.3. *The event order*

$$\forall E, F \in \mathcal{E}_v : E \leq F \text{ iff } \forall w \in \mathcal{S}[\ m_w(E) \leq m_w(F)].$$

Consider now an observable A in $\mathcal{O}$ and a Borel set Δ with characteristic function χ_Δ. Since χ_Δ is a Borel function, by Axiom M3, $\chi_\Delta(A)$ is an observable belonging to $\mathcal{O}$. Such an observable is also an event, because for any state w we have:

$$p(w, \chi_\Delta(A), \{0, 1\}) = p(w, A, \chi_\Delta^{-1}(\{0, 1\})) = p(w, A, \mathbb{R}) = 1.$$

As a consequence, any observable A determines the following *event-valued map*:

$$M_A : \mathcal{B}(\mathbb{R}) \mapsto \mathcal{E}_v, \text{ such that } \forall \Delta \in \mathcal{B}(\mathbb{R}) : M_A(\Delta) = \chi_\Delta(A).$$

Such a map is an *event-valued measure* on the Borel set structure. In other words, M_A is a σ-homomorphism of $\mathcal{B}(\mathbb{R})$ into $\mathcal{E}_v$, that satisfies the following conditions:

(i) $M_A(\emptyset) = \mathbf{0}$, $M_A(\mathbb{R}) = \mathbf{1}$;
(ii) $\forall \Delta, \Gamma \in \mathcal{B}(\mathbb{R})$: if $\Delta \cap \Gamma = \emptyset$, then $M_A(\Delta) \perp M_A(\Gamma)$;

(iii) Let $\{\Delta_i\}_{i\in I}$ be a countable set of real Borel sets such that, for $j \neq k, \Delta_j \cap \Delta_k = \emptyset$. Then:

$$M_A\left(\bigcup\{\Delta_i\}_{i\in I}\right) = \bigvee\{M_A(\Delta_i)\}_{i\in I}.$$

Axiom M6.

For an any event-valued measure M, there is an observable A in $\mathcal{O}$ such that $M = M_A$.

In other words, any event-valued measure M is associated to an observable A such that for any Borel set Δ: $M(\Delta) = \chi_\Delta(A)$.

Axiom M7. The partially ordered set of all events is isomorphic to the partially ordered set of all closed subspaces of a separable infinite dimensional Hilbert space over the complex numbers.

Let us now try to illustrate the intuitive meaning of these axioms.

Axiom M1 asserts that, for any choice of a particular state w and of a particular observable A, the map p gives rise to a probability measure defined on the set of all Borel sets of the real line.

As a consequence, any state w and any observable A uniquely determine a classical probability measure $\mu_{(A,w)}$ that is defined on the σ-complete Boolean algebra $\mathcal{B}(\mathbb{R})$.

The expected definition of $\mu_{(A,w)}$ will be:

$$\forall\Delta \in \mathcal{B}(\mathbb{R}),\ \mu_{(A,w)}(\Delta) = p(w, A, \Delta) = p_w((A, \Delta)).$$

Axiom M2 codifies a "principle of parsimony" (*entia non sunt multiplicanda praeter necessitatem!*). Observables and states that are statistically indistinguishable are identified.

Axiom M3 permits one to "create" new observables by a convenient application of Borel functions to observables that belong to $\mathcal{O}$. Let f be a Borel function. Then, the inverse image $f^{-1}(\Delta) = \{x \in \mathbb{R} : f(x) \in \Delta\}$ of a Borel set Δ is a Borel set. Now, for any observable A, Axiom M3 and Axiom M2(i) guarantee the existence and the uniqueness of the observable $f(A)$ such that for any Δ:

$$p(w, f(A), \Delta) = p(w, A, f^{-1}(\Delta)).$$

For instance, consider the Borel function $f(x) = x^2$, and let A be an observable in $\mathcal{O}$. Then, the observable $A^2 = f(A)$ also belongs to $\mathcal{O}$ and we have: $p(w, A^2, \Delta) = p(w, A, \{x : x^2 \in \Delta\})$.

Axiom M4 permits one to "create" new states by using σ-convex combinations of states belonging to $\mathcal{S}$.

Axiom M5 represents a quite strong postulate (which will generally fail in the case of unsharp QT). This axiom asserts a σ-orthocompleteness and a σ-additivity principle. Suppose we have a countable set of questions (A_n, Δ_n)

such that for any pair of distinct questions (A_i, Δ_i) and (A_j, Δ_j) and for any state w:

$$p_w((A_i, \Delta_i)) + p_w((A_j, \Delta_j)) \leq 1.$$

Then, there exists a question (A, Δ) that behaves as a sort of supremum of the questions (A_i, Δ_i). The probability-value $p_w(A, \Delta)$ is just the sum over all probability-values $p_w(A_n, \Delta_n)$.

What is the role of Axiom M6? Axioms M1-M5 guarantee that the set $\mathcal{O}$ of all observables is in one-to-one correspondence with the set $\mathcal{M}(\mathcal{O})$ of all event-valued measures M_A that are determined by an observable A. A natural question arises: does $\mathcal{M}(\mathcal{O})$ *exhaust* the set of *all* possible event-valued measures defined on $\mathcal{E}_v$? One can easily realize that a negative answer to this question is compatible with the Axioms M1-M5. However, according to Mackey, there are very plausible physical reasons to assume that *any* event-valued measure on $\mathcal{E}_v$ is determined by an observable. This is just the content of Axiom M6.

Axiom M7 represents the most critical axiom of Mackey's approach. Because of its apparent *ad hoc*-character, such a principle has not generally been accepted as a reasonable axiom. In fact, even Mackey noticed:

> Ideally one would like to have a list of physically plausible assumptions from which one could deduce Axiom M7.

We will return to this question in Chapter 3. In the present Chapter we will not refer again to Axiom M7.

It is worthwhile noticing that Mackey's Axioms M1-M6 correspond to a "weak" part of QT, where neither the dynamical evolution of quantum systems (Schrödinger equation) nor the critical *collapse of the wave-function* (von Neumann-Lüders postulate) are taken into consideration. Accordingly, we will also term such an axiomatization a *minimal abstract version of sharp QT*.

2.2. Events

As we have seen, in the framework of Mackey's axiomatization, events are defined as a special kind of observable. Unlike "normal" observables (whose possible values generally range over the whole $\mathbb{R}$) an event can be described as a *dichotomous observable* that only admits two possible values: a (logically) *positive value* (corresponding to the case where the event does occur) and a (logically) *negative value* (corresponding to the case where the event does not occur). Conventionally, one can identify the positive value with the number 1, while the negative value will be represented by the number 0.

The following two Lemmas, contained in Mackey's formulation, describe some characteristic properties of events.[2]

[2]See also (Gudder, 1979; Pták and Pulmannová, 1991).

LEMMA 2.2.1. *Let E be an event. For any state $w \in \mathcal{S}$, for any Borel set Δ:*

$$p(w, E, \Delta) = \begin{cases} 0 & \textit{if } \{0,1\} \subseteq \Delta^c, \\ 1 & \textit{if } \{0,1\} \subseteq \Delta, \\ p(w, E, \{1\}) & \textit{if } 1 \in \Delta\,, 0 \notin \Delta, \\ 1 - p(w, E, \{1\}) & \textit{if } 1 \notin \Delta\,, 0 \in \Delta. \end{cases}$$

LEMMA 2.2.2. *Let A be an observable. Then, A is an event iff $A = A^2$.*

Is the set of all possible events always nonempty? We have already seen that the answer to this question is yes. Take any observable A in $\mathcal{O}$ and take any Borel set Δ with characteristic function χ_Δ. Since χ_Δ is a Borel function, by Axiom M3, $\chi_\Delta(A)$ is an observable belonging to $\mathcal{O}$. Such an observable is also an event, satisfying the relation:

$$p(w, \chi_\Delta(A), \{0,1\}) = p(w, A, \chi_\Delta^{-1}(\{0,1\})) = p(w, A, \mathbb{R}) = 1.$$

As expected, there is a natural correspondence between questions and events. Take the question (A, Δ) and consider the event $\chi_\Delta(A)$. For any state w we have:

$$p(w, A, \Delta) = p(w, \chi_\Delta(A), \{1\}).$$

As a consequence, we say that the question (A, Δ) is represented by the event $\chi_\Delta(A)$. Conversely, any event E represents the question $(E, \{1\})$, because any event E turns out to have the form $\chi_{\{1\}}(E)$ as shown by the following lemma.

LEMMA 2.2.3. *For every event E:*

$$E = \chi_{\{1\}}(E).$$

Thus, an observable A is an event iff there exists a question (B, Δ) in $\mathcal{Q}$ such that $A = \chi_\Delta(B)$.

Notice, that generally the correspondence between the set of all questions and the set of all events is many-to-one. As an example, think of two different observables A and B. The two questions $(A, \emptyset)$ and $(B, \emptyset)$ are different. However:

$$\chi_\emptyset(A) = \chi_\emptyset(B).$$

As a consequence of Lemma 2.2.1, for any event E, the probability distribution $\mu_{(E,w)}$ is completely determined by the value $\mu_{(E,w)}(\{1\})$.

We will now investigate the structural properties of the set $\mathcal{E}_v$ of all events.

Let us first prove that the relation $\leq$ (we have defined on $\mathcal{E}_v$) is a partial order. According to Definition 2.1.3, $E \leq F$ iff all states assign to E a probability-value that is less than or equal to the probability-value assigned to F.

LEMMA 2.2.4. *Relation $\leq$ is a partial order.*

PROOF. That $\leq$ is reflexive and transitive is trivial. Antisymmetry follows from Lemma 2.2.1 and from Axiom M2. □

The event-poset $\langle \mathcal{E}_v , \leq \rangle$ turns out to be bounded. The minimum element $\mathbf{0}$ and the maximum element $\mathbf{1}$ can be defined as follows:

DEFINITION 2.2.5. *The minimum and the maximum*

(i) $\mathbf{0} = \chi_{\emptyset}(A)$, where A is any observable;
(ii) $\mathbf{1} = \chi_{\mathbb{R}}(A)$, where A is any observable.

Notice that both $\chi_{\emptyset}(A)$ and $\chi_{\mathbb{R}}(A)$ are observables in $\mathcal{O}$ and they are independent of the choice of the observable A.

One immediately realizes that $\mathbf{0}$ and $\mathbf{1}$ are the minimum and the maximum events with respect to the partial order $\leq$. This is because:

$$\forall w \in \mathcal{S} : m_w(\mathbf{0}) = p(w, A, \emptyset) = 0; \;\; m_w(\mathbf{1}) = p(w, A, \mathbb{R}) = 1,$$

so that for all events E:

$$m_w(\mathbf{0}) \leq m_w(E) \leq m_w(\mathbf{1}), \;\text{ i.e.,}$$
$$\mathbf{0} \leq E \leq \mathbf{1}.$$

As a consequence, the structure $\langle \mathcal{E}_v , \leq , \mathbf{0} , \mathbf{1} \rangle$ is a bounded poset.

Are there any interesting operations that can be defined on this event-poset on the basis Axioms M1-M6?

A unary operation $'$, which turns out to be an orthocomplement, can be naturally defined. Let us first consider the following Borel function f:

$$\forall x \in \mathbb{R} : f(x) = 1 - x.$$

Then, for any observable A, the observable $f(A)$ exists in $\mathcal{O}$.

LEMMA 2.2.6. *Consider the Borel function $f(x) = 1 - x$. Then, for any event E:*

(i) *$f(E)$ is an event;*
(ii) *$f(E) = \chi_{\{0\}}(E)$;*
(iii) *$\forall w \in \mathcal{S} : p(w, f(E), \{1\}) = 1 - p(w, E, \{1\})$.*

PROOF. The proof is straightforward. □

The orthocomplement $'$ can be now defined as follows:

DEFINITION 2.2.7. *The orthocomplement*

Consider the event-poset $\langle \mathcal{E}_v , \leq , \mathbf{0} , \mathbf{1} \rangle$. For any $E \in \mathcal{E}_v$,

$$E' := f(E), \text{ where } f \text{ is the Borel function } f(x) = 1 - x.$$

As a consequence, we obtain for any event E and for any state w:

$$m_w(E') = 1 - m_w(E).$$

LEMMA 2.2.8. *The operation $'$ is an orthocomplement. In other words, $'$ satisfies the following conditions for any $E, F \in \mathcal{E}_v$:*

(i) $E = E''$;
(ii) $E \leq F$ *implies* $F' \leq E'$;
(iii) $E \wedge E' = \mathbf{0}$;
(iv) $E \vee E' = \mathbf{1}$.

PROOF. The proof of (i) and (ii) is trivial. Condition (iv) follows from (iii). We prove (iii).
Let F be an event such that $F \leq E$ and $F \leq E'$. We want to show that $F = \mathbf{0}$. Since $p(w, E, \{1\}) + p(w, E', \{1\}) = 1$, by hypothesis we obtain: $\forall w \in \mathcal{S}$: $p(w, F, \{1\}) + p(w, E, \{1\}) \leq 1$, $p(w, F, \{1\}) + p(w, E', \{1\}) \leq 1$. Thus, by Axiom M5, there is an observable $A \in \mathcal{O}$ and a Borel set $\Delta \in \mathcal{B}(\mathbb{R})$ such that $\forall w \in \mathcal{S}$:

$$p(w, A, \Delta) = p(w, F, \{1\}) + p(w, E', \{1\}) + p(w, E, \{1\}).$$

Hence, $p(w, A, \Delta) = p(w, F, \{1\}) + 1$. Consequently: $\forall w \in \mathcal{S} : p(w, F, \{1\}) = 0$. Therefore, $F = \mathbf{0}$. □

As a consequence, the structure $\langle \mathcal{E}_v\,, \leq, ', \mathbf{0}, \mathbf{1} \rangle$ is a bounded involution poset, which is also an orthoposet. Since the noncontradiction and the excluded middle principles hold, we can conclude that all the events that are described in Mackey's theory are *sharp*.

As we already know (see Section 1.1) an orthogonality relation $\perp$ can be naturally defined in any bounded involution poset. In the case of the event-poset, we will obtain for any $E, F \in \mathcal{E}_v$:

$$E \perp F \text{ iff } \forall w \in \mathcal{S}\, [m_w(E) \leq 1 - m_w(F)] \text{ iff } \forall w \in \mathcal{S}\, [m_w(E) + m_w(F) \leq 1]\,.$$

In other words, two events are orthogonal iff the sum of their probability-values, assigned by any state, does not exceed 1. In fact, one can prove that the event-poset is also a σ-orthocomplete orthomodular poset.

THEOREM 2.2.9. *Consider the event-poset* $\langle \mathcal{E}_v\,, \leq\,, '\,, \mathbf{0}\,, \mathbf{1} \rangle$.

(i) *Let* $\{E_n\}_{n \in I}$ *be a countable set of pairwise orthogonal events (i.e.,* $E_i \perp E_j$, *if* $i \neq j$*). Then, the supremum* $\bigvee \{E_n\}_{n \in I}$ *exists in* $\mathcal{E}_v$.
(ii) *The orthomodular property holds. In other words, for any pair of events* E, F*:* $E \leq F$ *implies* $F \wedge E'$ *and* $E \vee (F \wedge E')$ *exist and*

$$E \leq F \text{ implies } F = E \vee (F \wedge E').$$

PROOF. (i) We prove that $\mathcal{E}_v$ is σ-orthocomplete. Let $\{E_n\}_{n \in I}$ be a countable set of pairwise orthogonal events. Then, by Axiom M5, there is an observable A and a Borel set $\Delta \in \mathcal{B}(\mathbb{R})$ such that $\forall w \in \mathcal{S}$:

$$p(w, A, \Delta) = \sum_{n \in I} p(w, E_n, \{1\}).$$

Thus, $\forall w \in \mathcal{S}$:

$$p(w, \chi_\Delta(A), \{1\}) = \sum_{n \in I} p(w, E_n, \{1\}). \qquad (*)$$

Let us denote $\chi_\Delta(A)$ (the unique event that satisfies equality (*)) by $\sum_{n\in I} E_n$. Clearly, $\sum_{n\in I} E_n$ is an upper bound of $\{E_n\}_{n\in I}$. We want to show that $\sum_{n\in I} E_n$ is the least upper bound of $\{E_n\}_{n\in I}$, i.e., $\sum_{n\in I} E_n = \bigvee \{E_n\}_{n\in I}$. Let G be an event such that $\forall n \in I$: $E_n \leq G$. We have to show that $\sum_{n\in I} E_n \leq G$.
Now, $\forall n \in I$: $E_n \perp G'$. Thus, by Axiom M5, there exists an observable $B \in \mathcal{O}$ and a Borel set $\Gamma \in \mathcal{B}(\mathbb{R})$ such that $\forall w \in \mathcal{S}$:

$$p(w, B, \Gamma) = \sum_{n\in I} p(w, E_n, \{1\}) + p(w, G', \{1\}) \leq 1.$$

Hence,

$$\sum_{n\in I} p(w, E_n, \{1\}) \leq 1 - p(w, G', \{1\}) = p(w, G, \{1\}).$$

Thus, $\sum_{n\in I} E_n \leq G$. Therefore, $\sum_{n\in I} E_n$ is the least upper bound of $\{E_n\}_{n\in I}$.
(ii) We prove that $\mathcal{E}_v$ is orthomodular. Let E, F be two events such that $E \leq F$. We have to prove that $F = E \vee (F \wedge E')$. By hypothesis, $E \perp F'$. By (i), $E \vee F'$ exists and is equal to $E + F'$. Now, by the de Morgan law, $E \leq E + F' = E \vee F' = (F \wedge E')'$. Hence, $E \perp (F \wedge E')$. Then, by (i), $E \vee (F \wedge E')$ exists and is equal to $E + (F \wedge E')$. Thus, $\forall w \in \mathcal{S}$, we have:

$$\begin{aligned} p(w, E + (F \wedge E'), \{1\}) &= p(w, E, \{1\}) + p(w, F \wedge E', \{1\}) \\ &= p(w, E, \{1\}) + 1 - p(w, E + F', \{1\}) \\ &= p(w, E, \{1\}) + 1 - p(w, E, \{1\}) - p(w, F', \{1\}) \\ &= 1 - p(w, F', \{1\}) \\ &= p(w, F, \{1\}). \end{aligned}$$

Consequently, $\forall w \in \mathcal{S}$, $m_w(F) = m_w(E + (F \wedge E'))$. Hence, $F = E \vee (F \wedge E')$. □

As a consequence, in particular, the supremum $E \vee F$ exists for any pair of orthogonal events E and F. From a logical point of view, we can say that the disjunction between two events is defined, provided that the events are *logically incompatible* (the first event implies the negation of the second event).

What happens in the general case, when one is dealing with a pair of events that are not necessarily orthogonal? In other words, can we conclude from the axioms that the event-poset is always a lattice? The answer to this question is negative. As shown by counterexamples, Mackey's axioms are not sufficient to guarantee that the set of all events is closed under $\wedge$ and $\vee$. The simplest such example was discovered by Janowitz. It involves 18 elements, and is therefore called $\mathcal{J}_{18}$. It is the union of four 8-element Boolean subalgebras, B_1, B_2, B_3 and B_4 such that $B_1 \cap B_2$, $B_2 \cap B_3$, $B_3 \cap B_4$, and $B_4 \cap B_1$ are distinct 4-element subalgebras.[3] The Greechie diagram is

[3]See (Greechie, 1971).

given in Figure 2.2.1. The reader is invited to study the poset corresponding

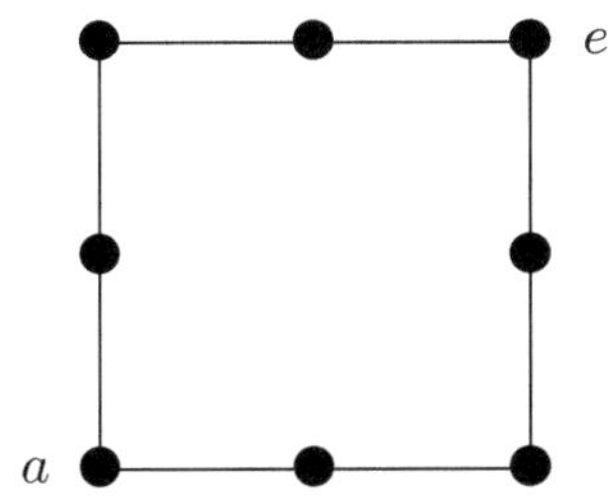

FIGURE 2.2.1. $\mathcal{J}_{18}$: the smallest OMP that is not an OML

to this diagram to the extent that it becomes clear why $a \vee e$ does not exist (perhaps referring to the discussion of Greechie diagrams given after Figure 1.1.3).

In the following, we will call any σ-orthocomplete orthomodular poset an *event algebra*.[4]

The following lemma sums up some basic properties of events, states and observables, in the framework of Mackey's systems.

LEMMA 2.2.10. *Given a system of events, states and observables satisfying Mackey's Axioms M1-M6 we have the following.*

(1) *The set of all events has the structure of an event algebra:*

$$\langle \mathcal{E}_v\,, \leq\,, '\,, \mathbf{0}\,, \mathbf{1} \rangle .$$

(2) *Any state w determines a map*

$$m_w : \mathcal{E}_v \mapsto [0,1].$$

Such a map is a probability measure on the event algebra. In other words:

(i) $m_w(\mathbf{0}) = 0,\ m_w(\mathbf{1}) = 1$;

(ii) *For any countable $\{E_n\}_{n \in I}$ of pairwise orthogonal events:*

$$m_w(\bigvee \{E_n\}_{n\in I}) = \sum_n m_w(E_n).$$

(3) *The set $\mathcal{S}$ of states is in one-to-one-correspondence with set $\mathcal{S}(\mathcal{E}_v)$ of all probability measures m_w that are determined by a state w.*

[4]As mentioned in the Introduction, *event algebras* have been often called *quantum logics* or even *logics* (see for example, (Gudder, 1979; Pták and Pulmannová, 1991; Dvurečenskij, 1993)). A stronger abstract axiomatization of sharp QT has been proposed by the so called "Geneva School" (Jauch, Piron, Aerts, and others). The axioms assumed in the framework of this approach guarantee a lattice-structure for the events (propositions) of a quantum system. See (Jauch, 1968; Piron, 1976; Aerts, 1982; Aerts, 1984).

(4) *The set* $\mathcal{S}$ *of states is* order determining. *In other words, for any pair of events* E *and* F*:*

$$\forall w \in \mathcal{S}\,[m_w(E) \le m_w(F)] \textit{ implies } E \le F.$$

(5) *The set* $\mathcal{S}$ *of states is* σ-convex. *In other words, for any countable set of states,* $\{w_n\}_{n\in I}$ *and for any countable set* $\{\lambda_n\}_{n\in I}$ *such that* $\sum_n \lambda_n = 1$*, there is a state* w *in* $\mathcal{S}$ *such that for any event* E*:*

$$m_w(E) = \sum_n \lambda_n m_{w_n}(E).$$

(6) *Any observable* A *determines an* event-valued map

$$M_A : \mathcal{B}(\mathbb{R}) \mapsto \mathcal{E}_v, \textit{ such that, } \forall \Delta \in \mathcal{B}(\mathbb{R}) : M_A(\Delta) = \chi_\Delta(A).$$

From the intuitive point of view, $M_A(\Delta)$ *represents the event that a measurement of the observable* A *gives a value belonging to* Δ*. Such a map is an* event-valued measure *on the Borel set structure.*

(7) *The set* $\mathcal{O}$ *of all observables is in one-to-one correspondence with the set* $\mathcal{M}(\mathcal{O})$ *of all event-valued measures* M_A *that are determined by an observable* A*.*

(8) *The set* $\mathcal{M}(\mathcal{O})$ *of all event-valued measures* M_A *that are determined by an observable* A *is in one-to-one correspondence with the set of all possible event-valued measures.*

2.3. Event-state systems

As we have seen, according to Mackey's axiomatization, the notions of *state*, *observable* and *probability* are assumed as primitive. *Events* are then defined as special observables.

In this framework, Mackey has analyzed the behavior of the pair $(\mathcal{E}_v, \mathcal{S})$, where $\mathcal{E}_v$ is the set of all events, while $\mathcal{S}$ is the set of states of a given state-observable probability system $(\mathcal{S}, \mathcal{O}, p)$ (governed by his axioms). We will call such a pair a *Mackey event-state system.* As a consequence of Lemma 2.2.10, one immediately obtains that any Mackey event-state system $(\mathcal{E}_v, \mathcal{S})$ satisfies the following properties:

- $\mathcal{E}_v$ has the structure of an event algebra;
- any state w uniquely determines a probability measure m_w defined on the event-algebra;
- $\mathcal{S}$ is order determining;
- $\mathcal{S}$ is σ-convex.

These results have suggested the possibility of trying to invert Mackey's procedure.[5] Instead of beginning with *states*, *observables* and *probability* as fundamental, one assumes only the notion of *event* as the fundamental concept and then defines both the notions of *state* and of *observable* based on *events*.

Consider an event algebra $\langle \mathcal{E}_v\,,\, \le\,,\, '\,,\, \mathbf{0}\,,\, \mathbf{1}\rangle$.

[5] See (Gudder, 1979; Zierler, 1961; Varadarajan, 1985).

DEFINITION 2.3.1. *State*
A *state of* $\mathcal{E}$ is a map

$$s : \mathcal{E}_v \to [0, 1],$$

that is a probability measure on the event algebra. In other words:

(1) $s(\mathbf{0}) = 0,\ s(\mathbf{1}) = 1$;
(2) For any countable $\{E_n\}_{n \in I}$ of pairwise orthogonal events:

$$s(\bigvee \{E_n\}_{n\in I}) = \sum_n s(E_n).$$

DEFINITION 2.3.2. *Observable*
An *observable of* $\mathcal{E}_v$ is an event-valued measure M on the Borel sets. In other words, M is a σ-homomorphism of $\mathcal{B}(\mathbb{R})$ into $\mathcal{E}_v$, that satisfies the following conditions:

(1) $M(\emptyset) = \mathbf{0},\ M(\mathbb{R}) = \mathbf{1}$;
(2) $\forall \Delta, \Gamma \in \mathcal{B}(\mathbb{R})$: if $\Delta \cap \Gamma = \emptyset$, then $M(\Delta) \perp M(\Gamma)$;
(3) If $\{\Delta_i\}_{i \in I}$ is a countable set of real Borel sets such that $\Delta_j \cap \Delta_k = \emptyset$, whenever $j \neq k$, then:

$$M\left(\bigcup \{\Delta_i\}_{i\in I}\right) = \bigvee \{M(\Delta_i)\}_{i\in I}.$$

The intuitive meaning of observables in this framework is quite clear. Any observable M associates to any Borel set Δ a particular event, which will be naturally interpreted as the following state of affairs: "the value of the observable M lies in the Borel set Δ". As expected, it is required that the map M preserves some structural properties of the σ-Boolean algebra of all Borel sets.

By abstracting from Mackey's systems and following an approach developed by Gudder, we will introduce a general notion that we will call *Gudder event-state system.*[6]

DEFINITION 2.3.3. *Gudder event-state system*
A *Gudder event-state system* is a pair $(\mathcal{E}_v, \mathcal{S})$, where:

(1) $\mathcal{E}_v$, the set of events, has the structure of an event algebra (a σ-orthocomplete orthomodular poset) :

$$\langle \mathcal{E}_v\, , \leq , {}' , \mathbf{0} , \mathbf{1} \rangle ;$$

(2) $\mathcal{S}$ is a nonempty set of states of $\mathcal{E}_v$;
(3) $\mathcal{S}$ is order determining for the event-algebra. In other words, for any pair of events E and F,

$$\forall s \in \mathcal{S}\, [s(E) \leq s(F)] \text{ implies } E \leq F.$$

[6]See (Gudder, 1979; Beltrametti and Cassinelli, 1981; Pták and Pulmannová, 1991).

(4) $\mathcal{S}$ is σ-convex. In other words, for any countable set of states, $\{s_n\}_{n\in I}$ and for any countable set $\{\lambda_n\}_{n\in I}$ of nonnegative real numbers such that $\sum_n \lambda_n = 1$, there is a state s in $\mathcal{S}$ such that for any event E:

$$s(E) = \sum_n \lambda_n s_n(E).$$

The state s (called a *convex combination* or *mixture* of the states s_n) will be indicated by $\sum_n \lambda_n s_n$.

One can prove that Mackey's notion of *state-observable probability system* and Gudder's notion of *event-state system* are equivalent.

THEOREM 2.3.4.

(i) *Any Mackey state-observable probability system* $\mathbf{S} = (\mathcal{S},\, \mathcal{O},\, p)$ *uniquely determines a Gudder event-state system*

$$\mathbf{S}^{\blacktriangledown} = (\mathcal{E}_v^{\blacktriangledown}, \mathcal{S}^{\blacktriangledown}),$$

where:

- $\mathcal{E}_v^{\blacktriangledown}$ = *the set* $\mathcal{E}_v$ *of all events of* $(\mathcal{S},\, \mathcal{O},\, p)$*;*
- $\mathcal{S}^{\blacktriangledown}$ = *the set* $\mathcal{S}(\mathcal{E}_v)$ *of all probability measures* m_w *that are determined by the states* w *of* $(\mathcal{S},\, \mathcal{O},\, p)$*.*

(ii) *Any Gudder event-state system* $\mathbf{S} = (\mathcal{E}_v,\, \mathcal{S})$ *uniquely determines a Mackey state-observable probability system*

$$\mathbf{S}^{\blacktriangle} = (\mathcal{S}^{\blacktriangle},\, \mathcal{O}^{\blacktriangle},\, p^{\blacktriangle}),$$

where:

- $\mathcal{S}^{\blacktriangle} = \mathcal{S}$*;*
- $\mathcal{O}^{\blacktriangle}$ = *the set of all observables* M *of* $(\mathcal{E}_v,\, \mathcal{S})$*;*
- $p^{\blacktriangle}(s, M, \Delta) = s(M(\Delta))$*, for any state* s*, any observable* M *and any Borel set* Δ*.*

(iii) *For any Mackey system* $\mathbf{S}$*,* $\mathbf{S}$ *is equal (up to isomorphism) to* $\mathbf{S}^{\blacktriangledown\blacktriangle}$*.*

(iv) *For any Gudder system* $\mathbf{S}$*,* $\mathbf{S}$ *is equal (up to isomorphism) to* $\mathbf{S}^{\blacktriangle\blacktriangledown}$*.*

PROOF.
(i) By Lemma 2.2.4, Lemma 2.2.8 and Theorem 2.2.9, $\mathcal{E}_v^{\blacktriangledown}$ is a σ -orthocomplete orthomodular poset. As a consequence of Lemma 2.2.10, $\mathcal{S}^{\blacktriangledown}$ is an order determining and σ-convex set of probability measures on $\mathcal{E}_v^{\blacktriangledown}$.
(ii) The validity of Axiom M1 follows from the fact that every $s \in \mathcal{S}^{\blacktriangle}$ is a probability measure on $\mathcal{E}_v$ and every $M \in \mathcal{O}^{\blacktriangle}$ is an $\mathcal{E}_v$-valued measure.
The validity of Axiom M2(i) follows from the fact that $\mathcal{S}$ is an order determining set of probability measures on $\mathcal{E}_v$. It remains to prove the validity of Axiom M2(ii). Let $s, t \in \mathcal{S}^{\blacktriangle}$. Suppose $\forall M \in \mathcal{O}^{\blacktriangle} \forall \Delta \in \mathcal{B}(\mathbb{R})$: $p(s, M, \Delta) = p(t, M, \Delta)$. Then, $s(M(\Delta)) = t(M(\Delta))$. Now, any element E of $\mathcal{E}_v$ determines an $\mathcal{E}_v$-valued measure M_E such that $M_E(\{1\}) = E$ and $M_E(\{0\}) = E'$. Thus, by hypothesis, $\forall E \in \mathcal{E}_v$: $s\,(M_E(\{1\})) = t\,(M_E(\{1\}))$. Therefore, $\forall E \in \mathcal{E}_v : s(E) = t(E)$. Hence, $s = t$.

The proof of the validity of Axiom M3 is straightforward, because $\mathcal{O}^{\blacktriangledown}$ is the set of *all* possible observables of $(\mathcal{E}_v, \mathcal{S})$.
The proof of the validity of Axiom M4 follows from the σ-convexity of $\mathcal{S}$.
We prove the validity of Axiom M5.
Let $\{M_n\}_{n\in I}$ be a countable set of $\mathcal{E}_v$-valued measures in $\mathcal{O}^{\blacktriangle}$ and let $\{\Delta_n\}_{n\in I}$ be a countable set of Borel sets such that $\forall s \in \mathcal{S}\, \forall i, j \in I (i \neq j)$:

$$p(s, M_i, \Delta_i) + p(s, M_j, \Delta_j) \leq 1,$$

i.e.,

$$s(M_i(\Delta_i)) + s(M_j(\Delta_j)) \leq 1.$$

Thus, $\forall i, j \in I(i \neq j)$, $M_i(\Delta_i) \perp M_j(\Delta_j)$. Since $\mathcal{E}_v$ is σ-orthocomplete, the sup of $\{M_n(\Delta_n)\}_{n\in I}$ exists in $\mathcal{E}_v$. Let $F := \bigvee \{M_n(\Delta_n)\}_{n\in I}$. Consider the $\mathcal{E}_v$-valued measure M_F such that $M_F(\{1\}) = F$ and $M_F(\{0\}) = F'$. It turns out that $p(s, M_F, \{1\}) = s\left(M_F\{1\}\right) = s(F) = s\left(\bigvee \{M_n(\Delta_n\}_{n\in I}\right) = \sum_{n\in I} s\left(M_n(\Delta_n)\right) = \sum_{n\in I} p(s, M_n, \Delta_n)$.
The validity of Axiom M6 is straightforward.
The proofs of (iii) and (iv) are also straightforward; the argument essentially uses Axiom M6.

□

As a consequence of this theorem, we can say that the theory we have called the *minimal abstract version of sharp QT* is equivalently characterized both by the class of all Mackey state-observable probability systems and by the class of all Gudder event-state systems. In the following we will prefer to use the framework of event-state systems.

2.4. Event-state systems and preclusivity spaces

So far we have mainly studied how states "act" on events, inducing a particular algebraic structure on the set $\mathcal{E}_v$. There is also an inverse "interaction:" the set of all events induces a *preclusivity space* on the set of all states.

Let us first recall the abstract definition of *preclusivity space*. In graph theory, this notion would translate as having an undirected graph with neither loops nor multiple edges.

DEFINITION 2.4.1. *Preclusivity space*
A *preclusivity space* is a system $(U, \perp)$, where

- U (called the *universe*) is a nonempty set of objects;
- $\perp$ is an irreflexive and symmetric relation defined on U. In other words:
  (i) $\forall x \in U:$ not $x \perp x$;
  (ii) $\forall x, y \in U: x \perp y$ implies $y \perp x$.

In our quantum theoretical applications, we will assume as universe the set $\mathcal{S}$ of all states of a Gudder event-state system $(\mathcal{E}_v, \mathcal{S})$. Hence, from an intuitive point of view, the universe will represent here a set of micro-objects prepared in different states. A preclusivity relation $\perp$ can then be defined by referring to the set $\mathcal{E}_v$ of all events.

DEFINITION 2.4.2. *The preclusivity relation between states*
Given two states $s, t \in \mathcal{S}$,

$$s \perp t \text{ iff } \exists E \in \mathcal{E}_v \left[s(E) = 1 \text{ and } t(E) = 0 \right].$$

In other words, two states are *preclusive* iff they are strongly distinguished by at least one event, which is certain for the first state and impossible for the second state.

One can easily check that $\perp$ is a preclusivity relation:

- $\perp$ is irreflexive (for any state s: not $s \perp s$).
  Since any s is a probability measure on $\mathcal{E}_v$, there exists no event E such that $s(E)$=1 and $s(E) = 0$.
- $\perp$ is symmetric ($s \perp t$ implies $t \perp s$).
  If $s \perp t$, then there exists an event E such that $s(E) = 1$ and $t(E) = 0$. Consequently: $t(E') = 1$ and $s(E') = 0$, so that $t \perp s$.

Every preclusivity space has a natural "twin space" $(\mathcal{S}, \mathcal{R})$, which is a *similarity space*.

DEFINITION 2.4.3. *Similarity space*
A *similarity space* is a system $(U, \mathcal{R})$, where

- U (called the *universe*) is a nonempty set of objects;
- $\mathcal{R}$ is a reflexive and symmetric relation defined on U. In other words:

  (i) $\forall x \in U : x\mathcal{R}x$;
  (ii) $\forall x, y \in U : x\mathcal{R}y$ implies $y\mathcal{R}x$.

The "twin" similarity space of the preclusivity space $(\mathcal{S}, \perp)$ is the space $(\mathcal{S}, \mathcal{R})$, where the similarity relation $\mathcal{R}$ is the negation of the preclusivity relation $\perp$. In other words:

$$\forall x, y \in U : x\mathcal{R}y \text{ iff not } x \perp y.$$

Apparently, the similarity relation $\mathcal{R}$ will have the following meaning:

$$s\mathcal{R}t \text{ iff there is no event } E \text{ such that: } s(E) = 1 \text{ and } t(E) = 0.$$

In other words, s and t are *similar* iff they cannot be strongly distinguished by any event.

We use the following abbreviations:

$$s \perp X \text{ for } \forall t \in X(s \perp t); \quad s\mathcal{R}X \text{ for } \forall t \in X(s\mathcal{R}t).$$

$$X \perp Y \text{ for } \forall t \in X(t \perp Y); \quad X\mathcal{R}Y \text{ for } \forall t \in X(t\mathcal{R}Y).$$

In quantum contexts, the similarity relation (which represents the negation of the orthogonality relation $\perp$) is usually indicated by $\not\perp$. While $\not\perp$ is reflexive and symmetric, it is not generally transitive.

Consider now the power set $\mathcal{P}(\mathcal{S})$ of the set of all states $\mathcal{S}$. The preclusivity relation $\perp$ permits one to define on $\mathcal{P}(\mathcal{S})$ a unary operation $^\perp$ (called the *preclusive complement*), which turns out to be a weak complement. For any set X of states:

$$X^\perp := \{x \in \mathcal{S} : \forall t \in X(s \perp t)\}.$$

The preclusive complement $^\perp$ satisfies the following properties for any sets X, Y of states:

- $X \subseteq X^{\perp\perp}$;
- $X \subseteq Y$ implies $Y^\perp \subseteq X^\perp$;
- $X \cap X^\perp = \emptyset$.

At the same time, the strong double negation principle ($X^{\perp\perp} \subseteq X$) and the excluded middle principle ($X \cup X^\perp = \mathcal{S}$) will generally fail.

Consider now the map $^{\perp\perp} : \mathcal{P}(\mathcal{S}) \mapsto \mathcal{P}(\mathcal{S})$ such that:

$$X \rightarrowtail X^{\perp\perp}, \text{ for any } X \subseteq \mathcal{S}.$$

One can easily check that this map is a *closure operator* satisfying the following conditions:

(i)	$\emptyset^{\perp\perp} = \emptyset$	(normalization)
(ii)	$X \subseteq X^{\perp\perp}$	(increasing)
(iii)	$X^{\perp\perp} = X^{\perp\perp\perp\perp}$	(idempotence)
(iv)	$X \subseteq Y \quad \text{implies} \quad X^{\perp\perp} \subseteq Y^{\perp\perp}$	(monotonicity)

Consider then the set $\mathbb{C}(\mathcal{P}(\mathcal{S}))$ of all *closed* elements of the power set of $\mathcal{S}$. By definition, we have:

$$X \in \mathbb{C}(\mathcal{P}(\mathcal{S})) \text{ iff } X = X^{\perp\perp}.$$

The elements of $\mathbb{C}(\mathcal{P}(\mathcal{S}))$ will be called *closed* sets of states. As we will see, in the following chapters, such sets will play a very significant role both in quantum theory and in quantum logic.

A characteristic property of the closed sets of a preclusivity space is described by the following lemma.

LEMMA 2.4.4. *If $(U, \mathcal{R})$ is a similarity space associated with a preclusivity space $(U, \perp)$, and if X is any subset of U, then, X is closed iff X satisfies the following condition:*

$$\forall x[x \in X \textit{ iff } \forall y\mathcal{R}x\exists z\mathcal{R}y(z \in X)].$$

PROOF. By definition of preclusive complement the following equivalences hold:

$$\begin{aligned} X = X^{\perp\perp} &\text{ iff } \forall x[x \in X \text{ iff } x \perp X^{\perp}] \\ &\text{ iff } \forall x[x \in X \text{ iff } \forall y(y \perp X \text{ implies } y \perp x)] \\ &\text{ iff } \forall x[x \in X \text{ iff } \forall y(\text{not } y \perp x \text{ implies not } y \perp X)] \\ &\text{ iff } \forall x[x \in X \text{ iff } \forall y \mathcal{R} x \exists z y (z \in X)]. \end{aligned}$$

□

The following theorem gives important information about the algebraic structure of $\mathbb{C}(\mathcal{P}(\mathcal{S}))$.

THEOREM 2.4.5. (Birkhoff, 1967)
The structure

$$\left\langle \mathbb{C}(\mathcal{P}(\mathcal{S}))\,,\subseteq\,,^{\perp}\,,\emptyset\,,\mathcal{S}\right\rangle$$

is a complete bounded ortholattice, where for any family $\{X_i\}_{i\in I} \subseteq \mathbb{C}(\mathcal{P}(\mathcal{S}))$:

- *the meet* $\bigwedge\{X_i\}_{i\in I}$ *exists and coincides with* $\bigcap\{X_i\}_{i\in I}$;
- *the join* $\bigvee\{X_i\}_{i\in I}$ *exists and coincides with* $(\bigcup\{X_i\}_{i\in I})^{\perp\perp}$;
- *the preclusive complement* $^{\perp}$ *is an orthocomplement.*

Generally the lattice $\mathbb{C}(\mathcal{P}(\mathcal{S}))$ fails to be distributive. As we will see in the next chapter, interesting counterexamples to such distributivity can be constructed in the framework of Hilbert space QT.

Now, we have focused two special structures:

- the σ-orthocomplete orthomodular poset

$$\langle \mathcal{E}_v\,,\leq\,,'\,,\mathbf{0}\,,\mathbf{1}\rangle$$

based on the set of all events;

- the complete ortholattice

$$\left\langle \mathbb{C}(\mathcal{P}(\mathcal{S}))\,,\subseteq\,,^{\perp}\,,\emptyset\,,\mathcal{S}\right\rangle$$

based on the set of all closed sets of states.

A natural question arises: is there any structural relation between the two structures? If we want to obtain some interesting connections between $\mathcal{E}_v$ and $\mathbb{C}(\mathcal{P}(\mathcal{S}))$, we shall require some further conditions.

To this aim, we will first introduce a special map, called **Yes**, that will associate to any event E the set of all states that assign to E probability-value 1. In other words:

$$\mathbf{Yes} : \mathcal{E}_v \mapsto \mathcal{P}(\mathcal{S}), \text{ where } \forall E \in \mathcal{E},$$

$$\mathbf{Yes}(E) := \{s \in \mathcal{S} : s(E) = 1\}.$$

From an intuitive point of view, the set $\mathbf{Yes}(E)$ represents a kind of *extension* of the event E. If we regard states as *possible worlds* (in the sense of Kripkean semantics), then $\mathbf{Yes}(E)$ can be thought of as the set of all the worlds that *certainly verify* the event E. By adopting a standard semantic jargon, we can say that $\mathbf{Yes}(E)$ represents the *proposition* that is associated to the event E.[7] In the following, we will also call $\mathbf{Yes}(E)$ the *Yes-domain* or the *positive domain* of E.

The map **Yes** is clearly order preserving. In other words, $\forall E, F \in \mathcal{E}_v$,

$$E \leq F \text{ implies } \mathbf{Yes}(E) \subseteq \mathbf{Yes}(F).$$

However, **Yes** is not generally injective. It may happen that two different events have one and the same Yes-domain. This would be the case, for instance, in the following situation:

(i) $\forall s : s(E) = 1$ iff $s(F) = 1$;
(ii) $\exists s : s(E) \neq s(F)$.

An example of an event-state system satisfying conditions (i) and (ii) is based on the orthomodular poset given in Figure 2.4.1. This figure is the

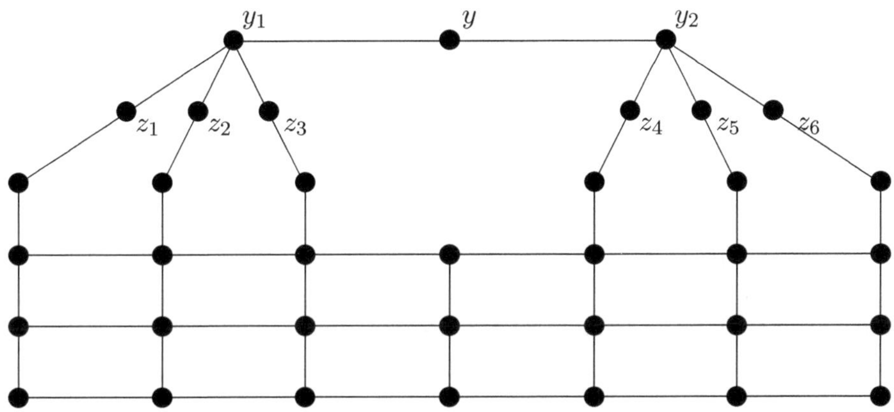

FIGURE 2.4.1. The Greechie diagram of $\mathcal{GGM}_{410}$

Greechie diagram of an orthomodular poset which admits an order determining set of states, yet the possible values of $s(y)$ are precisely $[\frac{1}{3}, 1]$ so that the possible values of $s(y')$ are $[0, \frac{2}{3}]$.[8] It follows that $\mathbf{Yes}(y')$ is empty, so that $\mathbf{Yes}(y')=\mathbf{Yes}(\mathbf{0})$. Figure 2.4.2 provides a state that takes on different values at y' and $\mathbf{0}$. Here are the details.

As we have learned in Section 1.1, each line in the Greechie diagram of an orthomodular poset $\mathcal{B}$ consists of a maximal pairwise orthogonal set of atoms and any such set is a line. Then, the set $\mathcal{L}$ of lines is a set of subsets of the set X of all atoms. It is easy to see that the states on a finite orthomodular poset $\mathcal{B}$ are completely determined by their values on the atoms of $\mathcal{B}$, and therefore can be read off from the Greechie diagram

[7]In the framework of Kripkean semantics, the set of all possible worlds that verify a given sentence is usually called the *proposition* associated to the sentence. See the Introduction to Part II. See also (Goldblatt, 1993).

[8]See (Gerelle, Greechie and Miller, 1974).

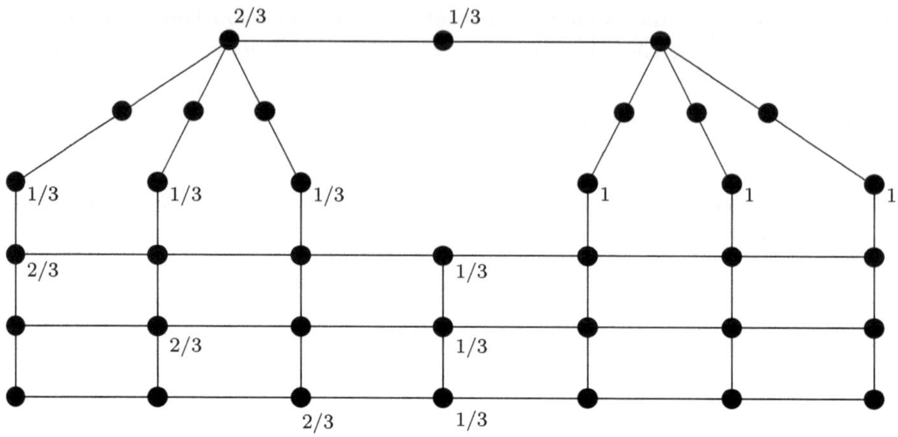

FIGURE 2.4.2. The state s_0 on $\mathcal{GGM}_{410}$

$(X\,,\mathcal{L})$ of $\mathcal{B}$. That is, the states on $\mathcal{B}$ are in one-to-one correspondence with the functions $s : X \to [0,1]$ such that, $\forall L \in \mathcal{L}$,

$$\sum_{x \in L} s(x) = 1.$$

Now consider the set X of all atoms of $\mathcal{GGM}_{410}$. Let s be *any* state on $\mathcal{GGM}_{410}$. Let $\lambda := \sum_{x \in X} s(x)$ and $\xi := \sum_{i=1}^{6} s(z_i)$. Noting that the points that are not labeled z_i consist of the disjoint union of eight lines, so that

$$\lambda = 8 + \xi.$$

Summing over the three 7-element horizontal lines, the six lines labeled x_i, z_i, y_i, and the line $\{y_1, y, y_2\}$ yields

$$\lambda + 3s(y_i) + 3s(y_2) = 3 + 6 + 1$$

after the overcount on y_1 and y_2 is considered. Thus,

$$\lambda = 10 - 3(s(y_1) + s(y_2)) = 10 - 3(1 - s(y)) = 7 + 3s(y).$$

It follows that $8 + \xi = 7 + 3s(y)$ so that $3s(y) = 1 + \xi \geq 1$ and $s(y) \geq \frac{1}{3}$. This forces

$$s(y') \leq \frac{2}{3} < 1$$

in $\mathcal{GGM}_{410}$ so that $\mathbf{Yes}(y') = \mathbf{Yes}(0) = \emptyset$ so that (i) is satisfied vacuously.

Figure 2.4.2 exhibits a state s_0 on $\mathcal{GGM}_{410}$. To see this, simply note that the image of each point is in $[0,1]$ and the sum of over each line is 1. This state takes on the value $\frac{1}{3}$ at y so that $s_0(y') = \frac{2}{3} \neq 0 = s_0(0)$; therefore (ii) holds also. In this figure the numbers next to a point indicate the value of the state at that point. Only nonzero values are indicated.

In the above discussion we have assumed that the set of states that we were dealing with was the set of all states. But the proof works for *any* order determining σ-convex set of states on $\mathcal{GGM}_{410}$ because, for any such set $\mathcal{S}$ of states $s(y) : s \in \mathcal{S} \subseteq [\frac{1}{3}, 1]$. This construction and other constructions restricting the values that states may take on at certain elements, while maintaining other conditions are discussed in (Gerelle et al., 1974).

The counterexample we have illustrated proves that the map **Yes** is not generally injective. More importantly, $\mathbf{Yes}(E)$ is not generally a closed set of states, i.e., $\mathbf{Yes}(E) \notin \mathbb{C}(\mathcal{P}(\mathcal{S}))$.

Is it possible to make **Yes** injective and $\mathbf{Yes}(E)$ a closed set?

To make **Yes** injective it is sufficient to require that the set $\mathcal{S}$ satisfies a special property, usually called *richness.*

DEFINITION 2.4.6. Let $(\mathcal{E}_v, \mathcal{S})$ be an event-state system. The set $\mathcal{S}$ is called *rich for* $\mathcal{E}_v$ iff $\forall E, F \in \mathcal{E}_v$:

$$\mathbf{Yes}(E) \subseteq \mathbf{Yes}(F) \text{ implies } E \leq F.$$

In other words, the set of states is rich for the set of all events, whenever the event order (which is determined by the set of states) turns out to be completely determined by the behavior of the states with respect to the certain probability-value 1. Needless to stress, the richness property is stronger than the order-determining property. For this reason, it has sometimes been called *strongly-order determining* in the literature.[9] One immediately sees that, if the set of states is rich, then the map **Yes** is injective. For, suppose $\mathbf{Yes}(E) = \mathbf{Yes}(F)$. Then, by richness, $E \leq F$ and $F \leq E$; hence, $E = F$ (by the antisymmetry of $\leq$). How can we make $\mathbf{Yes}(E)$ closed (for any event E)?

To this aim, let us first introduce the notion of *carrier of a state.*

DEFINITION 2.4.7. *Carrier of a state*
Let $(\mathcal{E}_v, \mathcal{S})$ be an event-state system. An event E is called the *carrier* of a state s iff the following conditions are satisfied:

(i) $s(E) = 1$,
(ii) $\forall F \in \mathcal{E}_v$: $s \in \mathbf{Yes}(F)$ implies $E \leq F$.

Apparently, the carrier of a state s is the smallest element of $\mathcal{E}_v$ to which s assigns value 1. Generally, it is not the case that every state has a carrier. However, one can easily show that if the carrier of a state exists, then it is unique. When existing, the carrier of the state s will be denoted by $car(s)$.

From an intuitive point of view, we could say that $car(s)$ represents the *characteristic property* of the physical object described by s (a kind of *individual concept* in Leibniz' sense).

A situation where any state has a carrier corresponds to a *richness property* of the set of all events: each state is "characterized" by a special event.

LEMMA 2.4.8. *Let $(\mathcal{E}_v, \mathcal{S})$ be an event-state system. Suppose that every state in $\mathcal{S}$ has a carrier in $\mathcal{E}_v$. Then, $\forall s, t \in \mathcal{S}$: $s \perp t$ iff $car(s) \perp car(t)$.*

---

[9]See, for instance, (Beltrametti and Cassinelli, 1981).

PROOF. Suppose $s \perp t$. Then, $\exists E \in \mathcal{E}_v$ such that $s(E) = 1$ and $t(E) = 0$. Thus, $car(s) \leq E$ and $car(t) \leq E'$ and therefore $car(s) \perp car(t)$. Conversely, suppose $car(s) \perp car(t)$. Now, $t(car(t)) = 1$. Thus, $t(car(s)') = 1$, so that $t(car(s)) = 0$. Since $s(car(s)) = 1$, we conclude that $s \perp t$. □

We call *normal* any event-state system $(\mathcal{E}_v, \mathcal{S})$ that satisfies the following conditions:

(i) The set $\mathcal{S}$ of all states is rich for the set $\mathcal{E}_v$ of all events.
(ii) For any state $s \in \mathcal{S}$, the carrier of s exists.

Event-state systems that are normal turn out to satisfy some important properties, as shown by the following two theorems.

THEOREM 2.4.9. *Let $(\mathcal{E}_v, \mathcal{S})$ be a normal event-state system. Then, $\forall E \in \mathcal{E}_v$,*

$$E = \bigvee \{car(s) : s \in \mathbf{Yes}(E)\}.$$

PROOF. Let E be an element of $\mathcal{E}_v$. Since $\bigvee \emptyset = \mathbf{0}$, we may assume that $E \neq \mathbf{0}$. Then, $\mathbf{Yes}(E) \neq \emptyset$, because $\mathcal{S}$ is rich. For each $s \in \mathbf{Yes}(E)$, we have $car(s) \leq E$. Let F be an element of $\mathcal{E}_v$ such that, $\forall s \in \mathbf{Yes}(E)$, $car(s) \leq F$. Now, $s(car(s)) = 1$. Thus, if $s \in \mathbf{Yes}(E)$, then $s(F) = 1$. Hence, $\mathbf{Yes}(E) \subseteq \mathbf{Yes}(F)$; and therefore $E \leq F$, since $\mathcal{S}$ is rich. □

THEOREM 2.4.10. *Let $(\mathcal{E}_v, \mathcal{S})$ be a normal event-state system. Then, $\forall E \in \mathcal{E}$,*

$$\mathbf{Yes}(E') = \mathbf{Yes}(E)^{\perp}.$$

PROOF. If $E = \mathbf{0}$, then $\mathbf{Yes}(E') = \mathcal{S} = \emptyset^{\perp} = \mathbf{Yes}(E)^{\perp}$. Thus, we can suppose $E \neq \mathbf{0}$.
(i) To prove $\mathbf{Yes}(E') \subseteq \mathbf{Yes}(E)^{\perp}$, suppose $s \in \mathbf{Yes}(E')$. Then, $s(E) = 0$. Suppose $t \in \mathbf{Yes}(E)$. Then, $t(E) = 1$. Hence, $s \perp t$.
(ii) To prove $\mathbf{Yes}(E)^{\perp} \subseteq \mathbf{Yes}(E')$, suppose $s \in \mathbf{Yes}(E)^{\perp}$. Then, $\forall t \in \mathbf{Yes}(E)$: $s \perp t$. We have to show that $s(E') = 1$. By Lemma 2.4.8, $car(s) \perp car(t)$, for every $t \in \mathbf{Yes}(E)$. By Theorem 2.4.9 and the de Morgan laws, $E' = (\bigvee \{car(t) : t \in \mathbf{Yes}(E)\})' = \bigwedge \{car(t)' : t \in \mathbf{Yes}(E)\}$. Thus, $car(s) \leq E'$ and therefore, $s(E') = 1$, since $s(car(s)) = 1$. □

As a consequence of Theorem 2.4.10, we obtain that for any $E \in \mathcal{E}_v$, the set $\mathbf{Yes}(E)$ is closed.

COROLLARY 2.4.11. *Let $(\mathcal{E}_v, \mathcal{S})$ be a normal event-state system. Then, $\forall E \in \mathcal{E}_v$,*

$$\mathbf{Yes}(E) = (\mathbf{Yes}(E))^{\perp\perp}.$$

PROOF. By Theorem 2.4.10: $\mathbf{Yes}(E'') = (\mathbf{Yes}(E))^{\perp\perp}$. Hence, by strong double negation of the orthocomplement $'$, $\mathbf{Yes}(E) = (\mathbf{Yes}(E))^{\perp\perp}$. □

From the semantic point of view, this means that, for normal event-state systems, any proposition associated to an event is a closed set (i.e., belongs to the ortholattice $\mathbb{C}(\mathcal{P}(\mathcal{S})))$. Thus, by using Lemma 2.4.4 we have the following result in a normal event-state system:

a state s certainly verifies an event E iff for any state t similar to s there exists a state u similar to t that certainly verifies E.

As we will see in Part II, this property will play an important role in the possible world semantics for quantum logic.

Let us finally ask whether the map **Yes** preserves the existing meets and joins of the poset $\mathcal{E}_v$. The answer is given by the following theorem.

THEOREM 2.4.12. *Let $(\mathcal{E}_v, \mathcal{S})$ be a normal event-state system. Then,* **Yes** *is an ortho-embedding of $\mathcal{E}_v$ into $\mathbb{C}(\mathcal{P}(\mathcal{S}))$ that preserves all existing meets and joins in $\mathcal{E}$.*

PROOF. That **Yes** is an ortho-embedding follows from Theorem 2.4.10 and from the fact that $\mathcal{S}$ is rich. We now prove that **Yes** preserves all existing meets and joins in $\mathcal{E}_v$. It is sufficient to show that **Yes** preserves all existing meets in $\mathcal{E}_v$. Let $\{E_i\}_{i\in I}$ be any set of elements of $\mathcal{E}_v$ such that the meet $\bigwedge\{E_i\}_{i\in I}$ exists in $\mathcal{E}_v$. We show that $\mathbf{Yes}\left(\bigwedge\{E_i\}_{i\in I}\right) = \bigcap\{\mathbf{Yes}(E_i)\}_{i\in I}$. The proof of the inclusion $\mathbf{Yes}\left(\bigwedge\{E_i\}_{i\in I}\right) \subseteq \bigcap\{\mathbf{Yes}(E_i)\}_{i\in I}$ is trivial. Let us prove that $\bigcap\{\mathbf{Yes}(E_i)\}_{i\in I} \subseteq \mathbf{Yes}\left(\bigwedge\{E_i\}_{i\in I}\right)$. Let us suppose that $s \in \bigcap\{\mathbf{Yes}(E_i)\}_{i\in I}$; then, $\forall i \in I$: $s(E_i) = 1$ and therefore $car(s) \leq E_i$. Hence, $car(s) \leq \bigwedge\{E_i\}_{i\in I}$. Thus, $s\left(\bigwedge\{E_i\}_{i\in I}\right) = 1$, since $s(car(s)) = 1$. □

Thus, for normal event-state systems, the map **Yes** is an embedding of the σ-orthocomplete orthomodular poset $\langle \mathcal{E}_v, \leq, ', \mathbf{0}, \mathbf{1}\rangle$ into the complete ortholattice $\left\langle \mathbb{C}(\mathcal{P}(\mathcal{S})), \subseteq, ^{\perp}, \emptyset, \mathcal{S}\right\rangle$.

One immediately sees that the map **Yes** is not generally surjective. In order to make the map surjective it is necessary and sufficient to require that the orthomodular poset $\mathcal{E}_v$ is a complete lattice.

THEOREM 2.4.13. *Let $(\mathcal{E}_v, \mathcal{S})$ be a normal event-state system. Then, the map* **Yes** *is surjective iff $\mathcal{E}_v$ is a complete lattice.*

Sketch of the proof. (i) Sufficiency is straightforward.
(ii) Suppose $\mathcal{E}_v$ is a complete lattice and let X be an element of $\mathbb{C}(\mathcal{P}(\mathcal{S}))$. We prove that $X = \mathbf{Yes}\left(\bigwedge_{s\in X^\perp} car(s)'\right)$.
(iia) For $X \subseteq \mathbf{Yes}\left(\bigwedge_{s\in X^\perp} car(s)'\right)$, suppose $t \in X$. Then, $\forall s \in X^\perp$: $s \perp t$. Thus, by Lemma 2.4.8, $\forall s \in X^\perp$: $car(t) \perp car(s)$. Hence, $car(t) \leq \bigwedge_{s\in X^\perp} car(s)'$. Thus, $t\left(\bigwedge_{s\in X^\perp} car(s)'\right) = 1$ since $t(car(t)) = 1$.
(iib) For $\mathbf{Yes}\left(\bigwedge_{s\in X^\perp} car(s)'\right) \subseteq X$, suppose $t \in \mathbf{Yes}\left(\bigwedge_{s\in X^\perp} car(s)'\right)$. By Theorem 2.4.12, $t \in \bigcap_{s\in X^\perp} \mathbf{Yes}(car(s)')$. To prove that $t \in X = X^{\perp\perp}$, we let $u \in X^\perp$ and show that $t \perp u$. Since, $\forall s \in X^\perp$, $t \in \mathbf{Yes}(car(s)')$ so that $t(car(s)') = 1$ and $t(car(s)) = 0$, we have $t(car(u)) = 0$. Therefore $t \perp u$, since $u(car(u)) = 1$.

□

From the semantic point of view, a situation where the map **Yes** is surjective leads to a kind of *extensional collapse*: the structure of all *events* is isomorphic to the structure of all possible *propositions*. We will call *supernormal* the event-state systems that satisfy this strong condition. As we will see in the next chapter, orthodox QT gives rise to event-state systems of this kind.

CHAPTER 3

Back to Hilbert space

We will now return to orthodox QT. We will compare the abstract event-state systems, studied in the previous chapter, with the concrete examples that emerge in the framework of Hilbert space structures.

As we have seen in Section 1.3, according to von Neumann's axiomatization, the "mathematical environment" for any quantum system $\mathfrak{S}$ is represented by a convenient complex separable Hilbert space $\mathcal{H}$. As a consequence, all the basic physical notions concerning $\mathfrak{S}$ will have a counterpart that "lives" in $\mathcal{H}$. According to Axiom vN3, any (sharp) event (which may occur to $\mathfrak{S}$) is represented by a projection P of $\mathcal{H}$. At the same time, by Axiom vN2, any pure or mixed state of $\mathfrak{S}$ is represented by a density operator ρ of $\mathcal{H}$. By the one-to-one correspondence between the set $\Pi(\mathcal{H})$ of all projections and the set $\mathcal{C}(\mathcal{H})$ of all closed subspaces, events turn out to be equivalently represented by the closed subspaces of $\mathcal{H}$.

Now we want to investigate more closely the basic properties of such concrete events and states.

3.1. Events as closed subspaces

We will first study the structural properties of the set $\mathcal{C}(\mathcal{H})$ of all closed subspaces of $\mathcal{H}$.

As established in Section 1.2, the set $\mathcal{C}(\mathcal{H})$ gives rise to a complete orthomodular lattice, which fails to be distributive.

Let us consider the following structure:

$$\langle \mathcal{C}(\mathcal{H}) \,, \subseteq \,, \,' \,, \mathbf{0} \,, \mathbf{1} \rangle \,,$$

where:

- $\subseteq$ is the set-theoretic inclusion;
- $\forall X \in \mathcal{C}(\mathcal{H}) : \ X' = \{\psi \in \mathcal{H} : \forall \varphi \in X \, (\langle \psi \mid \varphi \rangle = 0)\}$, where $\langle \psi \mid \varphi \rangle$ is the inner product of ψ and φ;
- $\mathbf{0}$ and $\mathbf{1}$ are the null space (the singleton set consisting of the null vector) and the total space $\mathcal{H}$, respectively.

THEOREM 3.1.1. (Kalmbach, 1983; Beltrametti and Cassinelli, 1981) $\langle \mathcal{C}(\mathcal{H}) \,, \subseteq \,, \,' \,, \mathbf{0} \,, \mathbf{1} \rangle$ *is a complete orthomodular lattice, such that for any set* $\{X_i\}_{i \in I} \subseteq \mathcal{C}(\mathcal{H})$ *(where I is any set of indices):*

(i) *the meet* $\bigwedge \{X_i\}_{i \in I}$ *exists and coincides with* $\bigcap \{X_i\}_{i \in I}$*;*

(ii) *the join* $\bigvee \{X_i\}_{i\in I}$ *is the smallest closed subspace in* $\mathcal{H}$ *including* $\bigcup \{X_i\}_{i\in I}$.

We will call a *Hilbert lattice* any ortholattice based on the set of all closed subspaces of a given Hilbert space of dimension at least two.

One can easily show that *all* Hilbert lattices are nondistributive. Consider any Hilbert lattice. Take three 1-dimensional closed subspaces X, Y, Z that are pairwise nonorthogonal and that are included in a 2-dimensional closed subspace. We will have:

- $X \wedge (Y \vee Z) = X$;
- $(X \wedge Y) \vee (X \wedge Z) = \mathbf{0}$.

Interestingly enough, as anticipated in Section 1.1, the concrete compatibility relation defined for observables in a Hilbert space $\mathcal{H}$ (see Definition 1.3.3) is fully captured by the abstract compatibility relation defined for any orthomodular lattice (see Definition 1.1.16).

THEOREM 3.1.2. *Suppose* M *and* N *are two observables in Hilbert space* QT. *The two following conditions are equivalent:*

- M *and* N *are compatible observables. In other words, for any pair of Borel sets* Γ *and* Δ,
$$M(\Gamma)N(\Delta) = N(\Delta)M(\Gamma);$$
- *for any pair of Borel sets* Γ *and* Δ, *the two events* $M(\Gamma)$ *and* $N(\Delta)$ *are compatible in the Hilbert lattice* $\mathcal{C}(\mathcal{H})$. *In other words:*
$$M(\Gamma) = (M(\Gamma) \wedge N(\Delta)') \vee (M(\Gamma) \wedge N(\Delta)).$$

Some interesting properties of Hilbert lattices are described by the following theorem.

THEOREM 3.1.3. (Halmos, 1951)
Consider any Hilbert lattice $\mathcal{C}(\mathcal{H})$.

(i) *for any set* $\{X_i\}_{i\in I} \subseteq \mathcal{C}(\mathcal{H})$: $\bigvee \{X_i\}_{i\in I} = \left(\bigcup \{X_i\}_{i\in I}\right)''$;
(ii) *Let* X *be a finite dimensional subspace in* $\mathcal{C}(\mathcal{H})$. *Then,* $\forall Y \in \mathcal{C}(\mathcal{H})$: $X \vee Y = X + Y := \{\psi + \varphi : \psi \in X, \varphi \in Y\}$;
(iii) $\forall X, Y \in \mathcal{C}(\mathcal{H})$: *if* X *and* Y *are orthogonal* ($X \subseteq Y'$), *then* $X \vee Y = X + Y$.

Hilbert lattices turn out to satisfy a number of special conditions that do not generally hold for the events of an abstract event-state system.

THEOREM 3.1.4. *Any Hilbert lattice* $\mathcal{C}(\mathcal{H})$ *is atomic, irreducible and satisfies the covering property. The atoms of the lattice are the 1-dimensional subspaces of the Hilbert space. Furthermore,* $\mathcal{C}(\mathcal{H})$ *is separable iff the Hilbert space* $\mathcal{H}$ *is separable.*[1]

[1]See (Beltrametti and Cassinelli, 1981; Kalmbach, 1983). Atomicity, irreducibility, the covering property, and separability have been defined in Section 1.1.

THEOREM 3.1.5. *Any Hilbert lattice $\mathcal{C}(\mathcal{H})$ is modular iff $\mathcal{H}$ is finite dimensional.*[2]

Summing up: **any Hilbert lattice is an atomistic, irreducible, complete orthomodular lattice that satisfies the covering property.**

3.2. Events as projections

As we have discussed earlier, there is a one-to-one correspondence between the set of all closed subspaces and the set of all projections in a Hilbert space. We will now discuss this correspondence in more detail.

According to the notation (assumed in Section 1.2), for any closed subspace X of a Hilbert space $\mathcal{H}$, the corresponding projection is denoted by P_X, while X_P denotes the closed subspace corresponding to the projection P.

We have seen that, for any closed subspace X and for any projection P, the following conditions hold:

$$X_{(P_X)} = X; \;\; P_{(X_P)} = P.$$

On this basis, the lattice-theoretic structure of $\mathcal{C}(\mathcal{H})$ can be transferred to the set $\Pi(\mathcal{H})$ of all projections of $\mathcal{H}$.

We obtain the following structure:

$$\left\langle \Pi(\mathcal{H}) \,,\, \leq \,,\, ' \,,\, \mathbf{0} \,,\, \mathbf{1} \right\rangle, \text{ where:}$$

- $\forall P_X, P_Y \in \Pi(\mathcal{H})$,
$$P_X \leq P_Y \text{ iff } X \subseteq Y;$$
- $\forall P_X \in \Pi(\mathcal{H})$,
$$(P_X)' = P_{X'}.$$
- $\mathbf{0}$ and $\mathbf{1}$ are the null and the identity operators, respectively;
- For any set $\{P_{X_i}\}_{i \in I} \subseteq \Pi(\mathcal{H})$,
$$\bigwedge \{P_{X_i}\}_{i\in I} = P_{\bigcap \{X_i\}_{i\in I}};$$
$$\bigvee \{P_{X_i}\}_{i\in I} = P_{\bigvee \{X_i\}_{i\in I}}.$$

The order relation $\leq$ can be also expressed in terms of the product operation, because: $P_X \leq P_Y$ iff $P_X P_Y = P_X$. The orthocomplement can be expressed in terms of the sum operation, because: $(P_X)' = \mathbf{1} - P_X$. If P_X and P_Y are two commuting projections ($P_X P_Y = P_Y P_X$), then the meet and the join of P_X and P_Y can be expressed in the following way:

$$P_X \wedge P_Y = P_X P_Y; \;\; P_X \vee P_Y = P_X + P_Y - P_X P_Y.$$

As a consequence, the set of all closed subspaces and the set of all projections in a Hilbert space give rise to two isomorphic complete orthomodular lattices. Owing to this isomorphism, the structure $\langle \Pi(\mathcal{H}) \,,\, \leq \,,\, ' \,,\, \mathbf{0} \,,\, \mathbf{1} \rangle$ also is called a *Hilbert lattice*.

[2]See (Beltrametti and Cassinelli, 1981). Modularity has been defined in Section 1.1.

3.3. Hilbert event-state systems

We will now study the behavior of states, in the framework of orthodox QT. As we have discussed earlier, the states of a quantum system $\mathfrak{S}$ are mathematically represented by *density operators* of the complex separable Hilbert space $\mathcal{H}$ associated to $\mathfrak{S}$. We recall that a density operator is a linear, bounded, positive, trace-class operator of trace 1. The class of all density operators of $\mathcal{H}$ is denoted by $D(\mathcal{H})$. A density operator ρ represents a *pure state* (*maximal information* about $\mathfrak{S}$) iff there is a unit vector ψ such that ρ is the projection $P_{[\psi]}$, where $[\psi]$ is the 1-dimensional closed subspace determined by ψ.

The following theorem asserts that any density operator ρ can be represented as a convex combination of pure states.

THEOREM 3.3.1. *For every density operator $\rho \in D(\mathcal{H})$, there is a countable orthonormal set $\{\psi_i\}_{i\in I}$ of vectors of $\mathcal{H}$ and a countable set $\{\lambda_i\}_{i\in I}$ of positive real numbers (called* weights *of ρ) such that*

$$\rho = \sum_{i\in I} \lambda_i P_{[\psi_i]} \text{ and } \sum_{i\in I} \lambda_i = 1.$$

THEOREM 3.3.2. *Let ρ be a density operator. Then, ρ represents a pure state iff $\rho^2 = \rho$ (i.e., ρ is idempotent).*

Any density operator ρ determines a map $m_\rho : \Pi(\mathcal{H}) \mapsto \mathbb{R}$ such that $\forall P \in \Pi(\mathcal{H})$:

$$m_\rho(P) = \mathrm{Tr}(\rho P).$$

One can show that m_ρ is a probability measure on $\Pi(\mathcal{H})$. In other words:

- $m_\rho(\mathbf{0}) = 0,\ \ m_\rho(\mathbf{1}) = 1$ and
- for any countable set $\{P_n\}_{n\in I}$ of pairwise orthogonal projections:

$$m_\rho(\bigvee \{P_n\}_{n\in I}) = \sum_n \{m_\rho(P_n)\}_{n\in I}.$$

Consider now the set

$$\mathcal{S}(\mathcal{H}) := \{m_\rho : \rho \in D(\mathcal{H})\}.$$

This set contains precisely all those probability measures on $\Pi(\mathcal{H})$ that are determined by density operators.

At first sight, nothing guarantees that *all* probability measures defined on $\Pi(\mathcal{H})$ are determined by a density operator.

One can prove that, if the dimension of $\mathcal{H}$ is at least three, then every probability measure on $\Pi(\mathcal{H})$ has the form m_ρ, for some $\rho \in D(\mathcal{H})$. This is the content of a celebrated theorem proved by Gleason in 1957.

THEOREM 3.3.3. Gleason's Theorem
Let $\mathcal{H}$ be a separable Hilbert space of dimension at least 3. Then, for every probability measure s on $\Pi(\mathcal{H})$, there exists a unique density operator $\rho \in D(\mathcal{H})$ such that $\forall P \in \Pi(\mathcal{H})$: $s(P) = m_\rho(P)$.[3]

We have focused upon the following:

- a special set of *events* $\Pi(\mathcal{H})$;
- a special set of *states*, identified with the set $\mathcal{S}(\mathcal{H})$ of all probability measures defined on $\Pi(\mathcal{H})$ and determined by density operators.

Consider the pair $(\Pi(\mathcal{H}), \mathcal{S}(\mathcal{H}))$. One can prove that this pair represents an event-state system in the sense of Definition 2.3.3.

THEOREM 3.3.4. *For any separable Hilbert space $\mathcal{H}$, the pair $(\Pi(\mathcal{H})\,\mathcal{S}(\mathcal{H}))$ is an event-state system.*[4]

We call the pair $(\Pi(\mathcal{H}), \mathcal{S}(\mathcal{H}))$ a *Hilbert event-state system.* Needless to say, instead of $\Pi(\mathcal{H})$, one can equivalently refer to the isomorphic structure $\mathcal{C}(\mathcal{H})$. Accordingly, the pair $(\mathcal{C}(\mathcal{H}), \mathcal{S}(\mathcal{H}))$ also is called a *Hilbert event-state system.*

THEOREM 3.3.5. *For any separable Hilbert space $\mathcal{H}$, the event-state system $(\mathcal{C}(\mathcal{H}), \mathcal{S}(\mathcal{H}))$ is normal. In other words, $\mathcal{S}(\mathcal{H})$ is rich for $\mathcal{C}(\mathcal{H})$ and every probability measure $m_\rho \in \mathcal{S}(\mathcal{H})$ has a carrier in $\mathcal{C}(\mathcal{H})$.*

PROOF.

(i) That $\mathcal{S}(\mathcal{H})$ is rich depends on the fact that for any $X \in \mathcal{C}(\mathcal{H})$ and for any $\psi \in \mathcal{H}$,

$$m_{P_{[\psi]}}(X) = 1 \text{ iff } \psi \in X.$$

(ii) We show that every probability measure m_ρ has a carrier in $\mathcal{C}(\mathcal{H})$. Let ρ be a density operator (possibly a pure state). By Theorem 3.3.1, $\rho = \sum_{i\in I} \lambda_i P_{[\psi_i]}$, where $\{\psi_i\}_{i\in I}$ is a countable orthonormal set of vectors of $\mathcal{H}$ and $\{\lambda_i\}_{i\in I}$ is a countable set of weights of ρ. Since $\mathcal{S}(\mathcal{H})$ is σ-convex, we obtain that $\forall X \in \mathcal{C}(\mathcal{H}) : m_\rho(X) = \sum_{i\in I} \lambda_i m_{P_{[\psi_i]}}(X)$. We want to show that the carrier of m_ρ is the event $\bigvee_{i\in I}[\psi_i]$. Let us first prove that $m_\rho(\bigvee_{i\in I}[\psi_i]) = 1$. Since m_ρ satisfies σ-additivity, we have:

$$m_\rho(\bigvee_{i\in I}[\psi_i]) = \sum_{i\in I} m_\rho([\psi_i]) = \sum_{i\in I} \mathrm{Tr}(\rho P_{[\psi_i]}) = \sum_{i\in I} \lambda_i = 1.$$

To show that $\bigvee_{i\in I}[\psi_i]$ is the carrier of m_ρ, it remains to show that:

$$\forall X \in \mathcal{C}(\mathcal{H}), m_\rho(X) = 1 \text{ implies } \bigvee_{i\in I}[\psi_i] \subseteq X.$$

[3]See (Gleason, 1957; Varadarajan, 1985; Dvurečenskij, 1993). Gleason's Theorem can be generalized also to Hilbert spaces over the quaternions.

[4]See, for instance, (Beltrametti and Cassinelli, 1981; Pták and Pulmannová, 1991).

Suppose $m_\rho(X) = 1$. Then,

$$1 = m_\rho(X) = \mathrm{Tr}(\rho P_X) = \mathrm{Tr}\sum_{i\in I}\lambda_i P_{[\psi_i]}P_X = \sum_{i\in I}\lambda_i \mathrm{Tr}(P_{[\psi_i]}P_X).$$

Since $\sum_{i\in I}\lambda_i = 1$, we obtain: $\forall i \in I : \mathrm{Tr}(P_{[\psi_i]}P_X) = 1$. Whence, $\forall i \in I : [\psi_i] \subseteq X$, and $\bigvee_{i\in I}[\psi_i] \subseteq X$.

□

As a consequence, one immediately obtains that also any event-state system $(\Pi(\mathcal{H}), \mathcal{S}(\mathcal{H}))$ is normal.

Let us now turn to the preclusivity space $(\mathcal{S}(\mathcal{H}), \perp)$ induced by a Hilbert event-state system $(\Pi(\mathcal{H}), \mathcal{S}(\mathcal{H}))$.

Consider the map **Yes** that associates to any projection P of $\Pi(\mathcal{H})$ the set of all states in $\mathcal{S}(\mathcal{H})$ assigning to X probability-value 1 (see the previous chapter). Since $(\Pi(\mathcal{H}), \mathcal{S}(\mathcal{H}))$ is normal (Theorem 3.3.5), the map **Yes** turns out to be an ortho-embedding of $\Pi(\mathcal{H})$ into the complete ortholattice $\mathbb{C}(\mathcal{P}(\mathcal{S}(\mathcal{H})))$ of all closed sets of $(\mathcal{S}(\mathcal{H}), \perp)$ (Theorem 2.4.12). By Theorem 3.1.1, the orthomodular lattice $\Pi(\mathcal{H})$ is complete. Thus, by Theorem 2.4.13, the map **Yes** is surjective. Accordingly, the event structure $\Pi(\mathcal{H})$ turns out to be isomorphic to $\mathbb{C}(\mathcal{P}(\mathcal{S}(\mathcal{H})))$. As a consequence, $(\Pi(\mathcal{H}), \mathcal{S}(\mathcal{H}))$ is a supernormal event-state system.

3.4. From abstract orthoposets of events to Hilbert lattices

As we have seen, in orthodox QT, the class of all events gives rise to a Hilbert lattice, which turns out to satisfy a number of special properties (it is an atomic, irreducible, complete orthomodular lattice that satisfies the covering property). Now, an important question arises: are these conditions sufficient for an orthomodular orthoposet $\mathcal{E}$ of events to be isomorphic to (or embeddable into) a classical Hilbert lattice? More generally, is it possible to *capture lattice-theoretically* the structure of Hilbert lattices?

This question has been investigated for at length in the *quantum logical* literature. The basic aim was to prove a kind of *representation theorem* that could reasonably replace the very strong assumption represented by Mackey's Axiom M7.[5] As we have discussed earlier, because of its apparent *ad hoc* character, such a principle has been never accepted as a reasonable axiom by the quantum logic community.

An important partial answer to the *representation problem* has been given by a theorem proved by Piron in 1964. In order to understand the content of this theorem, it will be first expedient to introduce some mathematical definitions.

In Chapter 1 we have defined the notion of *pre-Hilbert space*. We will now introduce a weaker concept, the category of all *generalized pre-Hilbert spaces*.

[5]The axiom asserts that: the partially ordered set of all events is isomorphic to the partially ordered set of all closed subspaces of a separable infinite dimensional complex Hilbert space.

Unlike the case of pre-Hilbert spaces, the division ring of a generalized pre-Hilbert space is not necessarily the real numbers or the complex numbers or the quaternions.

DEFINITION 3.4.1. *Generalized pre-Hilbert space*
A *generalized pre-Hilbert space* is a system

$$\langle \mathcal{V}, \theta, \langle . \mid . \rangle, \mathbf{D} \rangle,$$

where:

- $\mathcal{V}$ is a vector space over a division ring $\mathbf{D}$;
- θ is an involutive antiautomorphism on $\mathbf{D}$;
- $\langle . \mid . \rangle$ (to be interpreted as an inner product) is a *Hermitean form* on $\mathcal{V}$.[6]

Consider now a generalized pre-Hilbert space $\langle \mathcal{V}, \theta, \langle . \mid . \rangle, \mathbf{D} \rangle$. Let X be any subset of $\mathcal{V}$ and let $X' := \{\psi \in \mathcal{V} : \forall \varphi \in X[\langle \psi \mid \varphi \rangle = 0]\}$. A subspace X of $\mathcal{V}$ is called $\langle . \mid . \rangle$*-closed* iff $X = X''$. The set of all $\langle . \mid . \rangle$-closed subspaces of $\mathcal{V}$ will be denoted by $\mathcal{C}(\mathcal{V})$. Notice that if $\mathcal{V}$ is a Hilbert space (with inner product $\langle . \mid . \rangle$), then the class of all $\langle . \mid . \rangle$-closed subspaces of $\mathcal{V}$ coincides with the class of all closed subspaces of $\mathcal{V}$. We will denote by $\mathbf{0}$ and $\mathbf{1}$ the $\langle . \mid . \rangle$-closed sets such that: $\mathbf{0}$ is the singleton consisting of the null vector, while $\mathbf{1}$ is the whole space $\mathcal{V}$.

DEFINITION 3.4.2. *Generalized Hilbert space*
A *generalized Hilbert space* is a generalized pre-Hilbert space $\langle \mathcal{V}, \theta, \langle . \mid . \rangle, \mathbf{D} \rangle$ such that for any $\langle . \mid . \rangle$-closed subspace $X \in \mathcal{C}(\mathcal{V})$,

$$\mathcal{V} = X + X' := \{\psi + \varphi : \psi \in X, \varphi \in X'\}.$$

THEOREM 3.4.3. (Varadarajan, 1985)
Let $\langle \mathcal{V}, \theta, \langle . \mid . \rangle, \mathbf{D} \rangle$ *be a generalized Hilbert space. The structure*

$$\langle \mathcal{C}(\mathcal{V}), \subseteq, ', \mathbf{0}, \mathbf{1} \rangle$$

is an atomic, irreducible, complete orthomodular lattice with the covering property, where, for any set $\{X_i\}_{i \in I} \subseteq \mathcal{C}(\mathcal{V})$:

(i) *the meet* $\bigwedge \{X_i\}_{i \in I}$ *is the set-theoretic intersection* $\bigcap \{X_i\}_{i \in I}$;
(ii) *the join* $\bigvee \{X_i\}_{i \in I}$ *is the smallest* $\langle . \mid . \rangle$*-closed subspace in* $\mathcal{V}$ *including* $\bigcup \{X_i\}_{i \in I}$.

On this basis, one can prove the Piron Theorem (which represents a *weak representation theorem*).

---

[6]An *involutive antiautomorphism* of a division ring $\mathbf{D}$ is a map $\theta : D \mapsto D$ with the following properties $\forall a, b \in D$: (i) $\theta(a+b) = \theta(a) + \theta(b)$; (ii) $\theta(ab) = \theta(b)\theta(a)$; (iii) $\theta(\theta(a)) = a$. In the particular case where $\mathbf{D}$ is the complex field, θ coincides with the complex conjugation map.

A *Hermitean form* of $\mathcal{V}$ is a map $\langle . \mid . \rangle : V \times V \mapsto D$ that satisfies the following properties $\forall a, b \in D$, $\forall \psi_1, \psi_2, \psi, \varphi_1, \varphi_2, \varphi \in V$: (i) $\langle a\psi_1 + b\psi_2 \mid \varphi \rangle = a \langle \psi_1 \mid \varphi \rangle + b \langle \psi_2 \mid \varphi \rangle$ and $\langle \psi \mid a\varphi_1 + b\varphi_2 \rangle = \langle \psi \mid \varphi_1 \rangle \theta(a) + \langle \psi \mid \varphi_2 \rangle \theta(b)$; (ii) $\langle \psi \mid \varphi \rangle = \theta(\langle \varphi \mid \psi \rangle)$; (iii) $\langle \psi \mid \psi \rangle = 0$ implies that ψ is the null vector.

THEOREM 3.4.4. (Piron, 1976; Varadarajan, 1985) The Piron weak representation theorem
Let $\mathcal{B}$ be a complete, irreducible, atomic, orthomodular lattice satisfying the covering property. If $\mathcal{B}$ has at least four pairwise orthogonal elements, then $\mathcal{B}$ is isomorphic to the orthomodular lattice $\mathcal{C}(\mathcal{V})$ of all $\langle . \mid . \rangle$-closed subspaces of a generalized Hilbert space $\langle \mathcal{V}, \theta, \langle . \mid . \rangle, \mathbf{D} \rangle$. If $\mathbf{D}$ is either $\mathbb{R}$ or $\mathbb{C}$ or $\mathbb{Q}$ (the quaternions) and the antiautomorphism θ is continuous, then $\langle \mathcal{V}, \theta, \langle . \mid . \rangle, \mathbf{D} \rangle$ turns out to be a Hilbert space.

One also says that the lattice $\mathcal{B}$ that is isomorphic to the lattice of all closed subspaces is *coordinatized* by the generalized Hilbert space $\langle \mathcal{V}, \theta, \langle . \mid . \rangle, \mathbf{D} \rangle$.

Do the properties of the *coordinatized lattice* $\mathcal{B}$ of Theorem 3.4.4 restrict the choice of $\mathbf{D}$ either to the reals, the complexes or the quaternions - and therefore force the generalized Hilbert space to be an actual Hilbert space? Quite unexpectedly, Keller (1980) proved a negative result: there are lattices that satisfy all the conditions of Piron Theorem; at the same time, they are coordinatized by generalized Hilbert spaces over non-archimedean division rings. Keller's counterexamples have sometimes been interpreted as showing the definitive impossibility for the quantum logical approach to capture the Hilbert space mathematics. This impossibility was supposed to demonstrate the failure of the quantum logic approach in reaching its main goal: the "bottom-top" reconstruction of Hilbert lattices. Interestingly enough, such a negative conclusion has been contradicted by an important result proved by Solèr (1995):

Hilbert lattices can be characterized in a lattice-theoretic way.

We will now give a synthetic description of Solèr's main results.

DEFINITION 3.4.5. *The Solèr condition*
Let $\langle \mathcal{V}, \theta, \langle . \mid . \rangle, \mathbf{D} \rangle$ be an infinite dimensional generalized Hilbert space. We say that $\langle \mathcal{V}, \theta, \langle . \mid . \rangle, \mathbf{D} \rangle$ satisfies the *Solèr condition* iff there exists a set of vectors $\{\psi_i\}_{i \in \mathbb{N}}$ of $\mathcal{V}$ and a scalar c such that:

- $\forall i [\langle \psi_i \mid \psi_i \rangle = c]$;
- $\forall i, j [i \neq j$ implies $\langle \psi_i \mid \psi_j \rangle = 0]$.

In other words, the elements of the set $\{\psi_i\}_{i \in \mathbb{N}}$ of $\mathcal{V}$ (also called c-orthogonal set) are pairwise orthogonal; while the inner product of any element with itself is identically equal to c.

THEOREM 3.4.6. (Solèr, 1995) The Solèr strong representation theorem
Let $\langle \mathcal{V}, \theta, \langle . \mid . \rangle, \mathbf{D} \rangle$ be an infinite dimensional generalized Hilbert space over a division ring $\mathbf{D}$. The following conditions are equivalent:

(i) $\langle \mathcal{V}, \theta, \langle . \mid . \rangle, \mathbf{D} \rangle$ *satisfies the Solèr condition;*
(ii) $\langle \mathcal{V}, \theta, \langle . \mid . \rangle, \mathbf{D} \rangle$ *is a Hilbert space.*

As a consequence, the Solèr condition turns out to characterize Hilbert spaces in the class of all generalized Hilbert spaces. The important point is that one is dealing with a condition that admits a purely lattice-theoretic characterization, namely the so-called *angle bisecting condition* (Morash, 1973).

One can prove that every lattice of infinite length[7] that satisfies the angle bisecting condition (in addition to the conditions of the Piron Theorem) is isomorphic to a Hilbert lattice.

At first sight, it seems difficult to give an intuitive physical interpretation either for the Solèr condition or for the angle bisecting condition (whose formulation is quite long and complicated). Interestingly enough, Holland (1995) found another condition (called *ample unitary group condition*), which seems to be physically more attractive. One can show that the Solèr condition and the Holland condition are equivalent.

Before introducing the Holland condition, let us first recall that the notion of *unitary operator* can be also defined in the framework of generalized Hilbert spaces.

DEFINITION 3.4.7. *Unitary operator*
Let $\langle \mathcal{V}, \theta, \langle . \mid . \rangle, \mathbf{D} \rangle$ be a generalized Hilbert space. A *unitary operator* on $\langle \mathcal{V}, \theta, \langle . \mid . \rangle, \mathbf{D} \rangle$ is a map $U : V \mapsto V$ such that $\forall \psi, \varphi \in V$, $\forall a, b \in D$ the following conditions are satisfied:

(i) $U(a\psi + b\varphi) = aU(\psi) + bU(\varphi)$;
(ii) $\langle \psi \mid \varphi \rangle = \langle U\psi \mid U\varphi \rangle$.

In other words, a unitary operator of a generalized Hilbert space is a linear operator that preserves the inner product.

DEFINITION 3.4.8. *The ample unitary group condition*
A generalized Hilbert space $\langle \mathcal{V}, \theta, \langle . \mid . \rangle, \mathbf{D} \rangle$ satisfies the *ample unitary group condition* iff for any nonzero orthogonal vectors $\psi, \varphi \in V$, there is a unitary operator U and a scalar $c \in D$ such that $U(\psi) = c\varphi$.

Recall that in orthodox QT, the spontaneous time evolution of a quantum object (governed by the Schrödinger equation) is described by unitary operators. In a sense, we could say that, according to the ample unitary group condition, for any pair of pure states, either element of the pair can be regarded as a transformation of the other element, where the transformation seems to behave as a "quasi-time evolution."

THEOREM 3.4.9. (Holland, 1995)
Any infinite generalized Hilbert space $\langle \mathcal{V}, \theta, \langle . \mid . \rangle, \mathbf{D} \rangle$ satisfies the Solèr condition iff it satisfies the ample unitary group condition.

Both the Solèr condition and the ample unitary group condition have a major flaw: they essentially refer to the generalized Hilbert space machinery. Only the angle bisecting condition is purely lattice-theoretic. However, this condition is rather technical and by no means intuitive. To overcome this

[7]The *length* of a lattice $\mathcal{B}$ is defined to be the supremum, over all the chains of $\mathcal{B}$, of the numbers of elements in each chain minus 1.

difficulty, Aerts and Van Steirteghem (2000) have proposed a new lattice-theoretic condition, called *plane transitivity*. One can prove that the plane transitivity condition and the angle bisecting condition are equivalent. At the same time, from an intuitive point of view, the content of the plane transitivity condition turns out to be somewhat close to Holland's ample unitary group condition.

DEFINITION 3.4.10. *The plane transitivity condition*
Let $\mathcal{B}$ be an atomic orthomodular lattice. We say that $\mathcal{B}$ satisfies the *plane transitivity condition* iff for any two atoms $a, b \in \mathcal{B}$, there are two distinct atoms $a_1, b_1 \in \mathcal{B}$ and an isomorphism $h : B \mapsto B$ such that the following conditions are satisfied:

(i) $\forall c \in B$: $\mathbf{0} \leq c \leq a_1 \vee b_1$ implies $h(c) = c$;
(ii) $h(a) = b$.

Summing up:

THEOREM 3.4.11. *Let $\langle \mathcal{V}, \theta, \langle . \mid . \rangle, \mathbf{D} \rangle$ be an infinite dimensional generalized Hilbert space. The following conditions are equivalent:*

(i) $\langle \mathcal{V}, \theta, \langle . \mid . \rangle, \mathbf{D} \rangle$ *is a Hilbert space;*
(ii) $\langle \mathcal{V}, \theta, \langle . \mid . \rangle, \mathbf{D} \rangle$ *satisfies Solèr condition;*
(iii) $\langle \mathcal{V}, \theta, \langle . \mid . \rangle, \mathbf{D} \rangle$ *satisfies the ample unitary condition;*
(iv) *The orthomodular lattice of all $\langle . \mid . \rangle$-closed subspaces of $\langle \mathcal{V}, \theta, \langle ., . \rangle, \mathbf{D} \rangle$ satisfies the plain transitivity condition.*

As a consequence, one can show that every atomic, irreducible, complete orthomodular lattice $\mathcal{B}$ of infinite length, that satisfies the covering property and the plane transitivity condition is isomorphic to a Hilbert lattice $\mathcal{C}(\mathcal{H})$. Notice that the infinite length of the coordinatized lattice $\mathcal{B}$ implies that the coordinatizing generalized Hilbert space $\mathcal{H}$ is infinite dimensional. Furthermore, $\mathcal{B}$ is separable iff $\mathcal{H}$ is separable.

Let us now return to the class of all abstract event-state systems. Theorem 3.4.11 naturally suggests the following definition.

DEFINITION 3.4.12. *Solèr event-state system*
An event-state system $(\mathcal{E}, \mathcal{S})$ is called a *Solèr event-state system* iff the set of all events $\mathcal{E}$ has the structure of an atomic, irreducible, complete orthomodular lattice of infinite length, that satisfies the covering property and the plane transitivity condition.

As a consequence, one immediately obtains that all Solèr event-state systems $(\mathcal{E}, \mathcal{S})$ (such that $\mathcal{E}$ is separable) are supernormal (see Section 2.4).

All these results represent a satisfactory solution for the *representation problem* of the quantum logical approach to sharp QT: Solèr event-state systems represent a faithful abstract description of the basic structures of orthodox (sharp) QT. Mackey's critical Axiom M7 may now be replaced by an axiom that is not simply ad hoc.

CHAPTER 4

The emergence of fuzzy events in Hilbert space quantum theory

4.1. The notion of effect

The event-structures we have studied so far (both in the abstract and in the Hilbert space situation) are typically *sharp*: any event always satisfies the noncontradiction principle.

Let us now return to Hilbert space QT and to the concrete event-state systems $(\Pi(\mathcal{H}), \mathcal{S}(\mathcal{H}))$. An interesting question is the following: do the sets $\Pi(\mathcal{H})$ and $\mathcal{S}(\mathcal{H})$ correspond to an *optimal* possible choice of adequate mathematical representatives for the intuitive notions of *event* and of *state*, respectively?

Consider first the notion of *state*. Once $\Pi(\mathcal{H})$ is fixed, Gleason's Theorem guarantees that $\mathcal{S}(\mathcal{H})$ corresponds to an *optimal* notion of state: for, any probability measure defined on $\Pi(\mathcal{H})$ is determined by a density operator of $\mathcal{H}$ (provided the dimension of $\mathcal{H}$ is greater than or equal to 3).

Let us discuss then the notion of *event* and let us ask whether $\Pi(\mathcal{H})$ represents the largest set of operators assigned a probability-value, according to the Born rule. The answer to this question is negative.

One can easily recognize the existence of bounded linear operators E that are not projections and that satisfy the following condition:

$$\text{for any density operator } \rho, \ \operatorname{Tr}(\rho E) \in [0,1].$$

From an intuitive point of view, this means that such operators E "behave as possible events," because any state assigns to them a probability value.

An interesting example of this kind is represented by the operator $\frac{1}{2}\mathbb{1}$ (where $\mathbb{1}$ is the identity operator). One immediately realizes that $\frac{1}{2}\mathbb{1}$ is a linear bounded operator that is *not* a projection, because:

$$\frac{1}{2}\mathbb{1}\frac{1}{2}\mathbb{1} = \frac{1}{4}\mathbb{1} \neq \frac{1}{2}\mathbb{1}$$

(hence $\frac{1}{2}\mathbb{1}$ fails to be idempotent).

At the same time, for any density operator ρ we have:

$$\operatorname{Tr}(\rho\frac{1}{2}\mathbb{1}) = \frac{1}{2}.$$

Thus, $\frac{1}{2}\mathbb{I}$ seems to represent a totally *indeterminate* event, to which each state assigns probability $\frac{1}{2}$. Apparently, the event $\frac{1}{2}\mathbb{I}$ plays the role that, in fuzzy set theory, is played by the *semitransparent* fuzzy set $\frac{1}{2}\mathbf{1}$ such that for any object x of the universe:

$$\frac{1}{2}\mathbf{1}(x) = \frac{1}{2}.$$

This situation suggests that we liberalize the notion of *quantum event* and extend the set $\Pi(\mathcal{H})$ to a new set of operators. Following Ludwig (1983), the elements of this new set have been called *effects*.

DEFINITION 4.1.1. *Effects*
An *effect* of $\mathcal{H}$ is a bounded linear operator E that satisfies the following condition, for any density operator ρ:

$$\mathrm{Tr}(\rho E) \in [0, 1].$$

We denote by $\mathcal{E}(\mathcal{H})$ the set of all effects of $\mathcal{H}$.

Clearly, $\mathcal{E}(\mathcal{H})$ properly includes $\Pi(\mathcal{H})$. Because:

- any projection satisfies the definition of effect;
- there are examples of effects that are not projections (for instance the effect $\frac{1}{2}\mathbb{I}$, that is usually called *the semitransparent effect*).

By definition, effects turn out to represent a kind of maximal mathematical representative for the notion of quantum event, in agreement with the basic statistical rule of QT (the *Born rule*).

Unlike projections, effects represent quite general mathematical objects that describe at the same time events and states. Let E be any effect in $\mathcal{E}(\mathcal{H})$. The following conditions hold:

- E represents a sharp event ($\in \Pi(\mathcal{H})$) iff E is idempotent ($EE = E$);
- E is a density operator (representing a state) iff $\mathrm{Tr}(E) = 1$;
- E represents a pure state iff E is at the same time a projection and a density operator.

One can show:

LEMMA 4.1.2. *For any bounded linear operator E: $E \in \mathcal{E}(\mathcal{H})$ iff $\forall\psi \in \mathcal{H}$, $0 \leq \langle\psi|E\psi\rangle \leq \|\psi\|^2$*

As a consequence, effects can be alternatively defined as linear bounded operators E that satisfy the following condition:

$$\forall\psi \in \mathcal{H},\ 0 \leq \langle\psi|E\psi\rangle \leq \|\psi\|^2.$$

The notion of effect has been deeply investigated in the framework of the *unsharp approaches to QT* (first proposed by Ludwig and further developed by Kraus, Davies, Mittelstaedt, Busch, Lahti, Bugajski, Beltrametti, Cattaneo, Nisticò, Foulis, Bennett, Gudder, Pulmannová, Dvurečenskij, Riečan, Riečanova, Schroeck and many others including the authors).

As mentioned in the Introduction, one could say that moving to an unsharp approach represents an important step towards a kind of "second degree of fuzziness." In the framework of the sharp approach, any physical event E can be regarded as a kind of "clear" property. Whenever a state ρ assigns to E a probability-value different from 1 and 0, one can think that the semantic uncertainty involved in such a situation totally depends on the ambiguity of the state (first degree of fuzziness). In other words, even a pure state in QT does not represent *logically complete information* that is able to decide any possible physical event. In the unsharp approaches, instead, one takes into account also "genuinely ambiguous properties." This second degree of fuzziness may be regarded as depending on the accuracy of the measurement which tests the property, and also on the accuracy involved in the operational definition for the physical quantities which the property refers to.[1] There are different algebraic structures that can be induced on the set of all effects in a Hilbert space. We will investigate the basic properties of these structures in the following sections and in Chapter 5.

4.2. Effect-Brouwer Zadeh posets

One immediately realizes that the set $\mathcal{E}(\mathcal{H})$ can be naturally structured as a *regular poset*:

$$\langle \mathcal{E}(\mathcal{H}) \, , \, \leq \, , \, ' \, , \, \mathbf{0} \, , \mathbf{1} \rangle \, ,$$

where

(i) $\leq$ is the *natural* order determined by the set of all density operators. In other words:

$$E \leq F \text{ iff for any density operator } \rho \in D(\mathcal{H}), \mathrm{Tr}(\rho E) \leq \mathrm{Tr}(\rho F).$$

(i.e., any state assigns to E a probability-value that is less or equal than the probability-value assigned to F);

(ii) $E' = \mathbf{1} - E$ (where $-$ is the standard operator difference);

(iii) $\mathbf{0}$, $\mathbf{1}$ are the null projection ($\mathbb{O}$) and the identity projection ($\mathbb{1}$), respectively.

One can easily check that:

- $\leq$ is a partial order;
- $'$ is an involution;
- $\mathbf{0}$ and $\mathbf{1}$ are respectively the minimum and the maximum with respect to $\leq$;
- the regularity condition holds. In other words:

$$E \leq E' \text{ and } F \leq F' \text{ implies } E \leq F'.$$

The following condition holds.

LEMMA 4.2.1. *For all effects E, F:*

$$E \leq F \;\; \textit{iff} \;\; \forall \psi \in \mathcal{H} : \langle \psi | E \psi \rangle \leq \langle \psi | F \psi \rangle \, .$$

---

[1] The advantages of the effect approach from the point of view of the physical interpretation have been discussed in a large literature. See, for instance, (Davies, 1976; Busch, Lahti and Mittelstaedt, 1991; Busch, Grabowski and Lahti, 1995; Schroeck, 1996).

The following lemma sums up some standard results of operator theory.

LEMMA 4.2.2. *For any effect E the following conditions are satisfied:*

(i) $\forall\psi \in \mathcal{H}$: $\langle \psi | E\psi \rangle = 0$ *iff* $E\psi = \underline{0}$;
(ii) $\forall\psi \in \mathcal{H}$: $\langle \psi | E\psi \rangle = \|\psi\|^2$ *iff* $E\psi = \psi$;
(iii) $\forall\rho \in D(\mathcal{H})$: $m_\rho(E) = 1$ *iff* $\rho = \sum_{i\in I} \lambda_i P_{[\psi_i]}$, *where* $\forall i \in I$: $E\psi_i = \psi_i$;
(iv) $Ker(E) = Ran(E)'$, *where* $Ker(E) := \{\psi \in \mathcal{H} : E\psi = \underline{0}\}$ *and* $Ran(E)$ *is the range of* E.

The effect poset $\mathcal{E}(\mathcal{H})$ turns out to be properly fuzzy. The noncontradiction principle is violated: for instance the semitransparent effect $\frac{1}{2}\mathbb{I}$ satisfies the following condition:

$$\frac{1}{2}\mathbb{I} \wedge (\frac{1}{2}\mathbb{I})' = \frac{1}{2}\mathbb{I} \wedge \frac{1}{2}\mathbb{I} = \frac{1}{2}\mathbb{I} \neq \mathbf{0}.$$

This is one of the reasons why proper effects (those that are not projections) may be regarded as representing *unsharp physical events*. Accordingly, we will also call the involution operation of an effect-structure a *fuzzy complement*.

At the same time, the effect-poset fails to be a lattice. Unlike projections, some pairs of effects have no meet, as the following example shows (Gudder and Greechie, 1996).

EXAMPLE 4.2.3. Let us consider the following effects (in the matrix-representation) on the Hilbert space $\mathbb{R}^2$:

$$E = \begin{pmatrix} \frac{1}{2} & 0 \\ 0 & \frac{1}{2} \end{pmatrix} \quad F = \begin{pmatrix} \frac{3}{4} & 0 \\ 0 & \frac{1}{4} \end{pmatrix} \quad G = \begin{pmatrix} \frac{1}{2} & 0 \\ 0 & \frac{1}{4} \end{pmatrix}$$

It is not hard to see that $G \leq E, F$. Suppose, by contradiction, that $L = E \wedge F$ exists in $\mathcal{E}(\mathbb{R}^2)$. An easy computation shows that L must be equal to G. Let

$$M = \begin{pmatrix} \frac{7}{16} & \frac{1}{8} \\ \frac{1}{8} & \frac{3}{16} \end{pmatrix}.$$

Then, M is an effect such that $M \leq E, F$; however, $M \not\leq L$, which is a contradiction.

As we shall see in the following section, a lattice structure can be obtained by embedding $\mathcal{E}(\mathcal{H})$ into its *Mac Neille completion*.

The effect poset $\mathcal{E}(\mathcal{H})$ can be naturally extended to a richer structure, equipped with a new complement $^\sim$, that has an intuitionistic-like behavior. Such operation $^\sim$ is usually called the *Brouwer complement*.

DEFINITION 4.2.4. *The Brouwer complement*

$$\forall E \in \mathcal{E}(\mathcal{H}) : E^\sim = P_{Ker(E)}.$$

In other words, the Brouwer complement of E is the projection operator $P_{Ker(E)}$ whose range is $Ker(E)$, the kernel of E. By Lemma 4.2.2(iv),

$$E^{\sim} = P_{Ran(E)'}.$$

By definition, the Brouwer complement of an effect is always a projection. In the particular case, when E is a projection, it turns out that $E' = E^{\sim}$, in other words, the fuzzy and the intuitionistic complement collapse into one and the same operation.

The structure $\langle \mathcal{E}(\mathcal{H}), \leq, ', ^{\sim}, \mathbf{0}, \mathbf{1} \rangle$ turns out to be a particular example of a kind of abstract structure that has been termed *Brouwer Zadeh poset* (Cattaneo and Nisticò, 1986).

DEFINITION 4.2.5. *Brouwer Zadeh poset*
A *Brouwer Zadeh poset* (or *BZ-poset*) (BZP) is a structure

$$\langle B, \leq, ', ^{\sim}, \mathbf{0}, \mathbf{1} \rangle,$$

where

(i) $\langle B, \leq, ', \mathbf{0}, \mathbf{1} \rangle$ is a regular poset;

(ii) $^{\sim}$ is a unary operation that behaves like an intuitionistic complement:

  (iia) $a \wedge a^{\sim} = \mathbf{0}$;

  (iib) $a \leq a^{\sim\sim}$;

  (iic) $a \leq b$ implies $b^{\sim} \leq a^{\sim}$.

(iii) The following relation connects the fuzzy and the intuitionistic complement:

$$a^{\sim\prime} = a^{\sim\sim}.$$

Of course, any BZ-poset $\langle B, \leq, ', ^{\sim}, \mathbf{0}, \mathbf{1} \rangle$ where the two complements $'$ and $^{\sim}$ coincide, turns out to be an orthoposet.

DEFINITION 4.2.6. *Brouwer Zadeh lattice*
A *Brouwer Zadeh lattice* (briefly a *BZ-lattice*) (BZL) is a BZ-poset that is also a lattice.

One can prove:

THEOREM 4.2.7. (Cattaneo and Nisticò, 1986)
The structure

$$\langle \mathcal{E}(\mathcal{H}), \leq, ', ^{\sim}, \mathbf{0}, \mathbf{1} \rangle$$

is a Brouwer Zadeh poset.

PROOF. As we have already noticed, the structure $\langle \mathcal{E}(\mathcal{H}), \leq, ', \mathbf{0}, \mathbf{1} \rangle$ is a regular poset. Thus, to prove that $\langle \mathcal{E}(\mathcal{H}), \leq, ', ^{\sim}, \mathbf{0}, \mathbf{1} \rangle$ is a BZ-poset, we have to prove that conditions (ii) and (iii) of Definition 4.2.5 are satisfied.
(iii) $E^{\sim\sim} = E^{\sim\prime}$. The proof follows from the fact that $E^{\sim}$ is a projection and that the fuzzy and the intuitionistic complement collapse into the same operation whenever they are restricted to projections.

(iib) $E \leq E^{\sim\sim}$. The following holds: $E^{\sim\sim} = \left(P_{Ker(E)}\right)^{\sim} = \left(P_{Ker(E)}\right)' = P_{Ker(E)'}$. We want to show that for any $\psi \in \mathcal{H}$: $\langle\, \psi | E\psi \,\rangle \leq \langle\, \psi | P_{Ker(E)'}\psi \,\rangle$. Then, by Lemma 4.2.1 we can conclude that $E \leq E^{\sim\sim}$. Let ψ any vector of $\mathcal{H}$. By the projection theorem, $\psi = \psi_1 + \psi_2$, where $\psi_1 \in Ker(E) = Ran(E)'$ and $\psi_2 \in Ker(E)'$. Thus, $\langle\, \psi | E\psi \,\rangle = \langle\, \psi_1 + \psi_2 | E\psi_1 + E\psi_2 \,\rangle = \langle\, \psi_1 + \psi_2 | E\psi_2 \,\rangle = \langle\, \psi_1 | E\psi_2 \,\rangle + \langle\, \psi_2 | E\psi_2 \,\rangle$. Since $\psi_1 \in Ran(E)'$ and $E\psi_2 \in Ran(E)$, we obtain $\langle\, \psi_1 | E\psi_2 \,\rangle = 0$. Hence, $\langle\, \psi | E\psi \,\rangle = \langle\, \psi_2 | E\psi_2 \,\rangle$. Similarly, one can show that $\langle\, \psi | P_{Ker(E)'}\psi \,\rangle = \langle\, \psi_2 | \psi_2 \,\rangle$. By Lemma 4.1.2, $\langle\, \psi_2 | E\psi_2 \,\rangle \leq \langle\, \psi_2 | \psi_2 \,\rangle$. Consequently, $\langle\, \psi | E\psi \,\rangle \leq \langle\, \psi | P_{Ker(E)'}\psi \,\rangle = \langle\, \psi | E^{\sim\sim}\psi \,\rangle$.
(iic) $E \leq F$ implies $F^{\sim} \leq E^{\sim}$. Suppose $E \leq F$. To prove $F^{\sim} \leq E^{\sim}$ it is sufficient to show that $Ker(F) \subseteq Ker(E)$. Let $\psi \in Ker(F)$. Then, $F\psi = \underline{0}$. Consequently, $\langle\, \psi | E\psi \,\rangle = 0$ since $E \leq F$. By Lemma 4.2.2(i), we can conclude that $\psi \in Ker(E)$.
(iia) $E \wedge E^{\sim} = \mathbf{0}$. Let F be an effect such that $F \leq E,\ E^{\sim}$. By (iic) and (iii), we have $F^{\sim\sim} \leq E^{\sim\prime}$ and $F^{\sim\sim} \leq E^{\sim}$. Since $E^{\sim\prime}$ and $E^{\sim}$ are projections, we have: $E^{\sim\prime} \wedge E^{\sim} = \mathbf{0}$. Hence, $F^{\sim\sim} = \mathbf{0}$ and $F = \mathbf{0}$.

□

An interesting feature of the Brouwer Zadeh structures (BZ-posets and BZ-lattices) is the possibility to define two unary operations ν and μ, which turn out to behave as the modal operators *necessarily* and *possibly*, respectively.

DEFINITION 4.2.8. *The modal operators*
Let $\langle B\,, \leq\,, {}'\,, {}^{\sim}\,, \mathbf{0}\,, \mathbf{1}\rangle$ be a Brouwer Zadeh poset.

$$\nu(a) := a^{\prime\sim}; \quad \mu(a) := a^{\sim\prime}.$$

In other words, *necessity* is identified with the intuitionistic negation of the fuzzy negation, while *possibility* is identified with the fuzzy negation of the intuitionistic negation.

The modal operators ν and μ turn out to have a typical $\mathbf{S}_5$-like behavior.[2] For, the following conditions are satisfied:

- $\nu(a) \leq a$
  Necessarily a implies a.
- If $a \leq b$, then $\nu(a) \leq \nu(b)$
  If a implies b, then the necessity of a implies the necessity of b.
- $a \leq \nu(\mu(a))$
  a implies the necessity of its possibility.

[2]For the readers who are not familiar with modal logics, we recall that $\mathbf{S}_5$ represents a quite strong system in the family of all modal logics that are extensions of classical logics. The language of $\mathbf{S}_5$ contains the modal operators L (necessarily) and M (possibly), besides the classical connectives $\neg$ (not), $\curlywedge$ (and), $\curlyvee$ (or), $\rightarrow$ (if ... then), $\leftrightarrow$ (if and only if). Some important logical laws of $\mathbf{S}_5$ are the following (where $\alpha, \beta, \ldots$ represent any sentences of the language): 1) $L(\alpha \rightarrow \beta) \rightarrow (L\alpha \rightarrow L\beta)$; 2) $\alpha \rightarrow LM\alpha$; 3) $LL\alpha \leftrightarrow L\alpha$; 4) $LM\alpha \leftrightarrow M\alpha$. See, for instance, (Hughes and Cresswell, 1984).

- $\nu(\nu(a)) = \nu(a)$
  Necessity is equivalent to the necessity of the necessity.
- $\nu(\mu(a)) = \mu(a)$
  The necessity of a possibility is equivalent to the possibility.

Of course, in any BZ-poset $\langle B, \leq, ', ^\sim, \mathbf{0}, \mathbf{1}\rangle$ where the two complements $'$ and $^\sim$ coincide, we obtain a collapse of the modalities. In other terms, $\nu(a) = a = \mu(a)$.

Let us now return to concrete Brouwer Zadeh posets

$$\left\langle \mathcal{E}(\mathcal{H}), \leq, ', ^\sim, \mathbf{0}, \mathbf{1}\right\rangle,$$

and consider the necessity $\nu(E)$ of a given effect E (which may be either sharp or unsharp). One can easily prove the following lemma.

LEMMA 4.2.9.
(i) *E is a projection iff $E = \nu(E) = E'^{\sim} = P_{Ker(E')}$.*
(ii) *Let P be any projection.*

$$P \leq E \textit{ implies } P \leq \nu(E).$$

As a consequence, we can say that $\nu(E)$ represents a kind of "best sharp lower approximation of E." These important properties of both abstract and concrete Brouwer Zadeh structures are further analyzed in Section 4.4 and in Chapter 6.

4.3. Mac Neille completions

As we have seen, $\mathcal{E}(\mathcal{H})$ is not a lattice. A lattice structure, however, can be obtained by embedding $\mathcal{E}(\mathcal{H})$ into its *Mac Neille completion* $\overline{\mathcal{E}(\mathcal{H})}$.

Consider a generic bounded involution poset $\langle B, \leq, ', \mathbf{0}, \mathbf{1}\rangle$. For any nonempty subset X of B, let $l(X)$ and $u(X)$ represent the set of all lower bounds and the set of all upper bounds of X, respectively. In other words,

$$l(X) = \{a \in B : \forall b \in X[a \leq b]\};$$
$$u(X) = \{a \in B : \forall b \in X[b \leq a]\}.$$

Let $MC(B) := \{X \subseteq B \mid X = l(u(X))\}$.
It turns out that $X \in MC(B)$ iff $X = X^{\perp\perp}$, where

$$X^{\perp} := \{a \in B : \forall b \in X[a \leq b']\}.$$

Moreover, the structure

$$\overline{\mathcal{B}} = \left\langle MC(B), \subseteq, ^{\perp}, \{\mathbf{0}\}, B\right\rangle$$

is a complete bounded involution lattice (which is regular if $\mathcal{B}$ is regular), where $X \wedge Y = X \cap Y$ and $X \vee Y = (X \cup Y)^{\perp\perp}$. We call the lattice $\overline{\mathcal{B}}$ the *Mac Neille completion* of $\mathcal{B}$.
It turns out that $\mathcal{B}$ is embeddable into $\overline{\mathcal{B}}$, via the map $h : a \rightarrowtail (a]$, where $(a]$ is the *principal ideal* generated by a. In other words:

$$h(a) = (a] := \{b \in B : b \leq a\}.$$

Such an embedding preserves all *infima* and *suprema* existing in $\mathcal{B}$.

The Mac Neille completion of a bounded involution poset does not generally satisfy the noncontradiction principle ($a \wedge a' = \mathbf{0}$) and the excluded middle principle ($a \vee a' = \mathbf{1}$). In particular, the Mac Neille completion $\overline{\mathcal{E}(\mathcal{H})}$ of the regular poset $\mathcal{E}(\mathcal{H})$ is not an ortholattice.

Such a construction can be also extended to BZ-posets.

Let $\mathcal{B}_{BZ} = \langle B, \leq, ', ^{\sim}, \mathbf{0}, \mathbf{1} \rangle$ be a BZ-poset and let $\overline{\mathcal{B}}$ the Mac Neille completion of the regular bounded involution poset $\langle B, \leq, ', \mathbf{0}, \mathbf{1} \rangle$. For any nonempty subset X of B, let

$$X^{\sim} := \{a \in B : \forall b \in X [a \leq b^{\sim}]\}.$$

It turns out that the structure $\overline{\mathcal{B}_{BZ}} = \langle MC(B), \subseteq, ^{\perp}, ^{\sim}, \{\mathbf{0}\}, B \rangle$ is a complete BZ-lattice, which $\mathcal{B}$ can be embedded into, via the map h defined above (Giuntini, 1990).

4.4. Unsharp preclusivity spaces

In Chapter 2 we have studied how particular examples of preclusivity spaces naturally arise in the framework of abstract sharp QT. As a concrete application, any Hilbert event-state system $(\Pi(\mathcal{H}), \mathcal{S}(\mathcal{H}))$ gives rise to the following preclusivity space

$$(\mathcal{S}(\mathcal{H}), \perp),$$

where

- $\mathcal{S}(\mathcal{H})$ is the set of all states m_ρ, that are determined by density operators ρ of $\mathcal{H}$;
- the preclusivity relation $\perp$ is defined as follows: $\forall \rho, \sigma \in D(\mathcal{H})$,

$$m_\rho \perp m_\sigma \text{ iff } \exists P \in \Pi(\mathcal{H})[m_\rho(P) = 1 \text{ and } m_\sigma(P) = 0].$$

Let us now turn to the case of unsharp Hilbert space QT. As in the sharp case, any density operator ρ determines a function m_ρ such that $\forall E \in \mathcal{E}(\mathcal{H})$,

$$m_\rho(E) = \mathrm{Tr}(\rho E).$$

The set

$$\mathcal{S}(\mathcal{H}) := \{m_\rho : \rho \in D(\mathcal{H})\}$$

represents the set of all possible states that assign to any effect a probability-value.

Consider now the pair $(\mathcal{E}(\mathcal{H}), \mathcal{S}(\mathcal{H}))$. By generalizing the sharp situation, we call such a pair a *Hilbert effect-state system*.

We now investigate the preclusivity spaces that arise in the framework of such unsharp systems. In this particular case, the natural definition of the preclusivity relation is the following: $\forall \rho, \sigma \in D(\mathcal{H})$,

$$m_\rho \perp_{\mathcal{E}(\mathcal{H})} m_\sigma \text{ iff } \exists E \in \mathcal{E}(\mathcal{H})[m_\rho(E) = 1 \text{ and } m_\sigma(E) = 0].$$

In other words, two states are preclusive iff they are strongly distinguished by at least one effect.

LEMMA 4.4.1. $\forall E \in \mathcal{E}(\mathcal{H})\, \forall \rho \in D(\mathcal{H})\, [m_\rho(E) = 1$ *iff* $m_\rho(E'^{\sim}) = 1]$.

PROOF. Necessity follows from the fact that $E'^{\sim} \le E$.
Suppose that $m_\rho(E) = 1$. By Lemma 4.2.2(iii), $\rho = \sum_{i \in I} \lambda_i P_{[\psi_i]}$, where $\forall i \in I$: $E\psi_i = \psi_i$. Thus, $\forall i \in I$: $\psi_i \in Ker(E')$. Therefore, $E'^{\sim} P_{[\psi_i]} = P_{Ker(E')} P_{[\psi_i]} = P_{[\psi_i]}$. Hence, $m_\rho(E'^{\sim}) = \mathrm{Tr}\left(\sum_{i \in I} \lambda_i P_{[\psi_i]} E'^{\sim}\right) = \mathrm{Tr}\left(\sum_{i \in I} \lambda_i P_{[\psi_i]}\right) = 1$. □

Interestingly enough, the relations $\perp$ and $\perp_{\mathcal{E}(\mathcal{H})}$ turn out to be equivalent, as shown by the following lemma:

LEMMA 4.4.2. $\forall \rho, \sigma \in D(\mathcal{H})[m_\rho \perp_{\mathcal{E}(\mathcal{H})} m_\sigma$ *iff* $m_\rho \perp m_\sigma]$.

PROOF. Necessity is trivial. Suppose $m_\rho \perp_{\mathcal{E}(\mathcal{H})} m_\sigma$. Then, $\exists E \in \mathcal{E}(\mathcal{H})$ such that $m_\rho(E) = 1$ and $m_\sigma(E) = 0$. We want to show that $m_\rho \perp m_\sigma$. By hypothesis, $m_\rho(E) = \mathrm{Tr}(\rho E) = 1$. Thus, by Lemma 4.4.1, $m_\rho(E'^{\sim}) = 1$. Since $E'^{\sim} \le E$ and $m_\sigma(E) = 0$, we obtain $m_\sigma(E'^{\sim}) = 0$. Since $E'^{\sim}$ is a projection we can conclude that $m_\rho \perp m_\sigma$.

□

As a consequence, it also makes sense to refer to the preclusivity space $(\mathcal{S}(\mathcal{H}), \perp)$ in the unsharp case.

In Chapter 2, we have investigated the properties of the map **Yes** that associates to any event E the set of all states assigning to E probability-value 1. As noticed earlier, from the semantic point of view, the set $\mathbf{Yes}(E)$ can be regarded as the *proposition* associated to the event E: the set of all *possible worlds* that certainly verify the event. We have seen (Corollary 2.4.11) that in the case of *normal* event-state systems, for every event E, the set $\mathbf{Yes}(E)$ is a *closed subset* of the preclusivity space. In other words:

$$\mathbf{Yes}(E)^{\perp\perp} = \mathbf{Yes}(E).$$

Now, any Hilbert event-state system $(\Pi(\mathcal{H}), \mathcal{S}(\mathcal{H}))$ is normal. Hence, we obtain, $\forall P \in \Pi(\mathcal{H})$,

$$\mathbf{Yes}(P)^{\perp\perp} = \mathbf{Yes}(P).$$

What happens in the case of effects? The map **Yes** can be naturally defined for effects, as in the case of events. Hence, for any effect E, $\mathbf{Yes}(E)$ will contain all the states that assign to E probability-value 1. An effect-state system $(\mathcal{E}(\mathcal{H}), \mathcal{S}(\mathcal{H}))$ does not satisfy the characteristic properties of normal event-state systems. In fact, unlike the sharp case, the effect order is not completely determined by the behavior of all states with respect to the certain probability-value. For instance:

$$\mathbf{Yes}(\frac{3}{4}\mathbb{1}) = \mathbf{Yes}(\frac{1}{2}\mathbb{1}) = \emptyset; \quad \frac{3}{4}\mathbb{1} \not\le \frac{1}{2}\mathbb{1}.$$

In spite of this, one can prove that for any effect E, $\mathbf{Yes}(E)$ is a closed subset of the preclusivity space:

LEMMA 4.4.3. $\forall E \in \mathcal{E}(\mathcal{H})[\mathbf{Yes}(E)^{\perp\perp} = \mathbf{Yes}(E)]$.

PROOF. By Lemma 4.4.1, we have that

$$\forall E \in \mathcal{E}(\mathcal{H})\, \forall \rho \in D(\mathcal{H}) \left[m_\rho \in \mathbf{Yes}(E) \text{ implies } m_\rho \in \mathbf{Yes}(E'^{\sim}) \right].$$

At the same time, $\mathbf{Yes}(E'^{\sim}) \subseteq \mathbf{Yes}(E)$, because $E'^{\sim} \leq E$. Therefore, $\mathbf{Yes}(E'^{\sim}) = \mathbf{Yes}(E)$. Now, $E'^{\sim}$ is a projection. As a consequence, by Theorem 3.3.5 and by Corollary 2.4.11, $\mathbf{Yes}(E'^{\sim}) = \mathbf{Yes}(E'^{\sim})^{\perp\perp}$. Therefore, $\mathbf{Yes}(E) = \mathbf{Yes}(E)^{\perp\perp}$. □

As we have seen earlier, in the case of sharp Hilbert event-state systems (which are *supernormal*), the following structures turn out to be isomorphic:

- the complete ortholattice of all sharp events $\langle \Pi(\mathcal{H}), \subseteq, ', \mathbf{0}, \mathbf{1} \rangle$;
- the complete ortholattice $\langle \mathbb{C}(\mathcal{P}(\mathcal{S}(\mathcal{H}))), \subseteq, {}^{\perp}, \emptyset, \mathcal{S} \rangle$, based on the set of all closed sets of states of the preclusivity space $(\mathcal{S}(\mathcal{H}), \perp)$. By Lemma 4.4.2, this ortholattice coincides with the ortholattice of all closed sets of the preclusivity space $\left(\mathcal{S}(\mathcal{H}), \perp_{\mathcal{E}(\mathcal{H})}\right)$. As a consequence, we are entitled to use unambiguously the expression "closed set of states."

For any projection P, we let S^P indicate the corresponding closed set of states; while, for any closed set of states S, the corresponding projection is indicated by P^S.

As a consequence of this isomorphism, any effect E (which may be either sharp or unsharp) is associated, via the map **Yes**, to a privileged sharp event: the projection $P^{\mathbf{Yes}(E)}$ that corresponds to the closed set $\mathbf{Yes}(E)$. We call such a projection the *sharp representative* of E. Of course, instead of $P^{\mathbf{Yes}(E)}$, one could equivalently refer to the corresponding closed subspace $X_{P^{\mathbf{Yes}(E)}}$ (in $\mathcal{C}(\mathcal{H})$). It turns out that the sharp representative of any effect E coincides with the necessity $\nu(E)$ of E as the following lemma shows.

LEMMA 4.4.4. *For any effect $E \in \mathcal{E}(\mathcal{H})$,*

$$P^{\mathbf{Yes}(E)} = \nu(E) = E'^{\sim} = P_{Ker(E')} = P_{Ran(E')'}.$$

PROOF. The proof follows immediately from Lemma 4.4.1 and Lemma 4.2.2(iv). □

Of course, the sharp representative of a sharp effect is the effect itself:

$$\forall P \in \Pi(\mathcal{H}), P^{\mathbf{Yes}(P)} = P.$$

At the same time, it may happen that two different effects have one and the same sharp representative. For instance, the sharp representative of both $\frac{1}{2}\mathbb{I}$ and $\frac{3}{4}\mathbb{I}$ turn out to be the null projection $\mathbb{O}$ (the *impossible* sharp event).

We will use the following terminology (for any effect E):

- the closed set of states $\mathbf{Yes}(E)$ will be called the *positive proposition* (or also the *Yes-proposition*) of E.

- By *positive pure proposition* (or *Yes-pure proposition*) of E we will mean the subset of the sharp representative $X_{p\mathbf{Yes}(E)}$ of E, whose elements are pure states (unit vectors). Hence (by Lemma 4.2.2(ii)), we will obtain: a pure state ψ belongs to the positive pure proposition of E iff $\langle\psi|E\psi\rangle = 1$.

Both in the sharp and in the unsharp case, one can naturally introduce, besides the map **Yes**, the dual map **No**. The expected definition of **No** is the following, for any effect E of $\mathcal{E}(\mathcal{H})$ (which may be either sharp or unsharp):

$$\mathbf{No}(E) := \{m_\rho \in \mathcal{S}(\mathcal{H}) : m_\rho(E) = 0\}.$$

The set $\mathbf{No}(E)$ will be called the *negative proposition* (or the *No-proposition*) of E. Needless to say, $\mathbf{No}(E) = \mathbf{Yes}(E')$.

It follows immediately that the set $\mathbf{No}(E)$ is closed: for any effect E,

$$\mathbf{No}(E)^{\perp\perp} = \mathbf{No}(E).$$

Furthermore, one can easily show that:

$$\mathbf{No}(E) \perp \mathbf{Yes}(E).$$

At the same time, only sharp events (projections) generally satisfy the following important property:

$$\mathbf{No}(E) = \mathbf{Yes}(E)^{\perp}.$$

$$\mathbf{Yes}(E) = \mathbf{No}(E)^{\perp}.$$

This means that the positive proposition of a sharp effect determines the corresponding negative proposition (and vice versa). Such a perfect correspondence breaks down in the case of unsharp effects. It may happen that two different effects have one and the same positive proposition; at the same time the two corresponding negative propositions are different. For instance, take the effects $\mathbb{O}$ and $\frac{1}{2}\mathbb{I}$. We have:

$$\mathbf{Yes}(\mathbb{O}) = \mathbf{Yes}(\frac{1}{2}\mathbb{I}) = \emptyset;$$

$$\mathbf{No}(\mathbb{O}) = \mathcal{S}(\mathcal{H}); \;\; \mathbf{No}(\frac{1}{2}\mathbb{I}) = \emptyset.$$

Accordingly, in the framework of effect-state systems, it is interesting to introduce the notion of *pair-proposition*.

DEFINITION 4.4.5. *Pair-proposition*

For any effect E, the *pair-proposition* of E (also called the *Yes-No proposition* of E) is the pair $(\mathbf{Yes}(E), \mathbf{No}(E))$.

As we will see in Part II, pair-propositions will play an important role in the semantics for a particular form of *Brouwer Zadeh logic*.

CHAPTER 5

Effect algebras and quantum MV algebras

In Chapter 4 we have seen how the system of all effects in a Hilbert space gives rise to a "genuine" Brouwer Zadeh poset (which is neither an orthoposet nor a lattice). We will now study other interesting ways of structuring the set of all concrete effects. In particular, we will see how *effect algebras* and *quantum MV algebras* naturally emerge from effect-systems.

5.1. Effect algebras and Brouwer Zadeh effect algebras

We will first investigate effect algebras (also called *unsharp orthoalgebras*).[1] One is dealing with a particular kind of *partial structure*, equipped with a basic operation $\boxplus$ that is only defined for special pairs of elements. From an intuitive point of view, such an operation can be regarded as an *exclusive disjunction* (*aut*), defined for events that are *logically incompatible*.

The abstract definition of *effect algebra* is the following.

DEFINITION 5.1.1. *Effect algebra*
An *effect algebra* (EA) is a partial structure

$$\mathcal{A} = \langle A\,, \boxplus\,, \mathbf{0}\,, \mathbf{1}\rangle\,,$$

where $\boxplus$ is a partial binary operation on A, and $\mathbf{0}$ and $\mathbf{1}$ are special distinct elements of A. When $\boxplus$ is defined for a pair $a\,, b \in A$, we will write $\exists(a \boxplus b)$.

The following conditions hold:

(i) *Weak commutativity*
$\exists(a \boxplus b)$ implies $\exists(b \boxplus a)$ and $a \boxplus b = b \boxplus a$;

(ii) *Weak associativity*
$\exists(b \boxplus c)$ and $\exists(a \boxplus (b \boxplus c))$ implies $\exists(a \boxplus b)$ and $\exists((a \boxplus b) \boxplus c)$ and $a \boxplus (b \boxplus c) = (a \boxplus b) \boxplus c$;

(iii) *Strong excluded middle*
For any a, there exists a unique x such that $a \boxplus x = \mathbf{1}$;

(iv) *Weak consistency*
$\exists(a \boxplus \mathbf{1})$ implies $a = \mathbf{0}$.

An orthogonality relation $\perp$, a partial order relation $\leq$ and a generalized complement $'$ (which generally behaves as a fuzzy complement) can be defined in any effect algebra.

[1]See (Giuntini and Greuling, 1989; Dalla Chiara and Giuntini, 1994; Foulis and Bennett, 1994; Dvurečenskij and Pulmannová, 2000).

DEFINITION 5.1.2. Let $\langle A, \boxplus, \mathbf{0}, \mathbf{1}\rangle$ be an effect algebra and let $a, b \in A$.

(i) $a \perp b$ iff $a \boxplus b$ is defined in A.
(ii) $a \leq b$ iff $\exists c \in A$ such that $a \perp c$ and $b = a \boxplus c$.
(iii) The *generalized complement* of a is the unique element a' such that $a \boxplus a' = \mathbf{1}$.

THEOREM 5.1.3. *Let $\langle A, \boxplus, \mathbf{0}, \mathbf{1}\rangle$ be an effect algebra. For any $a, b \in A$ the following conditions hold:*

(i) $a \perp b$ *implies* $b \perp a$;
(ii) $a'' = a$;
(iii) $\mathbf{1}' = \mathbf{0}$ *and* $\mathbf{0}' = \mathbf{1}$;
(iv) $a \perp \mathbf{0}$ *and* $a \boxplus \mathbf{0} = a$;
(v) *If* $a \perp b$ *and* $a \boxplus b = \mathbf{0}$, *then* $a = b = \mathbf{0}$;
(vi) *If* $a \perp b$, *then* $a \perp (a \boxplus b)'$ *and* $b' = a \boxplus (a \boxplus b)'$;
(vii) $a \perp b$ *iff* $a \leq b'$;
(viii) *If* $a \leq b$ *then* $b = a \boxplus (a \boxplus b')'$

PROOF.

(i) The proof follows from weak commutativity.
(ii) By the strong excluded middle condition, $a \boxplus a' = \mathbf{1}$ and $a' \boxplus a'' = \mathbf{1}$. By weak commutativity, $a' \boxplus a = \mathbf{1}$; hence, by the strong excluded middle condition again, $a = a''$.
(iii) By the strong excluded middle condition, $\mathbf{1} \boxplus \mathbf{1}' = \mathbf{1}$. Thus, by weak commutativity and weak consistency, $\mathbf{1}' = \mathbf{0}$; therefore, by (ii), $\mathbf{0}' = \mathbf{1}'' = \mathbf{1}$.
(iv) $\mathbf{1} = \mathbf{1} \boxplus \mathbf{1}' = (a \boxplus a') \boxplus \mathbf{1}' = (a \boxplus a') \boxplus \mathbf{0} = (a' \boxplus a) \boxplus \mathbf{0} = a' \boxplus (a \boxplus \mathbf{0})$. Hence, by strong excluded middle: $a \boxplus \mathbf{0} = a'' = a$.
(v) Suppose $a \perp b$ and $a \boxplus b = \mathbf{0}$. Now, $a = a \boxplus \mathbf{0} = a \boxplus (a \boxplus b)$. Thus, $\mathbf{1} = (a \boxplus (a \boxplus b)) \boxplus a' = a \boxplus ((a \boxplus b) \boxplus a') = a \boxplus ((b \boxplus a) \boxplus a') = a \boxplus (b \boxplus (a \boxplus a')) = a \boxplus (b \boxplus \mathbf{1})$. Hence, $b \boxplus \mathbf{1}$ is defined, and by weak consistency, we obtain $b = \mathbf{0}$. Similarly, it follows that $a = \mathbf{0}$.
(vi) Suppose $a \perp b$. Then, $\mathbf{1} = (a \boxplus b) \boxplus (a \boxplus b)' = a \boxplus (b \boxplus (a \boxplus b)') = (b \boxplus (a \boxplus b)') \boxplus a = b \boxplus ((a \boxplus b)' \boxplus a) = b \boxplus (a \boxplus (a \boxplus b)')$. Thus, $a \perp (a \boxplus b)'$ and, by the strong excluded middle condition, $b' = a \boxplus (a \boxplus b)'$.
(vii) Suppose $a \perp b$. Then, by (vi), $b' = a \boxplus (a \boxplus b)'$. Thus, by definition of $\leq$, we obtain $a \leq b'$. Vice versa, suppose $a \leq b'$. Then, $\exists c \in A$ such that $a \perp c$ and $b' = a \boxplus c$. Thus, $\mathbf{1} = (a \boxplus c) \boxplus b = (c \boxplus a) \boxplus b = c \boxplus (a \boxplus b)$. Therefore, $a \boxplus b$ is defined, i.e., $a \perp b$.
(viii) Suppose $a \leq b$. Then, $a \leq b''$. By (vii), $a \perp b'$. Thus, by (vi), $b = b'' = a \boxplus (a \boxplus b')'$.

□

LEMMA 5.1.4. *Cancellation law*
Let $\langle A, \boxplus, \mathbf{0}, \mathbf{1}\rangle$ be an effect algebra, and let a, b, c be three elements of A such that $a \perp c$ and $b \perp c$. Then,

(i) $a \boxplus c = b \boxplus c$ *implies* $a = b$.
(ii) $a \boxplus c \leq b \boxplus c$ *iff* $a \leq b$.

PROOF.

(i) Suppose $a \boxplus c = b \boxplus c$. Then, $\mathbf{1} = (a \boxplus c) \boxplus (a \boxplus c)' = (b \boxplus c) \boxplus (a \boxplus c)'$. Thus, $a \boxplus (c \boxplus (a \boxplus c)') = b \boxplus (c \boxplus (a \boxplus c)') = \mathbf{1}$. By the strong excluded middle condition, we obtain $a = b$.

(ii) Suppose $a \boxplus c \leq b \boxplus c$. Then, $\exists d \in A$ such that $d \perp (a \boxplus c)$ and $b \boxplus c = (a \boxplus c) \boxplus d = a \boxplus (d \boxplus c) = (a \boxplus d) \boxplus c$. By (i), $b = a \boxplus d$, so that $a \leq b$. Suppose $a \leq b$. By Theorem 5.1.3(vii), $b = a \boxplus (a \boxplus b')'$. Consequently:

$$b \boxplus c = (a \boxplus (a \boxplus b')') \boxplus c = (a \boxplus c) \boxplus (a \boxplus b')'.$$

Thus, by definition of $\leq$, $a \boxplus c \leq b \boxplus c$.

□

The following theorem shows that any effect algebra gives rise to a bounded involution poset.

THEOREM 5.1.5. *Let* $\langle A, \boxplus, \mathbf{0}, \mathbf{1}\rangle$ *be an effect algebra and define* $\leq$ *and* $'$ *according to Definition 5.1.2. Then,* $\langle A, \leq, ', \mathbf{0}, \mathbf{1}\rangle$ *is a bounded involution poset.*

PROOF. (1) Let us first prove that $\leq$ is a partial order.
(1i) *Reflexivity* follows from Theorem 5.1.3(iv).
(1ii) *Antisymmetry*. Suppose $a \leq b$ and $b \leq a$. We have to prove that $a = b$. By hypothesis: $\exists c, d \in A$ such that $a \perp c$, $b \perp d$ and $b = a \boxplus c, a = b \boxplus d$. Then, $(a \boxplus \mathbf{0}) = a = b \boxplus d = (a \boxplus c) \boxplus d = a \boxplus (c \boxplus d)$. By cancellation law, $c \boxplus d = \mathbf{0}$. Thus, by Theorem 5.1.3(v), $c = d = \mathbf{0}$. Then, $a = b \boxplus d = b \boxplus \mathbf{0} = b$.
(1iii) *Transitivity*. Suppose $a \leq b$ and $b \leq c$. We have to prove that $a \leq c$. By hypothesis: $\exists d, e \in A$ such that $a \perp d$, $b \perp e$ and $b = a \boxplus d, c = b \boxplus e$. Then, $c = (a \boxplus d) \boxplus e = a \boxplus (d \boxplus e)$. Thus, $a \leq c$.
(2) That $\mathbf{0}$ is the *minimum* of the poset follows from Theorem 5.1.3(iv). That $\mathbf{1}$ is the *maximum* of the poset follows from the strong excluded middle condition.
(3) Let us prove that $'$ is an *involution*.
(3i) By Theorem 5.1.3(ii): $\forall a \in A$: $a'' = a$.
(3ii) Suppose $a \leq b$. We have to prove that $b' \leq a'$. By Theorem 5.1.3(vii), $a \perp b'$ and therefore, $b' \perp a$. Again, by Theorem 5.1.3(vii), $b' \leq a'$. □

The category of all effect algebras turns out to be (categorically) equivalent to the category of all *difference posets*, which have been first studied by Kôpka and Chovanec (1994) and further investigated by Pulmannová (1995) and others.

Effect algebras that satisfy the noncontradiction principle are called *orthoalgebras*. In such algebras, the involution $'$ becomes an orthocomplementation.

DEFINITION 5.1.6. (Foulis and Randall, 1981) *Orthoalgebras*
An *orthoalgebra* (OA) is an effect algebra $\langle A, \boxplus, \mathbf{0}, \mathbf{1}\rangle$ such that the following condition is satisfied:

$$\exists(a \boxplus a) \text{ implies } a = \mathbf{0} \quad (\textit{Strong consistency}).$$

In other words: **0** is the only element that is orthogonal to itself.

One can easily realize that orthoalgebras always determine an orthoposet. Let $\mathcal{A} = \langle A, \boxplus, \mathbf{0}, \mathbf{1}\rangle$ be an orthoalgebra. The structure

$$\langle A, \leq, ', \mathbf{0}, \mathbf{1}\rangle$$

(where $\leq$ and $'$ are the partial order and the generalized complement of $\mathcal{A}$) is an *orthoposet*. For, given any $a \in A$, the infimum $a \wedge a'$ exists and is equal to **0**; equivalently, the supremum $a \vee a'$ exists and is equal to **1**.

THEOREM 5.1.7. *Any orthoalgebra $\mathcal{A} = \langle A, \boxplus, \mathbf{0}, \mathbf{1}\rangle$ satisfies the following condition: if $a, b \in A$ and $a \perp b$, then $a \boxplus b$ is a minimal upper bound for a and b in $\mathcal{A}$.*

PROOF. Suppose $a \boxplus b$ is defined. Clearly, $a, b \leq a \boxplus b$. Let $a, b \leq c \leq a \boxplus b$, for a given c. We show that $c = a \boxplus b$. By hypothesis, $c = a \boxplus d$ and $c = b \boxplus e$, for some $d, e \in A$ such that $a \perp d$ and $b \perp e$. Furthermore, $a \boxplus b = c \boxplus f$, for some $f \in A$ such that $f \perp c$. We show that $f = \mathbf{0}$. The following holds:

$$a \boxplus b = c \boxplus f = (a \boxplus d) \boxplus f = a \boxplus (d \boxplus f), \text{ and}$$

$$a \boxplus b = c \boxplus f = (b \boxplus e) \boxplus f = b \boxplus (e \boxplus f) = (e \boxplus f) \boxplus b.$$

By the cancellation law (Lemma 5.1.4 (i)), $b = d \boxplus f$ and $a = (e \boxplus f)$. Since $a \boxplus b$ is defined, we obtain:

$$a \boxplus b = (e \boxplus f) \boxplus (d \boxplus f) = (e \boxplus d) \boxplus (f \boxplus f).$$

Thus, $f \boxplus f$ is defined. Since $\mathcal{A}$ is an orthoalgebra, by the strong consistency property, we obtain: $f = \mathbf{0}$.

□

COROLLARY 5.1.8. *Any orthoalgebra $\mathcal{A} = \langle A, \boxplus, \mathbf{0}, \mathbf{1}\rangle$ satisfies the following condition: for any $a, b \in A$ such that $a \perp b$, if the supremum $a \vee b$ exists, then $a \vee b = a \boxplus b$.*

Orthoalgebras and orthomodular posets turn out to be deeply connected. Any orthomodular poset $\langle A, \leq, ', \mathbf{0}, \mathbf{1}\rangle$ determines an orthoalgebra

$$\langle A, \boxplus, \mathbf{0}, \mathbf{1}\rangle,$$

where: $a \boxplus b$ is defined iff $a \leq b'$. Furthermore, when defined, $a \boxplus b = a \vee b$. At the same time, not every orthoalgebra is an orthomodular poset! (see Example 5.1.11).

DEFINITION 5.1.9. An orthoalgebra $\mathcal{A} = \langle A, \boxplus, \mathbf{0}, \mathbf{1} \rangle$ is called *orthocoherent* iff $\forall a, b, c \in A$,

$$a \perp b, a \perp c, b \perp c \text{ implies } a \perp (b \boxplus c).$$

THEOREM 5.1.10. (Dvurečenskij and Pulmannová, 2000)
Let $\mathcal{A} = \langle A, \boxplus, \mathbf{0}, \mathbf{1} \rangle$ be an orthoalgebra (with partial order $\leq$ and generalized complement $'$). The following conditions are equivalent:

(i) *$\langle A, \leq, ', \mathbf{0}, \mathbf{1} \rangle$ is an orthomodular poset;*
(ii) *$\forall a, b \in A[a \perp b$ implies $a \vee b$ exists];*
(iii) *$\mathcal{A}$ is orthocoherent.*

EXAMPLE 5.1.11. *The Wright triangle*
As shown by Wright (1990), the smallest orthoalgebra that is not orthocoherent is based on the set

$$WT = \{\mathbf{0}, \mathbf{1}, a, b, c, d, e, f, a', b', c', d', e', f'\},$$

where, apart the obvious cases, the partial sum $\boxplus$ is defined as follows:

$$a \boxplus b = d \boxplus e = c'$$
$$b \boxplus c = e \boxplus f = a'$$
$$c \boxplus d = f \boxplus a = e'$$
$$c \boxplus e = d', \ a \boxplus c = b', \ e \boxplus a = f'.$$

Let us check that this orthoalgebra $\langle WT, \boxplus, \mathbf{0}, \mathbf{1} \rangle$ is not orthocoherent. To see this, note that $a \perp c$, $a \perp e$ and $c \perp e$ but not $a \perp (c \boxplus e) = d'$ since $a \not\leq d$. Greechie diagrams (See Section 1.1 and 6.5) also work for orthoalgebras. One can represent the atoms of each maximal Boolean subalgebra of a finite orthoalgebra by points on a line. The Greechie diagram for the Wright triangle, $\mathcal{WT}$, is given in Figure 5.1.1. It is the smallest orthoalgebra that is not an orthomodular poset.

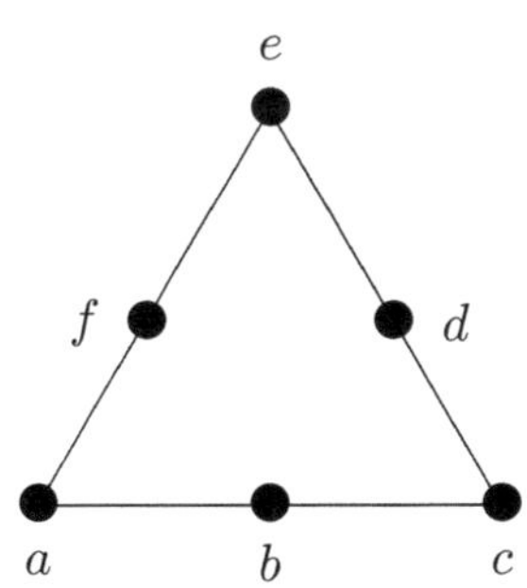

FIGURE 5.1.1. $\mathcal{WT}$: the smallest OA that is not an OMP

Genuine examples of effect algebras (which are not generally orthoalgebras) can be naturally obtained in the domain of fuzzy set systems.

EXAMPLE 5.1.12. *Effect algebras of fuzzy sets*
Let B be the set of all fuzzy subsets of a universe U (see Definition 1.1.27).

A partial operation $\boxplus$ can be defined on B. For any $f, g \in B$:

$$\exists(f \boxplus g) \text{ iff } \forall x \in U: \ f(x) + g(x) \leq 1,$$

where $+$ is the usual sum of real numbers. Furthermore:

$$\text{if } \exists(f \boxplus g), \text{ then } \quad f \boxplus g := f + g,$$

where:

$$\forall x \in U \{(f+g)(x) := f(x) + g(x)\}.$$

Let $\mathbf{1}$ be the classical characteristic function of the total set U, while $\mathbf{0}$ is the classical characteristic function of the empty set $\emptyset$.

The structure $\langle B, \boxplus, \mathbf{0}, \mathbf{1} \rangle$ is an effect algebra.
It turns out that the effect-algebra generalized complement $'$ coincides with the fuzzy complement. In other words:

$$\forall x \in U : f'(x) = 1 - f(x).$$

Furthermore, the effect-algebra partial order relation coincides with the natural partial order of B. In other words:

$$\forall x \in U \, [f(x) \leq g(x)] \text{ iff } \exists h \in B[f \perp h \text{ and } g = f \boxplus h].$$

The effect algebra $\langle B, \boxplus, \mathbf{0}, \mathbf{1} \rangle$ is not an orthoalgebra, because the strong consistency condition is violated by some genuine fuzzy sets (such as the *semitransparent* fuzzy set $\frac{1}{2}\mathbf{1}$ that assigns to any object x value $\frac{1}{2}$).

How can we induce the structure of an effect algebra on the set $\mathcal{E}(\mathcal{H})$ of all effects of the Hilbert space $\mathcal{H}$? As in the fuzzy-set case, it is sufficient to define the partial sum $\boxplus$ as follows:

$$\exists(E \boxplus F) \text{ iff } E + F \in \mathcal{E}(\mathcal{H}),$$

where $+$ is the usual sum-operator. Furthermore:

$$E \boxplus F := E + F, \text{ if } \exists(E \boxplus F).$$

It turns out that the structure $\langle \mathcal{E}(\mathcal{H}), \boxplus, \mathbb{O}, \mathbb{1} \rangle$ is an effect algebra (called *standard effect algebra* or *Hilbert effect algebra* (HEA)), where the generalized complement of any effect E is just $\mathbb{1} - E$. Furthermore, the effect-algebra order relation coincides with the *natural* order defined on $\mathcal{E}(\mathcal{H})$. In other words:

$$\forall \rho \in D(\mathcal{H}) \{\mathrm{Tr}(\rho E) \leq \mathrm{Tr}(\rho F) \text{ iff } \exists G \in \mathcal{E}(\mathcal{H})[E \perp G \text{ and } F = E \boxplus G]\}.$$

At the same time, this structure fails to be an orthoalgebra. For instance, the semitransparent effect $\frac{1}{2}\mathbb{1}$ gives rise to a counterexample to the strong consistency condition:

$$\frac{1}{2}\mathbb{1} \neq \mathbb{O} \text{ and } \frac{1}{2}\mathbb{1} \boxplus \frac{1}{2}\mathbb{1} = \frac{1}{2}\mathbb{1} + (\frac{1}{2}\mathbb{1})' = \mathbb{1}.$$

Interestingly enough, standard effect algebras represent particular examples of a special kind of effect algebras, that are called *interval effect algebras* (IA).

EXAMPLE 5.1.13. (Bennett and Foulis, 1997) *Interval effect algebras*
Let $\mathcal{G} = \langle G\,, +\,, 0\,, \leq\rangle$ be a partially ordered Abelian group.[2] Let u be an element of G such that $0 \leq u$ and let

$$G[0,u] := \{a \in G \,:\, 0 \leq a \leq u\}\,.$$

A partial operation $\boxplus$ can be defined on $G[0,u]$. The domain of $\boxplus$ is:

$$\{(a,b) \in G[0,u] \times G[0,u] \,:\, a+b \in G[0,u]\}\,.$$

Furthermore,

$$\text{if } \exists(a \boxplus b), \text{ then } a \boxplus b := a+b.$$

The structure

$$\mathcal{G}[0,u]_{eff} := \langle G[0,u]\,, \boxplus\,, 0\,, u\rangle$$

is an effect algebra such that $a' = u - a$.
Every effect algebra which is embeddable into the effect algebra of a partially ordered Abelian group is called an *interval effect algebra.*

The standard effect algebra $\mathcal{E}(\mathcal{H})$ of all effects of a Hilbert space is an interval effect algebra, because $\mathcal{E}(\mathcal{H}) = G[\,\mathbb{O}, \mathbb{I}]$, where $\mathcal{G}$ is the partially ordered Abelian group of all bounded operators of the Hilbert space $\mathcal{H}$.

Another canonical example of an interval effect algebra is given by the effect algebra that can be naturally induced on the real interval $[0,1]$.

EXAMPLE 5.1.14. *The* $[0,1]$*-effect algebra*
The $[0,1]$-effect algebra is the structure

$$\big\langle [0,1]\,, \boxplus\,, '\,, \mathbf{0}\,, \mathbf{1}\big\rangle\,,$$

where

- $\forall x,y \in [0,1] : \exists(x \boxplus y)$ iff $x+y \in [0,1]$; $x \boxplus y := x+y$, if $\exists(x \boxplus y)$.
- $\forall x \in [0,1] : x' = 1-x$;
- $\mathbf{0} = 0$;
- $\mathbf{1} = 1$.

One can easily realize that $\langle [0,1]\,, \boxplus\,, '\,, \mathbf{0}, \mathbf{1}\rangle$ is an example of an interval effect algebra, whose partial order relation $\leq$ coincides with the standard real order.

The states in an event-state system are assumed to be countable additive. But one may consider finitely additive states, even on an effect algebra. Thus, a state on an effect algebra $\mathcal{A} = \langle A\,, \boxplus\,, \mathbf{0}\,, \mathbf{1}\rangle$ is a map $s : A \mapsto [0,1]$ such that $s(\mathbf{0}) = 0$, $s(\mathbf{1}) = 1$, and $s(a \boxplus b) = s(a) + s(b)$, whenever $a \perp b$ in $\mathcal{A}$.

Also the notion of an order determining set of states, previously encountered in event-state systems, admits an expression in any effect algebra.

---

[2]See Definition 1.2.2.

Thus, an effect algebra $\mathcal{A}$ admits an *order determining set of states* in case there exists a set $\mathcal{S}$ of states on $\mathcal{A}$ such that, for any pair $a, b \in A$,

$$\forall s \in \mathcal{S}\,[s(a) \leq s(b)] \text{ implies } a \leq b.$$

It turns out that all effect algebras that admit an order determining set of states are interval algebras. This important result was proved in (Bennett and Foulis, 1997). Because of its importance to the theory we quote it as a theorem. The impact of this theorem is that all the effect algebras that admit an order determining set of states are intervals in a partially ordered Abelian group and the $\boxplus$ of the effect algebra becomes the restriction to an interval of a globally defined $+$ in an Abelian group.

THEOREM 5.1.15. Bennett and Foulis's representation theorem
Every effect algebra that admits an order determining set of states is an interval algebra.

So far we have focused two special algebraic structures that can be defined on the set of all effects in a Hilbert space:

- a Brouwer Zadeh poset $\langle \mathcal{E}(\mathcal{H}), \leq, {}', {}^{\sim}, \mathbf{0}, \mathbf{1} \rangle$;
- an effect algebra $\langle \mathcal{E}(\mathcal{H}), \boxplus, \mathbf{0}, \mathbf{1} \rangle$.

We have seen that these two structures have the same partial order relation $\leq$ and the same generalized complement $'$. This situation naturally suggests an abstract merging of the notions of *Brouwer Zadeh poset* and of *effect algebra*. The result will be the notion of *Brouwer Zadeh effect algebra*.

DEFINITION 5.1.16. (Cattaneo, 1997) *Brouwer Zadeh effect algebra*
A *Brouwer Zadeh effect algebra* (briefly, BZ-effect algebra) (BZEA) is a structure

$$\langle \mathcal{E}, \boxplus, {}^{\sim}, \mathbf{0}, \mathbf{1} \rangle,$$

such that:

(i) $\langle \mathcal{E}, \boxplus, \mathbf{0}, \mathbf{1} \rangle$ is an effect algebra;
(ii) $\langle \mathcal{E}, \leq, {}', {}^{\sim}, \mathbf{0}, \mathbf{1} \rangle$ is a Brouwer Zadeh poset, where $\leq$ and $'$ are the partial order relation and the generalized complement that are defined in the effect algebra $\langle \mathcal{E}, \boxplus, \mathbf{0}, \mathbf{1} \rangle$ (according to Definition 5.1.2).

As expected, one can show that the concrete structure $\langle \mathcal{E}(\mathcal{H}), \boxplus, {}^{\sim}, \mathbf{0}, \mathbf{1} \rangle$ is a Brouwer Zadeh effect algebra.

5.2. The Łukasiewicz operations

We will now investigate another kind of structure that naturally emerges from concrete effect systems. One is dealing with *quantum MV algebras* (QMV algebras): they are *weak variants* of MV algebras (which represent privileged abstractions from classical fuzzy set structures).

Before introducing the notion of QMV algebra, it will be useful to sum up some basic properties of MV algebras. As we have discussed, the fuzzy power set of a given set X gives rise to a de Morgan lattice, where the noncontradiction and the excluded middle principles are possibly violated. However, the operations we have considered so far (the lattice meet $\wedge$, the lattice join $\vee$, the fuzzy complement $'$) do not represent the only interesting fuzzy operations that can be defined. An important role is played by a new kind of conjunction and disjunction, which have been first investigated in the framework of Łukasiewicz' approach to many valued logics. These operations are usually called *Łukasiewicz operations*.

The definition of Łukasiewicz conjunction and disjunction in the framework of fuzzy set structures turns out to be quite natural. As we have already seen, fuzzy sets are generalized characteristic functions whose range is the real interval $[0, 1]$.

Of course, $[0, 1]$ is not closed under the ordinary real sum $+$ (we may have $x, y \in [0, 1]$ and $x + y \notin [0, 1]$). However, one can introduce a new operation $\oplus$, which is called *truncated sum*:

$$\forall x, y \in [0, 1] \, \{x \oplus y := \min(1, x + y)\} .$$

In other words, $x \oplus y$ is the ordinary sum $x + y$, whenever this sum belongs to the interval; otherwise $x \oplus y$ collapses into the maximum element 1.

One immediately realizes that $[0, 1]$ is closed under the operation $\oplus$. Now, we can use the truncated sum in order to define the Łukasiewicz disjunction between fuzzy sets (since no confusion is possible, it will be expedient to use the same symbol $\oplus$ both for the truncated sum and for the Łukasiewicz disjunction).

Let f, g be fuzzy subsets of a set X. The Łukasiewicz disjunction $\oplus$ is defined as follows:

$$\forall x \in X \, \{(f \oplus g)(x) := f(x) \oplus g(x) = \min(1, f(x) + g(x))\} .$$

On this basis, the Łukasiewicz conjunction $\odot$ can be defined, via de Morgan, in terms of $\oplus$ and $'$:

$$\forall x \in X \left\{(f \odot g)(x) := (f' \oplus g')'(x)\right\} .$$

As a consequence, one obtains:

$$(f \odot g)(x) = \max(0, f(x) + g(x) - 1).$$

From an intuitive point of view, the Łukasiewicz operations and the lattice operations represent different notions of conjunction and disjunction that can be used in a fuzzy situation. Consider two fuzzy sets f and g; they can be intuitively regarded as two *ambiguous properties*. The number $f(x)$ represents the "degree of certainty" according to which the object x satisfies the property f. A similar comment holds for g and $g(x)$. What does it mean that the object x satisfies the disjunctive property "f *or* g" with a given degree of certainty? If we interpret "or" as the lattice join, we assume the following choice: an object satisfies a disjunction according to a degree that corresponds to the maximum between the degrees of the two members. If we, instead, interpret "or" as the Łukasiewicz disjunction, we assume the following choice: the degrees of the members of the disjunction have to be summed in such a way that one never goes beyond the absolute certainty (the value 1). Of course, in the limit-case represented by crisp sets (i.e., classical characteristic functions) the Łukasiewicz disjunction and the lattice join will coincide. Suppose $x, y \in \{0, 1\}$, then $x \oplus y = \max(x, y)$.

From the definitions, one immediately obtains that the Łukasiewicz operations are not generally idempotent. It may happen:

$$a \oplus a \neq a \text{ and } a \odot a \neq a.$$

As noticed by Mundici (1992), this is a typical semantic situation that seems to be governed by the principle "repetita iuvant!" (repetitions are useful!). Of course repetitions are really useful in all physical circumstances that are accompanied by a certain *noise*. As a consequence, $\oplus$ and $\odot$ do not give rise to a lattice structure. At the same time, as with the lattice operations, they turn out to satisfy commutativity and associativity:

$$f \oplus g = g \oplus f; \quad f \odot g = g \odot f;$$

$$f \oplus (g \oplus h) = (f \oplus g) \oplus h; \quad f \odot (g \odot h) = (f \odot g) \odot h.$$

Unlike the fuzzy lattice operations, the Łukasiewicz conjunction and disjunction do satisfy both the excluded middle and the noncontradiction principle:

$$f \oplus f' = \mathbf{1}; \quad f \odot f' = \mathbf{0}.$$

Another important difference concerns the distributivity property. As opposed to the case of $\wedge$ and $\vee$ (which satisfy distributivity in the fuzzy set environment), it may happen:

$$f \odot (g \oplus h) \neq (f \odot g) \oplus (f \odot h);$$

$$f \oplus (g \odot h) \neq (f \oplus g) \odot (f \oplus h).$$

What can be said about the relationships between the Łukasiewicz operations and the lattice operations? Interestingly enough, the lattice operations turn out to be definable in terms of the fuzzy complement and of the Łukasiewicz operations. For, we have:

$$f \wedge g := (f \oplus g') \odot g;$$

$$f \vee g := (f \odot g') \oplus g.$$

5.3. MV algebras and QMV algebras

An interesting algebraic abstraction from fuzzy set structures can be obtained if we restrict our attention to the fuzzy complement, the lattice operations and the Łukasiewicz operations.

This gives rise to the abstract notion of an *MV algebra* (*multi-valued algebra*), first introduced by Chang, in order to provide an adequate semantic characterization for Łukasiewicz' many-valued logics.[3] MV algebras

[3] See (Chang, 1958; Chang, 1959).

represent a weakening of Boolean algebras, where the notion of *conjunction* (*disjunction*) is split into two different operations. The first kind of operation behaves like a Łukasiewicz conjunction (disjunction) and is generally nonidempotent; the second kind of operation is a lattice-meet (join). These algebras are also equipped with a generalized complement. In this framework, the lattice operations turn out to be defined in terms of the generalized complement and of the Łukasiewicz operations. Whenever the two conjunctions (resp., disjunctions) collapse into one and the same operation, one obtains a Boolean algebra.

Let us now give the formal definition of MV algebra.

DEFINITION 5.3.1. *MV algebra*[4]
An *MV algebra* is a structure

$$\mathcal{M} = \langle M\,, \oplus\,, {}'\,, \mathbf{0}\,, \mathbf{1}\rangle\,,$$

where $\oplus$ is a binary operation, $'$ is a unary operation and $\mathbf{0}$, $\mathbf{1}$ are special distinct elements satisfying the following conditions:

(MV1) $a \oplus b = b \oplus a$;
(MV2) $a \oplus (b \oplus c) = (a \oplus b) \oplus c$;
(MV3) $a \oplus a' = \mathbf{1}$;
(MV4) $a \oplus \mathbf{0} = a$;
(MV5) $a \oplus \mathbf{1} = \mathbf{1}$;
(MV6) $a'' = a$;
(MV7) $(a' \oplus b)' \oplus b = (b' \oplus a)' \oplus a$.

In any MV algebra $\mathcal{M} = \langle M\,, \oplus\,, {}'\,, \mathbf{0}\,, \mathbf{1}\rangle$, the Łukasiewicz conjunction $\odot$, the lattice operations $\wedge$ and $\vee$, the Łukasiewicz implication $\rightarrow_L$, the partial order relation $\leq$ can be defined as follows:

- $a \odot b := (a' \oplus b')'$;
- $a \wedge b := (a \oplus b') \odot b$;
- $a \vee b := (a \odot b') \oplus b$;
- $a \rightarrow_L b := a' \oplus b$;
- $a \leq b$ iff $a \wedge b = a$.

It is not difficult to see that $\forall a, b \in M$: $a \leq b$ iff $a \rightarrow_L b = a' \oplus b = \mathbf{1}$. Hence, the operation $\rightarrow_L$ represents a well behaved conditional.[5]

LEMMA 5.3.2. *Let $\mathcal{M} = \langle M\,, \oplus\,, {}'\,, \mathbf{0}\,, \mathbf{1}\rangle$ be an MV algebra. Consider the structure*

$$\langle M\,, \leq\,, {}'\,, \mathbf{0}\,, \mathbf{1}\rangle\,,$$

where $\leq$ is the partial order defined on $\mathcal{M}$. Such structure is a distributive bounded involution lattice, where $\wedge$ and $\vee$ represent the infimum and the

[4]See (Mangani, 1973; Cignoli, D'Ottaviano and Mundici, 2000).

[5]Generally, a binary operation $\rightarrow$ of a structure (which is at least a bounded poset) is considered a *well behaved conditional*, when: $a \leq b$ iff $a \rightarrow b = 1$, for any elements a and b. By assuming a natural logical interpretation, this means that the conditional $a \rightarrow b$ is "true" iff the "implication-relation" $a \leq b$ holds.

supremum, respectively. The noncontradiction principle ($a \wedge a' = \mathbf{0}$) and the excluded middle ($a \vee a' = \mathbf{1}$) are possibly violated.[6]

A privileged example of MV algebra can be defined by assuming as support the real interval $[0,1]$.

DEFINITION 5.3.3. *The $[0,1]$-MV algebra*
The $[0,1]$-*MV algebra* is the structure

$$\mathcal{M}^{[0,1]} = \langle [0,1]\,, \oplus\,, '\,, \mathbf{0}\,, \mathbf{1} \rangle,$$

where

- $\oplus$ is the truncated sum. In other words: $\forall x, y \in [0,1]\, \{x \oplus y = \min(1, x+y)\}$;
- $\forall x \in [0,1]\, \{x' = 1 - x\}$;
- $\mathbf{0} = 0$;
- $\mathbf{1} = 1$.

One can easily realize that $\mathcal{M}^{[0,1]}$ is a special example of MV algebra where:

- the partial order $\leq$ is a total order (coinciding with the natural real order);
- $x \wedge y = \min(x, y)$;
- $x \vee y = \max(x, y)$.

The following deep theorem is due to Chang:

THEOREM 5.3.4. (Chang, 1959) Chang representation theorem
Any MV algebra is the subdirect product of totally ordered MV algebras.[7]

Let us now return to fuzzy set systems. Consider the structure

$$\mathcal{M} = \langle M\,, \oplus\,, '\,, \mathbf{0}\,, \mathbf{1} \rangle,$$

where:

- M is the set of all fuzzy subsets of a given set X;

---

[6] See, for instance, (Cignoli et al., 2000).

[7] Let $\{\mathcal{M}_i\}_{i \in I}$ be a nonempty family of MV algebras. The *direct product of* $\{\mathcal{M}_i\}_{i \in I}$) is the structure $\prod_{i \in I} \mathcal{M}_i = \langle \prod_{i \in I} M_i\,, \oplus\,, '\,, \mathbf{0}\,, \mathbf{1} \rangle$ defined as follows: (i) $\prod_{i \in I} M_i$ is the set of all maps $f : I \to \bigcup_{i \in I} M_i$ such that $\forall i \in I$: $f(i) \in M_i$; (ii) $\oplus$ is a binary map on $\prod_{i \in I} M_i$ such that $\forall f, g \in \prod_{i \in I} M_i \,\forall i \in I$: $(f \oplus g)(i) = f(i) \oplus_i g(i)$, where $\oplus_i$ is the Łukasiewicz disjunction of the MV algebra $\mathcal{M}_i$; (iii) $'$ is a unary map on $\prod_{i \in I} M_i$ such that $\forall f \in \prod_{i \in I} M_i \,\forall i \in I$: $f'(i) = f(i)'^{i}$, where $'^{i}$ is the generalized complement of the MV algebra $\mathcal{M}_i$; (iv) $\mathbf{0}$ is the map from I into $\bigcup_{i \in I} M_i$ such that $\forall i \in I$: $\mathbf{0}(i) = \mathbf{0}_i$, where $\mathbf{0}_i$ is the minimum of the MV algebra $\mathcal{M}_i$. (v) $\mathbf{1}$ is the map from I into $\bigcup_{i \in I} M_i$ such that $\forall i \in I$: $\mathbf{1}(i) = \mathbf{1}_i$, where $\mathbf{1}_i$ is the maximum of the MV algebra $\mathcal{M}_i$.
It turns out that $\prod_{i \in I} \mathcal{M}_i$ is an MV algebra. For every $i \in I$, we can define the *projection map* $\pi_j : \prod_{i \in I} M_i \to M_j$ as follows: $\forall f \in \prod_{i \in I} M_i$, $\pi_j(f) = f(j)$. It turns out that the map π_j is a homomorphism of $\prod_{i \in I} \mathcal{M}_i$ onto $\mathcal{M}_j$.
An MV algebra $\mathcal{M}$ is a *subdirect product* of a family $\{\mathcal{M}_i\}_{i \in I}$iff there exists an embedding $h : M \to \prod_{i \in I} M_i$ such that $\forall i \in I$, the composite map $\pi_i \circ h$ is a homomorphism of $\mathcal{M}$ onto $\mathcal{M}_i$, i.e., h is an onto map that preserves the defining operations of MV algebras.

- $\oplus$ is the Łukasiewicz disjunction defined on M. In other words:
$$\forall f, g \in M \, \forall x \in X[(f \oplus g)(x) = \min(1, f(x) + g(x))];$$
- $'$ is the generalized complement defined on M. In other words:
$$\forall f \in M \, \forall x \in X[f'(x) = (1 - f)(x) = 1 - f(x)];$$
- $\mathbf{0}$ is the classical characteristic function of the empty set $\emptyset$;
- $\mathbf{1}$ is the classical characteristic function of the total set X.

One can easily realize that $\mathcal{M}$ is an MV algebra such that $\forall f, g \in M \, \forall x \in X$:

$$(f \wedge g)(x) = \min(f(x), g(x)); \quad (f \vee g)(x) = \max(f(x), g(x)).$$

As a consequence, the MV-partial order of $\mathcal{M}$ turns out to coincide with the standard fuzzy partial order.

A natural question arises: do fuzzy set structures play for MV algebras the same role that classical set structures have for Boolean algebras? As is well known, according to a celebrated "representation theorem" due to Stone, any Boolean algebra is isomorphic to a *concrete* Boolean algebra (based on a field of sets where the Boolean operations coincide with the set-theoretic intersection, union and relative complement). From the intuitive point of view, this means that any Boolean algebra can be thought of as a set-theoretic Boolean algebra, without any loss of generality. Such a nice situation can be only partially generalized to the case of MV algebras and fuzzy set MV algebras. Let us first introduce the notions of *simple* and *semisimple* MV algebra.

DEFINITION 5.3.5. *Simple MV algebra*
An MV algebra $\mathcal{M}$ is called *simple* iff $\forall a \neq \mathbf{0} \, \exists n \in \mathbb{N}$ such that $na := \underbrace{a \oplus \ldots \oplus a}_{n-times} = \mathbf{1}$.

DEFINITION 5.3.6. *Semisimple MV algebra*
An MV algebra $\mathcal{M}$ is called *semisimple* iff $\mathcal{M}$ is the subdirect product of simple MV algebras.

Unlike Boolean algebras, MV algebras are not necessarily semisimple.

Belluce (1986) has proved the following theorem (which represents a weak version of the Stone's Theorem for MV algebras).

THEOREM 5.3.7. *An MV algebra is semisimple iff it is isomorphic to a fuzzy-set MV algebra.*

As a consequence, MV algebras cannot be regarded as fully faithful abstractions from fuzzy-set structures. However, the class of all MV algebras and the class of all fuzzy-set MV algebras determine the same equational class, in other words, they satisfy the same class of equations.

Let us now return to the concrete effect-structure $\langle \mathcal{E}(\mathcal{H}), \boxplus, \mathbf{0}, \mathbf{1} \rangle$. The partial operation $\boxplus$ can be naturally extended to a total operation $\oplus$ that behaves similarly to a *truncated sum*. For any $E, F \in \mathcal{E}(\mathcal{H})$,

$$E \oplus F := \begin{cases} E + F & \text{if } \exists (E \boxplus F), \\ \mathbf{1} & \text{otherwise.} \end{cases}$$

Furthermore, let us define:

$$E' := 1\!\!\mathrm{I} - E.$$

The structure $\langle \mathcal{E}(\mathcal{H}), \oplus, ', \mathbf{0}, \mathbf{1} \rangle$ turns out to be "very close" to an MV algebra. However, something is missing: $\mathcal{E}(\mathcal{H})$ satisfies the first six axioms of the definition of an MV algebra; at the same time one can easily check that the final axiom (usually called "Łukasiewicz axiom") is violated. For instance, consider two nontrivial projections P, Q such that P is not orthogonal to Q' and Q is not orthogonal to P'. Then, by the definition of $\oplus$ given immediately above, we have that $P \oplus Q' = 1\!\!\mathrm{I}$ and $Q \oplus P' = 1\!\!\mathrm{I}$. Hence, $(P' \oplus Q)' \oplus Q = Q \neq P = (P \oplus Q')' \oplus P$.

As a consequence, the Łukasiewicz axiom must be conveniently weakened to obtain an adequate description of concrete effect structures. This can be done by means of the notion of *quantum MV algebra* (QMV algebra).[8]

As with MV algebras, QMV algebras are total structures having the following form: $\mathcal{M} = \langle M, \oplus, ', \mathbf{0}, \mathbf{1} \rangle$, where:

(i) $\mathbf{0}, \mathbf{1}$ represent the impossible and the certain object, respectively;
(ii) $'$ is the negation-operation;
(iii) $\oplus$ represents a disjunction (*or*) which is generally nonidempotent ($a \oplus a \neq a$).

A (generally nonidempotent) conjunction (*and*) is then defined via the de Morgan law:

$$a \odot b := (a' \oplus b')'.$$

On this basis, a pair consisting of an idempotent conjunction *et* and of an idempotent disjunction *vel* is then defined. As we have already discussed, in any MV algebra such idempotent operations behave as a lattice-meet and lattice-join, respectively. However, this is not the case for QMV algebras. As a consequence, in such a more general situation, we will denote the *et* operation by the symbol $\Cap$, while the *vel* will be indicated by $\Cup$. The definition of *et* and *vel* is as in the MV-case:

$$a \Cap b := (a \oplus b') \odot b$$

$$a \Cup b := (a \odot b') \oplus b.$$

DEFINITION 5.3.8. (Giuntini, 2000) *QMV algebra*
A *quantum MV algebra* (*QMV algebra*) (QMV) is a structure

$$\mathcal{M} = \langle M, \oplus, ', \mathbf{0}, \mathbf{1} \rangle,$$

where $\oplus$ is a binary operation, $'$ is a unary operation, and $\mathbf{0}, \mathbf{1}$ are special distinct elements of M. For any $a, b \in M$: $a \odot b := (a' \oplus b')'$, $a \Cap b := (a \oplus b') \odot b$, $a \Cup b := (a \odot b') \oplus b$. Assume that the following conditions hold:

[8] See (Giuntini, 1996).

(QMV1) $a \oplus b = b \oplus a$;
(QMV2) $a \oplus (b \oplus c) = (a \oplus b) \oplus c$;
(QMV3) $a \oplus a' = \mathbf{1}$;
(QMV4) $a \oplus \mathbf{0} = a$;
(QMV5) $a \oplus \mathbf{1} = \mathbf{1}$;
(QMV6) $a'' = a$;
(QMV7) $a \oplus [(a' \Cap b) \Cap (c \Cap a')] = (a \oplus b) \Cap (a \oplus c)$.

By Axioms (QMV3), (QMV1) and (QMV4), one immediately obtains that

$$\mathbf{0}' = \mathbf{1}.$$

The operations $\Cap$ and $\Cup$ of a QMV algebra $\mathcal{M}$ are generally noncommutative. As a consequence, they do not represent lattice-operations. It is not difficult to prove that $\Cap$ is commutative iff $\Cup$ is commutative iff (MV7) of Definition 5.3.1 holds. From this it easily follows that a QMV algebra $\mathcal{M}$ is an MV algebra iff $\Cap$ or $\Cup$ is commutative.

At the same time (as in the MV-case), we can define in any QMV algebra $\langle M, \oplus, ', \mathbf{0}, \mathbf{1}\rangle$ the following relation:

$$a \leq b \text{ iff } a \Cap b = a.$$

We will prove (Theorem 5.3.17) that the structure

$$\langle M, \leq, ', \mathbf{0}, \mathbf{1}\rangle$$

is a bounded involution poset.

The following lemmas describe some basic properties of QMV algebras. Note that we sometimes write $a \odot b \oplus c$ for $(a \odot b) \oplus c$. Let $\langle M, \oplus, ', \mathbf{0}, \mathbf{1}\rangle$ be any QMV algebra.

LEMMA 5.3.9. *The following conditions hold:*

(i) $\odot$ *is associative and commutative;*
(ii) $a \odot a' = \mathbf{0}$;
(iii) $a \odot \mathbf{1} = a$ *and* $a \odot \mathbf{0} = \mathbf{0}$;
(iv) $a \Cap \mathbf{1} = \mathbf{1} \Cap a = a$;
(v) $a \Cap \mathbf{0} = \mathbf{0} \Cap a = \mathbf{0}$;
(vi) $a \Cap a = a$;
(vii) $(a \Cap b)' = a' \Cup b'$ *and* $(a \Cup b)' = a' \Cap b'$;
(viii) $a \leq b$ *implies* $a = b \Cap a$;
(ix) $a \oplus b = \mathbf{0}$ *iff* $a \Cup b = \mathbf{0}$ *iff* $a = b = \mathbf{0}$;
(x) $a \odot b = \mathbf{1}$ *iff* $a \Cap b = \mathbf{1}$ *iff* $a = b = \mathbf{1}$.

PROOF. Straightforward. □

LEMMA 5.3.10.

(i) $a \oplus (b \Cap a') = a \oplus b$;
(ii) $a \oplus (a' \Cap b) = a \oplus b$;
(iii) $a \oplus (b \Cap (c \Cap a')) = (a \oplus b) \Cap (a \oplus c)$;
(iv) $(a \Cap b) \oplus (b \Cap a)' = \mathbf{1}$.

PROOF. Parts (i), (ii) and (iii) are proved by the following three groups of equations, respectively.

$$\begin{aligned}
a \oplus (b \Cap a') &= a \oplus [\mathbf{1} \odot (b \Cap a')] && \text{(Lemma 5.3.9(i), (iii))}\\
&= a \oplus (a \oplus a' \oplus a' \odot b') \odot (b \Cap a') && \text{((QMV3), (QMV5))}\\
&= a \oplus (a' \oplus (b' \Cup a)) \odot (b \Cap a') &&\\
&= a \oplus (a' \oplus (b \Cap a')') \odot (b \Cap a') && \text{(Lemma 5.3.9(vii), (QMV6))}\\
&= a \oplus (a' \Cap (b \Cap a')) &&\\
&= a \oplus ((a' \Cap \mathbf{1}) \Cap (b \Cap a')) && \text{(Lemma 5.3.9(iv))}\\
&= (a \oplus \mathbf{1}) \Cap (a \oplus b) && \text{(QMV7)}\\
&= a \oplus b. && \text{((QMV5), Lemma 5.3.9(iv))}
\end{aligned}$$

$$\begin{aligned}
a \oplus b &= (a \oplus b) \Cap (a \oplus \mathbf{1}) && \text{((QMV5), Lemma 5.3.9(iv))}\\
&= a \oplus ((a' \Cap b) \Cap (\mathbf{1} \Cap a')) && \text{(QMV7)}\\
&= a \oplus ((a' \Cap b) \Cap a')) && \text{(Lemma 5.3.9(iv))}\\
&= a \oplus (a' \Cap b). && \text{(Lemma 5.3.10(i))}
\end{aligned}$$

$$\begin{aligned}
a \oplus (b \Cap (c \Cap a')) &= a \oplus (b \oplus (c' \Cup a)) \odot (c \Cap a')\\
&\qquad\qquad \text{(Lemma 5.3.9(vii), (QMV6))}\\
&= a \oplus (b \oplus (c' \odot a') \oplus a) \odot (c \Cap a')\\
&= a \oplus ((a' \Cap b) \oplus a \oplus c' \odot a') \odot (c \Cap a')\\
&\qquad\qquad \text{((QMV1),(QMV2), Lemma 5.3.10(ii))}\\
&= a \oplus ((a' \Cap b) \oplus (c \Cap a')') \odot (c \Cap a')\\
&\qquad\qquad \text{(Lemma 5.3.9(vii))}\\
&= a \oplus ((a' \Cap b) \Cap (c \Cap a'))\\
&= (a \oplus b) \Cap (a \oplus c). \qquad\qquad \text{(QMV7)}
\end{aligned}$$

(iv) The proof follows from (ii) after applying Lemma 5.3.9(vii). □

The following lemma asserts an important connection between the *implication-relation* $a \leq b$ and the *truth* of $a' \oplus b$.

LEMMA 5.3.11. $\forall a, b \in M,\ a \leq b$ *implies* $a' \oplus b = \mathbf{1}$.

PROOF. Suppose $a \leq b$. Then, $a = a \Cap b$. By Lemma 5.3.9(vii) we have $a' = a' \Cup b'$. Hence, $a' \oplus b = (a' \Cup b') \oplus b = (a' \odot b) \oplus b' \oplus b = \mathbf{1}$. □

The converse of the above lemma does not generally hold in a QMV algebra. In fact, the property

$$\forall a, b \in M : \ a' \oplus b = \mathbf{1} \ \text{ implies } a \leq b$$

characterizes MV algebras in the class of all QMV algebras.

THEOREM 5.3.12. *Let $\mathcal{M}$ be a QMV algebra. The following conditions are equivalent:*

(i) *$\mathcal{M}$ is an MV algebra;*
(ii) $\forall a, b \in M$ [*if* $a' \oplus b = \mathbf{1}$, *then* $a \leq b$].

PROOF. To see that (i) implies (ii), let $\mathcal{M}$ be an MV algebra and suppose that $a' \oplus b = \mathbf{1}$. We want to show that $a = a \wedge b$. By Lemma 5.3.2, the operation $\wedge$ is commutative so that, $a \wedge b = b \wedge a = (b \oplus a') \odot a = 1 \odot a = a$. To see that (ii) implies (i), suppose (ii) is satisfied. To prove that $\mathcal{M}$ is an MV algebra it is sufficient to show that $\Cup$ is commutative. We first prove that, $\forall a, b \in \mathcal{M}$, $a \Cup b \leq b \Cup a$, i.e., $\forall a, b \in \mathcal{M}$, $(a \Cup b)' \oplus (b \Cup a) = \mathbf{1}$. But

$$\begin{aligned} (a \Cup b)' \oplus (b \Cup a) &= (a' \Cap b') \oplus (b \Cup a) && \text{(Lemma 5.3.9(vii)} \\ &= (a' \Cap b') \oplus (b' \Cap a')' && \text{((QMV6), Lemma 5.3.9(vii))} \\ &= \mathbf{1}. && \text{(Lemma 5.3.10(iv))} \end{aligned}$$

Interchanging a and b in the above argument provides the proof of the reverse inequality. □

As a corollary of this theorem, we obtain that no genuine QMV algebra (one that is not an MV algebra) admits a *conditional* operation $\rightarrow$, that is the *right adjoint* to the operation $\odot$. In other words, there exists no binary function $\rightarrow$ such that:

$$a \odot b \leq c \text{ iff } b \leq a \rightarrow c.$$

Hence, there exists no QMV algebra $\mathcal{M} = \langle M, \oplus, ', \mathbf{0}, \mathbf{1} \rangle$ that satisfies the following conditions:

(i) $\mathcal{M}$ is *not* an MV algebra;
(ii) $\langle M, \leq, \odot \rangle$ is a (commutative) autonomous poset;[9]
(iii) $a' = a \rightarrow \mathbf{0}$.

On this basis, MV algebras can be characterized as those QMV algebras that admit a conditional $\rightarrow$ (such that $a \odot b \leq c$ iff $b \leq a \rightarrow c$). Similarly, Boolean algebras can be characterized as those orthomodular lattices that admit a conditional $\rightarrow$ (such that $a \wedge b \leq c$ iff $b \leq a \rightarrow c$). Hence, the relationship between QMV algebras and MV algebras is, in this way, similar to the relationship between orthomodular lattices and Boolean algebras.

The following lemmas describe some other important properties of QMV algebras (in the proofs we will freely use associativity and commutativity of the operations $\oplus$ and $\odot$, without explicit reference to the axioms). Let $\langle M, \oplus, ', \mathbf{0}, \mathbf{1} \rangle$ be any QMV algebra.

LEMMA 5.3.13.

(i) $a \odot b \leq a$;

---

[9] A *commutative autonomous poset* (Rosenthal, 1990) is a structure $\langle P, \leq, \cdot, \rightarrow \rangle$ such that: (i) $\langle P, \leq \rangle$ is a poset; (ii) $\cdot$ is an associative and commutative binary operation; (iii) $\rightarrow$ is a binary operation such that $\forall a, b, c \in P[a \cdot b \leq c$ iff $b \leq a \rightarrow c]$.

(ii) $a \le a \oplus b$;
(iii) $a \Cap b \le b$.

PROOF. Parts (i), (ii) and (iii) are proved by the following three groups of equations, respectively.

$$\begin{aligned}
(a \odot b) \Cap a &= (a \odot b \oplus a') \odot a \\
&= (b \Cup a') \odot a \\
&= (a' \oplus (b' \Cap a))' \\
&= (a' \oplus b')' && \text{(Lemma 5.3.10(i))} \\
&= a \odot b.
\end{aligned}$$

$$\begin{aligned}
a \Cap (a \oplus b) &= (a \oplus \mathbf{0}) \Cap (a \oplus b) \\
&= a \oplus ((a' \Cap \mathbf{0}) \Cap (b \Cap a')) && \text{(QMV7)} \\
&= a \oplus \mathbf{0} \\
&= a.
\end{aligned}$$

$$\begin{aligned}
(a \Cap b) \Cap b &= ((a \Cap b) \oplus b') \odot b \\
&= (a \oplus b') \odot b && \text{(Lemma 5.3.10(i))} \\
&= a \Cap b.
\end{aligned}$$

□

It is worthwhile noticing (in the following Figure 5.3.1, for example) that generally:

$$a \Cap b \not\le a \text{ and } a \not\le a \Cup b.$$

LEMMA 5.3.14.

(i) $a \oplus (b \Cap (a \oplus c)') = (a \oplus b) \Cap (a \oplus a' \odot c')$;
(ii) $a \Cup (b \Cap a) = a$.

PROOF. Parts (i) and (ii) are proved by the following two groups of equations, respectively.

$$\begin{aligned}
a \oplus (b \Cap (a \oplus c)') &= a \oplus (b \Cap (a \oplus (c \Cap a'))') && \text{(Lemma 5.3.10(i))} \\
&= a \oplus (b \Cap (a \oplus (c \oplus a) \odot a')') \\
&= a \oplus (b \Cap ((a \oplus c) \Cup a)') \\
&= a \oplus (b \Cap ((a' \odot c') \Cap a')) \\
&= (a \oplus b) \Cap (a \oplus a' \odot c'). && \text{(Lemma 5.3.10(iii))}
\end{aligned}$$

$$\begin{aligned}
a \Cup (b \Cap a) &= a \odot (b' \Cup a') \oplus (b \Cap a) \\
&= (a' \oplus (b \Cap a))' \oplus (b \Cap a) \\
&= (a' \oplus b)' \oplus (b \Cap a) && \text{(Lemma 5.3.10(i))} \\
&= a \odot b' \oplus (b \oplus a') \odot a \\
&= a \Cup (a \odot b') \\
&= (a' \Cap (a' \oplus b))' \\
&= a. && \text{(Lemma 5.3.13(ii))}
\end{aligned}$$

□

LEMMA 5.3.15.

(i) $a \leq b$ *implies* $b = b \Cup a$;
(ii) $a \leq b$ *implies* $b' \leq a'$.

PROOF. (i) This follows from Lemma 5.3.14(ii) upon expanding $b \Cup a$ when $a = a \Cap b$.
(ii) This follows from (i). □

LEMMA 5.3.16.

(i) $a \leq b$ *implies* $\forall c \in M: \ a \oplus c \leq b \oplus c$;
(ii) $(a \Cap b) \Cap c = (a \Cap b) \Cap (b \Cap c)$;
(iii) $a \leq b$ *implies* $\forall c \in M: \ a \Cap c \leq b \Cap c$;
(iv) $a \leq b, c \leq d$ *imply* $a \oplus c \leq b \oplus d$.

PROOF. Parts (i), and (ii) are proved by the following two groups of equations, respectively.

$$\begin{aligned}
(a \oplus c) \Cap (b \oplus c) &= ((a \oplus c) \oplus b' \odot c')) \odot (b \oplus c) \\
&= ((a \oplus c) \oplus (b' \Cap a') \odot c') \odot (b \oplus c) \\
&&& \text{(Lemma 5.3.15(ii), HP)} \\
&= ((a \oplus c) \oplus (b' \oplus a) \odot a' \odot c') \odot ((b \Cup a) \oplus c) \\
&&& \text{(Lemma 5.3.15(i), HP)} \\
&= ((a \oplus c) \oplus (b' \oplus a) \odot a' \odot c') \odot (b \odot a' \oplus a \oplus c) \\
&= (a \oplus c) \Cap (b \odot a' \oplus a \oplus c) \\
&= a \oplus c. && \text{(Lemma 5.3.13(ii))}
\end{aligned}$$

$$\begin{aligned}
(a \Cap b) \Cap (b \Cap c) &= ((a \Cap b) \oplus (b' \Cup c')) \odot (b \Cap c) \\
&= ((a \Cap b) \oplus b' \odot c \oplus c') \odot (b \oplus c') \odot c \\
&= (((a \Cap b) \oplus c') \Cap (b \oplus c')) \odot c \\
&= ((a \Cap b) \oplus c') \odot c \quad \text{(Lemma 5.3.13(iii), Lemma 5.3.16(i))} \\
&= (a \Cap b) \Cap c.
\end{aligned}$$

Finally, (iii) follows from (ii) and (iv) follows from (i). □

On this basis, we can finally prove that any QMV algebra gives rise to a bounded involution poset.

THEOREM 5.3.17. *Let* $\langle M, \oplus, ', \mathbf{0}, \mathbf{1}\rangle$ *be a QMV algebra (with partial order* $\leq$*). Then,* $(M, \leq, ', \mathbf{0}, \mathbf{1})$ *is a bounded involution poset.*

PROOF. The relation $\leq$ is reflexive by Lemma 5.3.9(vi). It is antisymmetric by Lemma 5.3.9(viii). It is transitive by Lemma 5.3.16(ii). By Lemma 5.3.9(iv-v), $\mathbf{1}$ and $\mathbf{0}$ are, respectively, the maximum and the minimum of the poset $(M, \leq)$. By axiom (QMV6) and Lemma 5.3.15(ii), the operation $'$ is an involution. □

The smallest genuine QMV algebra is the "diamond" $\mathcal{M}_4$ (see Figure 5.3.1), where, apart from the obvious conditions, the operation $\oplus$ is defined as follows: $a \oplus a = a \oplus b = b \oplus b = \mathbf{1}$. As in the case of effect algebras, an important class of QMV algebras is given by *interval QMV algebras.*

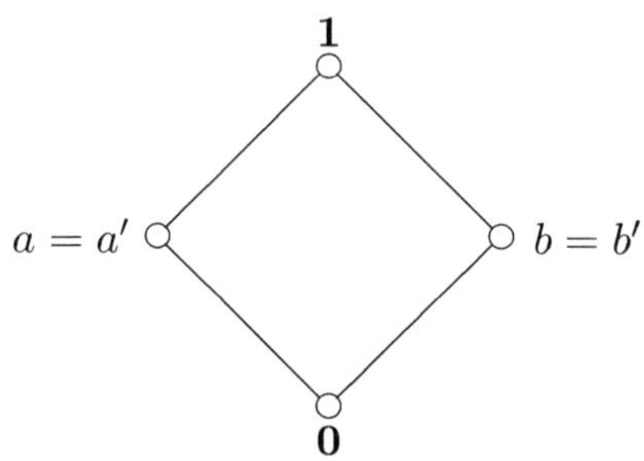

FIGURE 5.3.1. $\mathcal{M}_4$: the smallest QMV that is not an MV

EXAMPLE 5.3.18. *Interval QMV algebra*
Let $\mathcal{G} = \langle G, +, 0, \leq\rangle$ be a partially ordered Abelian group. Let u be an element in the positive cone of $\mathcal{G}$ and let

$$G[0, u] := \{a \in G \mid 0 \leq a \leq u\}.$$

Define on $G[0, u]$ the operations $\oplus$ and $'$ as follows:

$$a \oplus b := \begin{cases} a + b & \text{if } a + b \leq u, \\ u & \text{otherwise;} \end{cases}$$

$$a' := u - a.$$

The structure $\mathcal{G}[0, u] := \langle G[0, u], \oplus, ', 0, u\rangle$ is a QMV algebra.

Every QMV algebra that is embeddable into the QMV algebra of a partially ordered Abelian group is called an *interval QMV algebra.*

Another interesting class of QMV algebras arises from the category of all orthomodular lattices.

EXAMPLE 5.3.19. (Giuntini, 1996)
Take an orthomodular lattice $\langle B, \wedge, \vee, ', \mathbf{0}, \mathbf{1}\rangle$. Put $a \oplus b := a \vee b$. The structure $\langle B, \oplus, ', \mathbf{0}, \mathbf{1}\rangle$ turns out to be a QMV algebra, which is not, generally, an MV algebra since it need not be distributive.
Vice versa, take a QMV algebra $\mathcal{M} = \langle M, \oplus, ', \mathbf{0}, \mathbf{1}\rangle$ and consider the set M^I of all its idempotent elements (i.e., elements such that $a = a \oplus a$). This gives rise to an orthomodular lattice $\langle M^I, \leq, ', \mathbf{0}, \mathbf{1}\rangle$, where $\leq$ and $'$ are the partial order relation and the generalized complement of $\mathcal{M}$ restricted to M^I.

On this basis, one can say that QMV algebras represent at the same time nonidempotent generalizations (with respect to $\oplus$) of orthomodular lattices and noncommutative generalizations (with respect to $\Cap$) of MV algebras. Let us now return to discussing concrete effects ("living" in a Hilbert space $\mathcal{H}$). One can prove QMV algebras represent a "good abstraction" of these concrete effect-systems. Consider again the structure

$$\langle \mathcal{E}(\mathcal{H}), \oplus, ', \mathbf{0}, \mathbf{1}\rangle,$$

where $'$ and $\oplus$ are the operations defined above (the standard effect-involution and the truncated sum). Instead of proving directly that the QMV-axioms hold in the effect-structure $\langle \mathcal{E}(\mathcal{H}), \oplus, ', \mathbf{0}, \mathbf{1}\rangle$, we will follow here a more abstract procedure. After proving some general properties of QMV algebras, we will show that any effect algebra can be canonically transformed into a QMV algebra (whose total operation $\oplus$ extends the partial operation $\boxplus$ of the original effect algebra). As a consequence, we will obtain that any concrete effect algebra $\langle \mathcal{E}(\mathcal{H}), \boxplus, \mathbf{0}, \mathbf{1}\rangle$ can be canonically transformed into a QMV algebra $\langle \mathcal{E}(\mathcal{H}), \oplus, ', \mathbf{0}, \mathbf{1}\rangle$.

Such concrete structures $\langle \mathcal{E}(\mathcal{H}), \oplus, ', \mathbf{0}, \mathbf{1}\rangle$ will be also called *Hilbert QMV algebras* (or, alternatively, *standard QMV algebras*).

5.4. Quasi-linear QMV algebras and effect algebras

In Section 5.3 we have encountered the prototypical example of an MV algebra: the structure $\mathcal{M}^{[0,1]}$, based on the real unit interval $[0,1]$. Such a structure is also called the *standard MV algebra*. As we have seen, the order-relation of $\mathcal{M}^{[0,1]}$ coincides with the restriction to $[0,1]$ of the natural order of $\mathbb{R}$. Consequently, $\mathcal{M}^{[0,1]}$ is *linear* (*totally ordered*):

$$\forall a, b \in [0,1]\,\{\text{either } a \leq b \text{ or } b \leq a\}.$$

As one can readily see, a genuine QMV algebra cannot be totally ordered. However, there are some interesting QMV algebras (*in primis* the standard QMV algebras) that satisfy some weaker conditions. These conditions turn out to be equivalent to linearity in any MV algebra.

DEFINITION 5.4.1. *Quasi-linearity*
A *quasi-linear QMV algebra* (qLQMV) is a QMV algebra $\mathcal{M} = (M, \oplus, ', \mathbf{0}, \mathbf{1})$ that satisfies the following condition $\forall a, b \in M$:

$$a \not\leq b \text{ implies } a \Cap b = b.$$

Thus, for any quasi-linear QMV algebra the following relation holds for all elements a, b:

$$a \Cap b = \begin{cases} a & \text{if } a \leq b, \\ b & \text{otherwise.} \end{cases}$$

Interestingly enough, all interval QMV algebras (see Example 5.3.18) turn out to be quasi-linear. One can easily prove the following:

LEMMA 5.4.2. *Let $\mathcal{M}$ be a QMV algebra. The following conditions are equivalent:*

(i) *$\mathcal{M}$ is quasi-linear;*
(ii) *$\forall a, b \in M\ [a \oplus b \neq \mathbf{1}$ implies $a \leq b']$;*
(iii) *$\forall a, b, c \in M\ [a \oplus c = b \oplus c \neq \mathbf{1}$ implies $a = b]$.*

DEFINITION 5.4.3. *Weak linearity*
A QMV algebra $\mathcal{M}$ is called *weakly-linear* iff

$$\forall a, b \in M[a \Cap b = b \text{ or } b \Cap a = a].$$

It is easy to see that every quasi-linear QMV algebra is weakly linear, but not the other way around (Giuntini, 1995a) as the following counterexample shows (see Figure 5.4.1 and 5.4.2).

		$\oplus$
a	a	b'
a	b	b'
a	b'	$\mathbf{1}$
a'	a'	$\mathbf{1}$
a'	b	$\mathbf{1}$
a'	b'	$\mathbf{1}$
b	b	a'
b	a	b'
b	a'	$\mathbf{1}$
b'	b'	$\mathbf{1}$
b'	a	$\mathbf{1}$
b'	a'	$\mathbf{1}$

FIGURE 5.4.1. The operation $\oplus$ of $\mathcal{M}_{wl}$

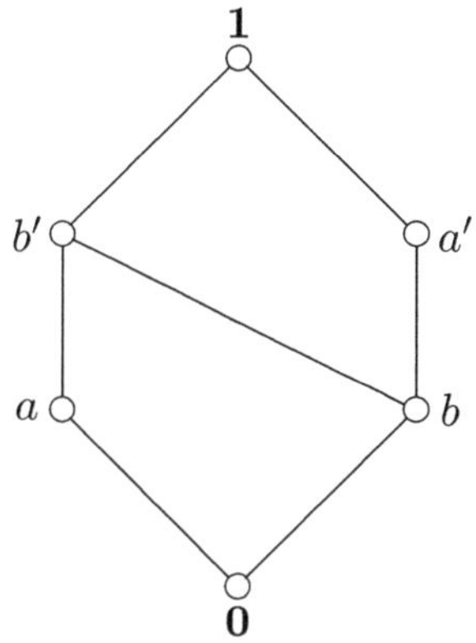

FIGURE 5.4.2. The Hasse diagram of $\mathcal{M}_{wl}$

The operation $\oplus$ of $\mathcal{M}_{wl}$ is defined, apart the obvious conditions, in Figure 5.4.1. One can easily show that $\mathcal{M}_{wl}$ is a weakly-linear QMV algebra. However, $\mathcal{M}_{wl}$ is not quasi-linear because $a \oplus a \neq \mathbf{1}$ but $a \Cap a' \neq a$.

A very strong relation connects the class of all effect algebras with the class of all quasi-linear QMV algebras: every effect algebra can be uniquely transformed into a quasi-linear QMV algebra and vice versa.

Let $\mathcal{B} = \langle B\,, \boxplus\,, \mathbf{0}\,, \mathbf{1}\rangle$ be an effect algebra. The operation $\boxplus$ can be extended to a total operation

$$\oplus : B \times B \mapsto B$$

in the following way:

$$a \oplus b := \begin{cases} a \boxplus b & \text{if } \exists(a \boxplus b), \\ \mathbf{1} & \text{otherwise.} \end{cases}$$

The resulting structure $\langle B\,, \oplus\,, '\,, \mathbf{0}\,, \mathbf{1}\rangle$ (where $'$ is the generalized complement defined in $\mathcal{B}$) will be denoted by $\mathcal{B}^{\blacktriangledown}$.

Vice versa, let $\mathcal{M} = \langle M\,, \oplus\,, '\,, \mathbf{0}\,, \mathbf{1}\rangle$ be a QMV algebra. Then, one can define a partial operation $\boxplus$ on M such that:

$$Dom(\boxplus) := \{(a, b) \in M \times M \,:\, a \leq b'\}\,;$$

$$a \boxplus b := a \oplus b, \text{ if } a \leq b'.$$

The resulting structure $\langle M\,, \boxplus\,, \mathbf{0}\,, \mathbf{1}\rangle$ will be denoted by $\mathcal{M}^{\blacktriangle}$.

THEOREM 5.4.4. (Gudder, 1995; Giuntini, 1995a)
Let $\mathcal{B} = \langle B\,, \boxplus\,, \mathbf{0}\,, \mathbf{1}\rangle$ be an effect algebra and let $\mathcal{M} = \langle M\,, \oplus\,, '\,, \mathbf{0}, \mathbf{1}\rangle$ be a QMV algebra.

(i) *$\mathcal{B}^{\blacktriangledown}$ is a quasi-linear QMV algebra;*
(ii) *$\mathcal{M}^{\blacktriangle}$ is an effect algebra;*
(iii) *$\mathcal{B}^{\blacktriangledown\blacktriangle} = \mathcal{B}$;*

(iv) $\mathcal{M}$ *is quasi-linear iff* $\mathcal{M}^{\blacktriangle\blacktriangledown} = \mathcal{M}$;
(v) $\mathcal{B}^{\blacktriangledown}$ *is the unique quasi-linear QMV algebra such that:*
(va) $\oplus$ *extends* $\boxplus$;
(vb) $[a \leq b$ *in* $\mathcal{B}^{\blacktriangledown}]$ *implies* $[a \leq b$ *in* $\mathcal{B}]$.

PROOF.
(i) Let $\mathcal{B} = \langle B, \boxplus, \mathbf{0}, \mathbf{1}\rangle$ be an effect algebra, with inherited partial order $\leq$ and generalized complement $'$ (see Definition 5.1.2). By Theorem 5.1.5, the structure $\langle B, \leq, ', \mathbf{0}, \mathbf{1}\rangle$ is a bounded involution poset. We want to show that $\mathcal{B}^{\blacktriangledown}$ is a quasi-linear QMV algebra. First, we prove that $\forall a, b \in B$:

$$a \leq b \text{ iff } a = a \Cap b := (a \oplus b') \odot b.$$

Suppose $a \leq b$. By contraposition, $b' \leq a'$. Thus, by Theorem 5.1.3 (vi)-(vii): $b' \perp a$ and $a' = b' \boxplus (b' \boxplus a)'$. Hence,

$$a = a'' = (b' \boxplus (b' \boxplus a)')' = (b' \oplus (b' \oplus a)')'$$
$$= (a \oplus b') \odot b = a \Cap b.$$

Suppose $a = a \Cap b := (a \oplus b') \odot b$. We want to show that $a \leq b$. Consider first the case in which $a \oplus b' = \mathbf{1}$. Then:

$$a = a \Cap b = \mathbf{1} \odot b = (\mathbf{0} \oplus b')' = (\mathbf{0} \boxplus b')' = b,$$

so that $a \leq b$. Now consider the case in which $a \oplus b' \neq \mathbf{1}$. Then, by definition, $a \boxplus b'$ is defined and $a \oplus b' = a \boxplus b'$. Hence, $a \perp b'$ and $a \leq b'' = b$ (by Theorem 5.1.3(vi)). It follows that:

$$a \Cap b = \begin{cases} a & \text{if } a \leq b, \\ b & \text{otherwise.} \end{cases}$$

This shows that if $\mathcal{B}^{\blacktriangledown}$ is a QMV algebra, then $\mathcal{B}^{\blacktriangledown}$ is quasi-linear. We will now prove that $\mathcal{B}^{\blacktriangledown}$ is a QMV algebra.

(i.1) (QMV1) $a \oplus b = b \oplus a$. By weak commutativity of effect algebras, $a \boxplus b$ is defined iff $b \boxplus a$ is defined. Hence, whenever $a \boxplus b$ is defined we have: $a \oplus b = a \boxplus b = b \boxplus a = b \oplus a$. When $a \boxplus b$ is not defined, then also $b \boxplus a$ is not defined, so that, by definition of $\oplus$, we obtain $a \oplus b = \mathbf{1} = b \oplus a$.

(i.2) (QMV2) $a \oplus (b \oplus c) = (a \oplus b) \oplus c$. Two cases are possible: (I) $a \boxplus (b \boxplus c)$ is defined, and (II) $a \boxplus (b \boxplus c)$ is not defined. In case (I), since $a \boxplus (b \boxplus c)$ is defined, we have $a \boxplus (b \boxplus c) = a \oplus (b \oplus c)$. By weak associativity, $a \boxplus b$ is defined, $(a \boxplus b) \boxplus c$ is defined and $(a \boxplus b) \boxplus c = a \boxplus (b \boxplus c)$. Since $(a \boxplus b) \boxplus c = (a \oplus b) \oplus c$, we obtain: $a \oplus (b \oplus c) = (a \oplus b) \oplus c$. In case (II), two subcases are possible: (II1) $b \boxplus c$ is defined and (II2) $b \boxplus c$ is not defined. Consider (II1), then $\mathbf{1} = a \oplus (b \boxplus c) = a \oplus (b \oplus c)$. Suppose, by contradiction, that $(a \oplus b) \oplus c \neq \mathbf{1}$. Then, $a \oplus b \neq \mathbf{1}$. Thus, $(a \boxplus b) \boxplus c$ is defined. Then, by weak associativity, $a \boxplus (b \boxplus c)$ is defined, contradiction. Consider (II2), then $b \boxplus c$ is not defined. Then, $b \oplus c = \mathbf{1}$. Hence, $a \oplus (b \oplus c) = a \oplus \mathbf{1} = \mathbf{1}$.

Suppose, by contradiction, that $(a \oplus b) \oplus c \neq \mathbf{1}$. Then, $(a \boxplus b) \boxplus c$ is defined. Again, by weak associativity, $a \boxplus (b \boxplus c)$ is defined, contradiction.

(i.3) The proof of (QMV3)-(QMV6) is straightforward.

(i.4) (QMV7) $a \oplus [(a' \Cap b) \Cap (c \Cap a')] = (a \oplus b) \Cap (a \oplus c)$. The following cases are possible: (1) $a \boxplus b$ and $a \boxplus c$ are not defined; (2) $a \boxplus b$ is not defined and $a \boxplus c$ is defined; (3) $a \boxplus b$ is defined and $a \boxplus c$ is not defined; (4) both $a \boxplus b$ and $a \boxplus c$ are defined. We will freely use the following facts:
$a \odot \mathbf{0} = \mathbf{0}$; $a \odot \mathbf{1} = \mathbf{1}$; $a \Cap \mathbf{1} = \mathbf{1} \Cap a = a$;
$a \Cap a = a$; $a \oplus a' = \mathbf{1}$; and $a \leq b$ implies $a = b \Cap a$.

Case (1). By hypothesis $a \boxplus b$ and $a \boxplus c$ are not defined. Then, $a \oplus b = \mathbf{1} = a \oplus c$. Consequently: $c \Cap a' = (c \oplus a) \odot a' = \mathbf{1} \odot a' = a'$. Thus, $rhs = (a \oplus b) \Cap (a \oplus c) = \mathbf{1} \oplus \mathbf{1} = \mathbf{1}$. Now, $lhs = a \oplus [(a' \Cap b) \Cap (c \Cap a')] = a \oplus [(a' \Cap b) \Cap a']$. Consider first the case: $a' \leq b$. Then, $(a' \Cap b) \Cap a' = a' \Cap a' = a'$, so that $lhs = a \oplus a' = \mathbf{1} = rhs$. Now consider the case: $a' \not\leq b$. Then, by quasi-linearity, $a' \Cap b = b$; thus $lhs = a \oplus [(a' \Cap b) \Cap a'] = a \oplus [b \Cap a'] = a \oplus [(b \oplus a) \odot a'] = a \oplus (\mathbf{1} \odot a') = a \oplus a' = \mathbf{1} = rhs$.

Case (2). By hypothesis: $a \oplus b = \mathbf{1}$, $a \oplus c = a \boxplus c$ and $a \leq c'$. Thus, $rhs = \mathbf{1} \Cap (a \boxplus c) = a \boxplus c$. Since $a \leq c'$, we also have: $c \leq a'$. Hence, $lhs = a \oplus [(a' \Cap b) \Cap c]$. Consider first the case $(a' \Cap b) \not\leq c$. Then, $(a' \Cap b) \Cap c = c$, so that $lhs = a \oplus c = a \boxplus c = rhs$. Now consider the case $(a' \Cap b) \leq c$. Then, $lhs = a \oplus (a' \Cap b)$. Two subcases are possible: (I) $a' \leq b$ and (II) $a' \not\leq b$. Consider (I): then $a' = a' \Cap b \leq c$. But by hypothesis $c \leq a'$ so that $c = a'$ and $lhs = a \oplus (a' \Cap b) = a \oplus a' = a \oplus c = rhs$. We prove now that (II) is impossible. Suppose, by contradiction, that $a' \not\leq b$. Then, $a' \Cap b = b$, so that $b = a' \Cap b \leq c$. By hypothesis, $c \leq a'$, so that $b \leq a'$ and $a \leq b'$. Consequently $a \boxplus b$ is defined, contradiction.

Case (3). By hypothesis, $a \oplus b = a \boxplus b$, $a \leq b'$ and $a \oplus c = \mathbf{1}$. Hence, $rhs = (a \boxplus b) \Cap \mathbf{1} = a \boxplus b$. Since $a \oplus c = \mathbf{1}$, we have $c \Cap a' = a'$. Thus, $lhs = a \oplus [(a \Cap b) \Cap a']$. By hypothesis $a \leq b'$, so that $b \leq a'$. Hence, $a' \Cap b = b$ and $lhs = a \oplus (b \Cap a') = a \oplus b = a \boxplus b = rhs$.

Case (4). By hypothesis, $a \oplus b = a \boxplus b$, $a \leq b'$ and $a \oplus c = a \boxplus c$, $a \leq c'$. Two subcases are possible: I) $a \boxplus b \leq a \boxplus c$ and II) $a \boxplus b \not\leq a \boxplus c$. Consider I): then, by Lemma 5.1.4(ii), $b \leq c$. Thus, $lhs = a \oplus [(a' \Cap b) \Cap (c \Cap a')] = a \oplus (b \Cap c) = a \oplus b = a \boxplus b = rhs$. Consider II): then we have: $rhs = (a \oplus b) \Cap (a \oplus c) = (a \boxplus b) \Cap (a \boxplus c) = a \boxplus c$ (by quasi-linearity). First, we show that $b \not\leq c$. Suppose, by contradiction, that $b \leq c$; then, by Lemma 5.1.4(ii), $a \boxplus b \leq a \boxplus c$, contradiction. Hence, $b \not\leq c$ and $b \Cap c = c$. Consequently (by quasi-linearity): $lhs = a \oplus [a' \Cap b) \Cap (c \Cap a')] = a \oplus (b \Cap c) = a \oplus c = rhs$.

(ii) Let $\mathcal{M} = \langle M, \oplus, ', \mathbf{0}, \mathbf{1} \rangle$ be a QMV algebra. We want to show that $\mathcal{M}^{\blacktriangle} = \langle M, \boxplus, \mathbf{0}, \mathbf{1} \rangle$ is an effect algebra.

(ii.1) $\boxplus$ is weakly commutative. Suppose $a \boxplus b$ is defined. By definition, $a \leq b'$, i.e., $a = a \Cap b'$ and $a \boxplus b = a \oplus b$. By Theorem 5.3.17, $\langle M, \leq, ', \mathbf{0}, \mathbf{1} \rangle$ is a bounded involution poset. Thus, $b \boxplus a$ is defined and $b \boxplus a = b \oplus a$. By commutativity of $\oplus$ (QMV1) we obtain: $a \boxplus b = b \boxplus a$.

(ii.2) $\boxplus$ is weakly associative. Suppose $a \boxplus (b \boxplus c)$ is defined. Then, $b \leq c'$ and $a \leq (b \boxplus c)' = (b \oplus c)'$. Thus, $a \boxplus (b \boxplus c) = a \oplus (b \oplus c)$. We want to show that $a \leq b'$ and $(a \boxplus b) = (a \oplus b) \leq c'$. By hypothesis $a \leq (b \oplus c)' = b' \odot c'$. By Lemma 5.3.13(i), $b' \odot c' \leq b'$ so that $a \leq b'$. It remains to show that $a \oplus b \leq c'$. We have: $b \leq c'$. Thus, by Lemma 5.3.15(i), $c' = c' \Cup b$. The following equalities hold:

$$\begin{aligned}
(a \oplus b) \Cap c' &= (a \oplus b) \Cap (c' \Cup b) & \\
&= (b \oplus a) \Cap (b \oplus c' \odot b') & \text{(definition of } \Cup) \\
&= b \oplus [(b' \Cap a) \Cap ((c' \odot b') \Cap b')] & \text{(QMV7)} \\
&= b \oplus [a \Cap ((c' \odot b') \Cap b')] & (a \leq b') \\
&= b \oplus [a \Cap (c' \odot b')] & (c' \odot b' \leq b') \\
&= b \oplus a & (a \leq (b \boxplus c)' = b' \odot c') \\
&= a \oplus b. &
\end{aligned}$$

Consequently $(a \oplus b) \leq c'$. Then, both $(a \oplus b) \boxplus c$ and $(a \boxplus b) \boxplus c$ are defined and equal to $(a \oplus b) \oplus c$. Whence, by (QMV2): $(a \oplus b) \oplus c = a \oplus (b \oplus c) = a \boxplus (b \boxplus c)$.

(ii.3) The strong excluded middle principle holds. Let $a \in M$. We have $a \leq a''$ so that $a \boxplus a'$ is defined and $a \boxplus a' = a \oplus a' = \mathbf{1}$. Let b be an element of M such that $a \boxplus b$ is defined and $a \boxplus b = \mathbf{1}$. Then, $a \oplus b = \mathbf{1}$ and, therefore, $a = a \Cap b' = (a \oplus b) \odot b' = \mathbf{1} \odot b' = b'$.

(ii.4) The weak consistency principle holds, trivially.

(iii) The proof of $\mathcal{B}^{\blacktriangledown\blacktriangle} = \mathcal{B}$ is trivial.

(iv) Necessity follows from (i). Suppose $\mathcal{M} = \langle M, \oplus, ', \mathbf{0}, \mathbf{1} \rangle$ is a quasi-linear QMV algebra. Let us denote the sum and the generalized complement of the QMV algebra $\mathcal{M}^{\blacktriangle\blacktriangledown}$ by $\oplus_{\blacktriangle\blacktriangledown}$ and by $'^{\blacktriangle\blacktriangledown}$, respectively. Clearly, $'^{\blacktriangle\blacktriangledown}$ coincides with $'$. Thus, to prove that $\mathcal{M} = \mathcal{M}^{\blacktriangle\blacktriangledown}$ it suffices to show that $\oplus = \oplus_{\blacktriangle\blacktriangledown}$. Let $a, b \in M$. Two cases are possible: (I) $a \oplus b = \mathbf{1}$ and (II) $a \oplus b \neq \mathbf{1}$.
In case (I), two subcases are possible: (I.1) $a \leq b'$; (I.2) $a \not\leq b'$. In case (I.1), we obtain: $a = b'$, so that $a \oplus b = a \boxplus a' = a \oplus_{\blacktriangle\blacktriangledown} a' = a \oplus_{\blacktriangle\blacktriangledown} b$, where $\boxplus$ is the partial sum of the effect algebra $\mathcal{M}^{\blacktriangle}$ determined by $\mathcal{M}$. In case (I.2), $a \boxplus b$ is not defined. Consequently $a \oplus_{\blacktriangle\blacktriangledown} b = \mathbf{1} = a \oplus b$. In case (II), $a \oplus b \neq \mathbf{1}$. Since $\mathcal{M}$ is quasi-linear, by Lemma 5.4.2 (ii), we obtain $a \leq b'$. Thus, $a \boxplus b$ is defined and $a \oplus b = a \boxplus b = a \oplus_{\blacktriangle\blacktriangledown} b$.

(v) Let $\mathcal{B} = \langle B, \boxplus, \mathbf{0}, \mathbf{1} \rangle$ be an effect algebra (with generalized complement $'$). Let $\mathcal{B}^{\blacktriangledown} = \langle B, \oplus, ', \mathbf{0}, \mathbf{1} \rangle$ be the QMV algebra determined by $\mathcal{B}$. Clearly, $\oplus$ extends $\boxplus$. As shown in the proof of (i), $a = a \Cap b$ iff $a \leq b$ in $\mathcal{B}$. Hence, $a \leq b$ in $\mathcal{B}^{\blacktriangledown}$ iff $a \leq b$ in $\mathcal{B}$. Suppose there exists a quasi-linear QMV algebra $\mathcal{B}^{o} = \langle B, \oplus^{o}, \mathbf{0}, \mathbf{1} \rangle$ that satisfies the following conditions: (v1) $\oplus^{o}$ extends $\boxplus$; (v2) $a \leq b$ in $\mathcal{B}^{o}$ implies $a \leq b$ in $\mathcal{B}$. We want to show that $\mathcal{B}^{o} = \mathcal{B}^{\blacktriangledown}$. It suffices to show that $\forall a, b \in B : a \oplus b = a \oplus^{o} b$. If $a \boxplus b$ is defined, then $a \oplus^{o} b = a \boxplus b = a \oplus b$. If $a \boxplus b$ is not defined, then $a \oplus b = \mathbf{1}$.

Suppose, by contradiction, that $a \oplus^o b \neq \mathbf{1}$. Since, $\mathcal{B}^o$ is quasi-linear, by Lemma 5.4.2, $a \leq b'$ in $\mathcal{B}^o$ so that, by (v2), $a \leq b'$ in $\mathcal{B}$, and consequently, $a \boxplus b$ is defined, contradiction. In each case $a \oplus b = a \oplus^o b$, completing the proof. $\square$

CHAPTER 6

Abstract axiomatic foundations of unsharp quantum theory

In the previous chapters we have studied the emergence of fuzzy events in the framework of Hilbert space QT. We now abstract from the Hilbert space formalism, paralleling the case of sharp QT, in order to obtain an abstract axiomatic foundation of unsharp QT.

Some general requirements, that should be satisfied by any abstract reconstruction of an unsharp version of QT, have been proposed by Cattaneo and Gudder (1999). Let us first sum up these basic principles.

I) The abstract structure of the unsharp quantum events should admit the concrete structure of all effects in a Hilbert space $\mathcal{H}$ as a particular model.

II) As in the sharp case, the theory should describe some special objects representing *observables* and *states*. In the Hilbert space model, observables should be exemplified by effect-valued measures (which are natural generalizations of event-valued measures). At the same time, states should correspond to the probability measures that are determined by density operators.

III) The theory should provide a formal distinction between *sharp* and *unsharp* effects. Such a distinction should capture, in the Hilbert space model, precisely the set of all projection operators as the sharp effects and the set of all proper effects as the unsharp effects.

IV) Any abstract effect E should be associated to two privileged sharp elements: its *neccessity* $\nu(E)$, representing the *best sharp lower approximation of* E, and its *possibility* $\mu(E)$, representing the *best sharp upper approximation of* E.

6.1. A minimal axiomatization of unsharp QT

The abstract axiomatization of unsharp QT that we are about to present is a modified version of the axiomatic systems proposed by Cattaneo-Garola-Nisticò and by Cattaneo-Gudder (in the spirit of Ludwig's approach).[1]

[1]See (Cattaneo, Garola and Nisticò, 1989; Cattaneo and Gudder, 1999). A different abstract approach to unsharp QT has been developed in a series of papers by Beltrametti and Bugajski (1995; 1997).

The basic notion is that of *effect-state system*: a convenient "unsharp variant" of both the notions of *state-observable probability system* and of *event-state system* (analyzed in Chapter 2).

As expected, the intended physical interpretation of an effect-state system $(\mathcal{E}, \mathcal{S})$ will be the following:

- The elements of $\mathcal{E}$, indicated by $E, F, G, \ldots\ldots$, represent *effects* tested by *dichotomous measuring macroscopic devices*. When interacting with a single sample of the physical object under investigation, such devices produce a certain definite macroscopic result. The response may be either *positive* ("yes") or *negative* ("no"), but not both.
- The elements of $\mathcal{S}$, indicated by $s, t, \ldots$, represent *states*. They are supposed to be physically realized by macroscopic apparatuses that prepare both individual samples and ensembles of identical non interacting physical objects under well defined and repeatable conditions. Mathematically, every state s represents a generalized probability function such that for any $E \in \mathcal{E}$:
$$s(E) \in [0, 1].$$
For any E, the number $s(E)$ represents the probability to obtain response "yes" for the effect E, when the physical object is prepared in the state corresponding to s.

As in Mackey's axiomatization, abstract unsharp QT is characterized by the class **K** of all effect-state systems $(\mathcal{E}, \mathcal{S})$ that satisfy an appropriate set of axioms. We first formally present the axioms (U1-U9); then we comment on them from an intuitive point of view and after we will prove some relevant theorems.

DEFINITION 6.1.1. *Effect-state system*
An *effect-state system* is a pair $(\mathcal{E}, \mathcal{S})$, where $\mathcal{E}$ represents a nonempty set of effects and $\mathcal{S}$ represents a nonempty set of states, satisfying the following Axioms (U1-U9).

Axiom U1. *Effects and states*

(i) $\forall E \in \mathcal{E}\ \forall s \in \mathcal{S}\, [s(E) \in [0, 1]]$;

(ii) $\mathcal{E}$ contains an effect $\mathbf{1}$, called the *certain* effect, such that $\forall s \in \mathcal{S}[s(\mathbf{1}) = 1]$.

Axiom U2. *Indistinguishability principle for states*
$\forall s, t \in \mathcal{S}$:
$$\text{if } \forall E \in \mathcal{E}[s(E) = t(E)], \text{ then } s = t.$$

Axiom U3. *Indistinguishability principle for effects*
$\forall E, F \in \mathcal{E}$:
$$\text{if } \forall s \in \mathcal{S}[s(E) = s(F)], \text{ then } E = F.$$

Axiom U4. *The excluded middle principle*
$\forall E \in \mathcal{E}\, \exists F \in \mathcal{E}$ such that:

$$\forall s \in \mathcal{S}[s(E) + s(F) = 1].$$

Axiom U5. *The convexity principle*
Let $\{s_n\}_{n \in I}$ be a countable set of states and let $\{\lambda_n\}_{n \in I}$ a countable set of nonnegative real numbers such that $\sum_n \lambda_n = 1$. Then, there exists a state $s \in \mathcal{S}$ such that $\forall E \in \mathcal{E}$:

$$s(E) = \sum_n \lambda_n s_n(E).$$

The state s is called a *convex combination* or *mixture* of the states s_n, where each λ_n represents the *weight* of the corresponding state s_n.

On the basis of these axioms, a partial order relation $\leq$, a generalized complement $'$, an orthogonality relation $\perp$ and a special effect $\mathbf{0}$ (the *impossible effect*) can be defined.

DEFINITION 6.1.2. *The effect order* $\leq$

$$\forall E, F \in \mathcal{E}[E \leq F \text{ iff } \forall s \in \mathcal{S}[s(E) \leq s(F)]].$$

DEFINITION 6.1.3. *The generalized complement* $'$

$$\forall E, F \in \mathcal{E}[F = E' \text{ iff } \forall s \in \mathcal{S}[s(E) + s(F) = 1]].$$

The operation $'$ is well defined by Axioms U4 and U3, which jointly guarantee the existence and the uniqueness of E' for any effect E.

DEFINITION 6.1.4. *The orthogonality relation*

$$\forall E, F \in \mathcal{E}[E \perp F \text{ iff } E \leq F'].$$

DEFINITION 6.1.5. *The impossible effect*

$$\mathbf{0} = \mathbf{1}'.$$

Axiom U6. *The partial sum*
Suppose we have two effects E, F of $\mathcal{E}$ such that $E \perp F$. Then:

$$\exists G \in \mathcal{E}\, \forall s \in \mathcal{S}[s(G) = s(E) + s(F)].$$

When existing, the effect G such that $s(G) = s(E) + s(F)$ is unique (by Axiom U3). As a consequence, we can define the following partial operation $\boxplus$:

$$Dom(\boxplus) := \{(E, F) \in \mathcal{E} : E \perp F\};$$
$$E \boxplus F := G \text{ iff } \forall s \in \mathcal{S}[s(G) = s(E) + s(F)].$$

One immediately obtains that, when defined, $\boxplus$ is commutative and associative. This permits us to write $E_1 \boxplus \cdots \boxplus E_n$, ignoring all brackets. If $n = 1$, we put $E_1 \boxplus \cdots \boxplus E_n := E_1$.

Axiom U7. *The weak σ-orthocompleteness principle*
Let $\{E_n\}_{n\in\mathbb{N}}$. be a sequence of effects such that $\forall n \in \mathbb{N}$ the partial sum $E_1 \boxplus \cdots \boxplus E_n$ is defined. Then, there exists an effect E such that

$$\forall s \in \mathcal{S} \left[s(E) = \lim_{n\to\infty} s(E_1 \boxplus \cdots \boxplus E_n) \right].$$

This effect E, which is unique (by Axiom U3), will be denoted by $\sum \{E_n\}_{n\in\mathbb{N}}$.

In this framework, the notion of *sharp effect* is defined in the expected way.

DEFINITION 6.1.6. *Sharp effect*
An effect E is called *sharp* iff the infimum $E \wedge E'$ (with respect to the partial order $\leq$) exists and is equal to $\mathbf{0}$.

We will indicate by $\mathcal{E}_s$ the set of all sharp elements of $\mathcal{E}$.

Axiom U8. *The sharply dominating principle*
For any effect E, there exists an effect F that represents a *sharp lower approximation of* E. In other words, F satisfies the following conditions:

(i) F is sharp (i.e., $F \in \mathcal{E}_s$);
(ii) $F \leq E$;
(iii) $\forall G \in \mathcal{E}_s$: if $G \leq E$, then $G \leq F$.

We will say that such an $(\mathcal{E}, \mathcal{S})$ is *sharply dominating.*[2]

By Axiom U3 the sharp lower approximation of an effect E is unique. Hence, we can define the following map:

$$\nu : \mathcal{E} \mapsto \mathcal{E}_s,$$

that associates to any effect E its sharp lower approximation $\nu(E)$.

Before introducing the last axiom, let us first define the notions of *carrier* of a state and of *sharply rich* effect-state system.

DEFINITION 6.1.7. *The carrier of a state*
An effect E is called a *carrier* of a state $s \in \mathcal{S}$ iff the following conditions are satisfied:

(i) $s(E) = 1$;
(ii) $\forall F \in \mathcal{E}$: $s(F) = 1$ implies $E \leq F$.

One can prove that, if it exists, the carrier of a state is unique.

[2]See (Gudder, 1998).

DEFINITION 6.1.8. *Sharply rich*
Let $(\mathcal{E}, \mathcal{S})$ be a pair satisfying the previous axioms. As in the discussion of Section 4.4, we define, for any effect E, $\mathbf{Yes}(E) := \{s \in \mathcal{S} : s(E) = 1\}$. $\mathcal{S}$ is called *sharply rich* for $\mathcal{E}$ iff $\forall E, F \in \mathcal{E}_s$:

$$\mathbf{Yes}(E) \subseteq \mathbf{Yes}(F) \text{ implies } E \leq F.$$

Axiom U9. *The sharply normal principle*

(i) For any state $s \in \mathcal{S}$, the carrier of s exists;
(ii) $\mathcal{S}$ is sharply rich for $\mathcal{E}$;
(iii) $\forall E \in \mathcal{E}$: $\mathbf{Yes}(E) = \mathbf{Yes}(\nu(E))$.

We will say that such an $(\mathcal{E}, \mathcal{S})$ is *sharply normal.*

In the next sections we will comment on the intuitive meaning of the axioms. At the same time we will prove and discuss some interesting consequences thereof.

6.2. The algebraic structure of abstract effects

Axiom U1 asserts that any state assigns to any effect a value in the interval $[0,1]$ and that the value of the certain effect $\mathbf{1}$ is 1. As in Mackey's axiomatization, Axioms U2 and U3 codify a principle of parsimony: states and events that are statistically indistinguishable are identified. Axioms U3 and U4 jointly guarantee, for any effect E, the existence and the uniqueness of an effect F such that the sum of the probabilities assigned by any state s to E and to F (respectively) is exactly 1. This justifies the definition of the generalized complement $'$. As in Mackey's axiomatization, Axiom U5 permits one to "create" new states by using convex combinations of states belonging to $\mathcal{S}$.

As we did in the sharp case, we will now investigate both the structural properties of effect-systems and the *interplay* between states and effects in systems satisfying these axioms.

We have already seen that the set of states permits one to define a partial order on the set of all effects: $E \leq F$ means that all states assign to E a probability-value that is less than or equal to the probability-value assigned to F.

On this basis, one can easily prove that the abstract set of effects gives rise to a regular poset (as, of course, happens in the case of a concrete structure based on $\mathcal{E}(\mathcal{H})$).

THEOREM 6.2.1. *The structure*

$$\langle \mathcal{E}, \leq, ', \mathbf{0}, \mathbf{1} \rangle$$

is a regular poset.

PROOF. The proof is straightforward. □

As in $\mathcal{E}(\mathcal{H})$ but unlike the sharp case, the poset $\mathcal{E}$ fails to be an orthoposet. One can easily realize that an abstract effect poset may contain a totally indeterminate effect to which any state s assigns probability $\frac{1}{2}$. Also in the abstract case, we call such a paradigmatic indeterminate effect, if it exists, *the semitransparent effect*, and we will denote it by $\frac{1}{2}\mathbf{1}$. Thus, $\forall E \in \mathcal{E}$,

$$E = \frac{1}{2}\mathbf{1} \text{ iff } \forall s \in \mathcal{S} : s(E) = \frac{1}{2}.$$

Of course, the existence of the semitransparent effect is only *compatible* with the axioms. However, when existing, $\frac{1}{2}\mathbf{1}$ is unique (Axiom U3).

As in the concrete case, the semitransparent effect gives rise to a violation of the noncontradiction principle:

$$\frac{1}{2}\mathbf{1} \wedge (\frac{1}{2}\mathbf{1})' = \frac{1}{2}\mathbf{1} \wedge \frac{1}{2}\mathbf{1} = \frac{1}{2}\mathbf{1} \neq \mathbf{0}.$$

As a consequence, generally, the set $\mathcal{E}_s$ of all sharp elements (which satisfy the noncontradiction principle) will be a proper subset of $\mathcal{E}$.

What can be said about the structural properties of the set of all sharp elements?

One can easily show that $\mathcal{E}_s$ is closed under the generalized complement $'$. Furthermore, $\mathbf{0}$ and $\mathbf{1}$ turn out to be trivially sharp.

Consider now the following structure

$$\langle \mathcal{E}_s \,, \leq_s \,, {}' \,, \mathbf{0} \,, \mathbf{1} \rangle \,,$$

where $\leq_s$ is the restriction of the partial order defined on $\mathcal{E}$ to the subset $\mathcal{E}_s$. From the definition, one immediately obtains that such a structure is an orthoposet (where the operation $'$ satisfies the noncontradiction principle).

As we have seen in Chapter 5, any concrete effect system $\mathcal{E}(\mathcal{H})$ gives rise to an effect algebra $\langle \mathcal{E}(\mathcal{H}) \,, \boxplus \,, \mathbf{0} \,, \mathbf{1} \rangle$. It is natural to ask: can such a situation be generalized to the case of our abstract effect systems? Axiom U6 permits us to give a positive answer to this question: also abstract effects have the structure of an effect algebra.

THEOREM 6.2.2. *The structure*

$$\langle \mathcal{E} \,, \boxplus \,, \mathbf{0} \,, \mathbf{1} \rangle \,,$$

is an effect algebra, where $\boxplus$ is the partial operation defined after Axiom U6.

PROOF. The proof is straightforward. □

We have also seen, in Chapter 5, that an orthogonality relation $\perp_A$, a partial order relation $\leq_A$, a generalized complement ${}'^A$ can be defined in any abstract effect algebra $\langle A \,, \boxplus \,, \mathbf{0} \,, \mathbf{1} \rangle$. The following conditions hold:

- $a \perp_A b$ iff $a \boxplus b$ is defined.
- $a \leq_A b$ iff $\exists c \in A$ such that $a \perp_A c$ and $b = a \boxplus c$.
- a'^A is the unique element such that $a \boxplus a' = \mathbf{1}$.

Interestingly enough, in the case of an effect algebra $\langle \mathcal{E}\,, \boxplus\,, \mathbf{0}\,, \mathbf{1}\rangle$ (based on a set $\mathcal{E}$ of abstract quantum effects), the operation ${}'^{\mathcal{E}}$ and the relations $\leq_{\mathcal{E}}$, $\perp_{\mathcal{E}}$ turn out to coincide respectively with the "natural" generalized complement, orthogonality and partial order that are defined on $\mathcal{E}$. For, one can easily show the following.

LEMMA 6.2.3.

(i) $E \perp_{\mathcal{E}} F$ *iff* $E \perp F$;
(ii) $E \leq_{\mathcal{E}} F$ *iff* $E \leq F$;
(iii) $E'^{\mathcal{E}} = E'$.

PROOF. The proof is straightforward. □

THEOREM 6.2.4. *Let* $\langle \mathcal{E}\,, \mathcal{S}\rangle$ *be a structure satisfying Axioms U1-U6. The following conditions are equivalent:*

(i) *Axiom U7 holds;*
(ii) *Let* $E_1, E_2, ...$ *be a sequence of effects such that* $\forall n \in \mathbb{N}\, \forall s \in \mathcal{S} : s(E_n) \leq s(E_{n+1})$. *Then, there exists an effect* E *such that:*

$$\forall s \in \mathcal{S}\left[s(E) = \lim_{n\to\infty} s(E_n)\right].$$

PROOF. (i) implies (ii) Let $E_1, E_2, ...$ be a sequence of effects such that $\forall n \in \mathbb{N}\, \forall s \in \mathcal{S} : s(E_n) \leq s(E_{n+1})$. We have to show that there exists an effect E such that: $\forall s \in \mathcal{S}\,[s(E) = \lim_{n\to\infty} s(E_n)]$. Let us define an infinite sequence $F_1, F_2, \ldots$ in the following way:

$$F_1 = E_1 \text{ and, for } n > 1,\ F_n = \left(E_n' \boxplus E_{n-1}\right)'.$$

It follows that $\forall n \in \mathbb{N}$, $F_1 \boxplus \cdots \boxplus F_n$ is defined and $F_1 \boxplus \cdots \boxplus F_n = E_n$. Thus, by Axiom U7, there exists an effect E such that

$$\forall s \in \mathcal{S}\left[s(E) = \lim_{n\to\infty} s(F_1 \boxplus \cdots \boxplus F_n) = \lim_{n\to\infty} s(E_n)\right].$$

(ii) implies (i). Let $E_1, E_2, ...$ be a sequence of effects such that $\forall n \in \mathbb{N}$ the partial sum $E_1 \boxplus \cdots \boxplus E_n$ is defined. We have to prove that there exists an effect E such that $\forall s \in \mathcal{S}\,[s(E) = \lim_{n\to\infty} s(E_1 \boxplus \cdots \boxplus E_n)]$. Define an infinite sequence $F_1, F_2, \ldots$ in the following way:

$$F_1 = E_1 \text{ and, for } n > 1,\ F_n = F_{n-1} \boxplus E_n.$$

By hypothesis, this sequence is well defined. Furthermore, $\forall n \in \mathbb{N}$: $F_n \leq F_{n+1}$ and $F_n = E_1 \boxplus \cdots \boxplus E_n$. By condition (ii), there exists an effect E such that

$$\forall s \in \mathcal{S}\left[s(E) = \lim_{n\to\infty} s\,(F_n) = \lim_{n\to\infty} s(E_1 \boxplus \cdots \boxplus E_n)\right].$$

□

Axiom U7 represents a *weak σ-orthocompleteness* principle. Unlike the sharp case (see Definition 2.3.1 and Definition 2.3.3) this axiom does not require that, for any sequence $\{E_n\}_{n\in I}$ of pairwise orthogonal events,

$$s(\bigvee \{E_n\}_{n\in I}) = \sum_n s(E_n).$$

As we have already seen, the strong σ-orthocompleteness principle is responsible for the validity of the noncontradiction principle (see the proof of Lemma 2.2.8). Hence, one could think that, in this abstract framework, the difference between the weak version and the strong version of the σ-orthocompleteness principle represents a *dividing line* between *sharp* and *unsharp* systems.[3]

At the same time, Axiom U7 permits one to prove that any effect algebra $\langle \mathcal{E}, \boxplus, \mathbf{0}, \mathbf{1}\rangle$, where $\mathcal{E}$ is the effect-set of a given effect-state system, is a *σ-effect algebra.*

DEFINITION 6.2.5. *σ-effect algebra*
A *σ-effect algebra* is an effect algebra $\langle A, \boxplus, \mathbf{0}, \mathbf{1}\rangle$ that satisfies the following condition. Suppose that $a_1, a_2, ...$ is a sequence of elements such that $\forall n \in \mathbb{N}$ the partial sum $a_1 \boxplus \cdots \boxplus a_n$ is defined. Then, there exists an element $\boxed{+}\{a_n\}_{n\in\mathbb{N}}$ such that:

$$\boxed{+}\{a_n\}_{n\in\mathbb{N}} = \bigvee \{a_1 \boxplus \cdots \boxplus a_n\}_{n\in\mathbb{N}}$$

(where $\bigvee$ is the supremum with respect to the effect order $\leq$).

LEMMA 6.2.6. *Suppose we have an effect-state system $(\mathcal{E}, \mathcal{S})$ and a sequence $E_1, E_2, ...$ of effects such that $\forall n \in \mathbb{N}$ the partial sum $E_1 \boxplus \cdots \boxplus E_n$ is defined. Then,*

$$\sum \{E_n\}_{n\in\mathbb{N}} = \boxed{+}\{E_n\}_{n\in\mathbb{N}},$$

where $\sum \{E_n\}_{n\in\mathbb{N}}$ is the element that exists by Axiom U7.

PROOF. Suppose that $(\mathcal{E}, \mathcal{S})$ is an effect-state system and that $E_1, E_2, ...$ is a sequence of effects such that $\forall n \in \mathbb{N}$ the partial sum $E_1 \boxplus .... \boxplus E_n$ is defined. Then, by Axiom U7, there exists a unique effect $\sum \{E_n\}_{n\in\mathbb{N}}$ such that

$$\forall s \in \mathcal{S} \left[s\left(\sum \{E_n\}_{n\in\mathbb{N}}\right) = \lim_{n\to\infty} s(E_1 \boxplus ... \boxplus E_n)\right].$$

Since $\forall s \in \mathcal{S}$, the number $s\left(\sum \{E_n\}_{n\in\mathbb{N}}\right)$ is the limit of the monotone nondecreasing bounded sequence $\{s(E_1 \boxplus .... \boxplus E_n)\}_{n\in\mathbb{N}}$, we obtain that $s\left(\sum \{E_n\}_{n\in\mathbb{N}}\right)$ is the least upper bound of $\{s(E_1 \boxplus .... \boxplus E_n)\}_{n\in\mathbb{N}}$. Consequently,

$$\sum \{E_n\}_{n\in\mathbb{N}} = \boxed{+}\{E_n\}_{n\in\mathbb{N}}.$$

[3] Actually, the σ-orthocompleteness in the Proof of Lemma 2.2.8 was applied only to a set with two elements so we could have, had we chosen to, attained the validity of the noncontradiction principle from an assumption of only finite orthoadditivity. But, feeling the need for countably additive states, we have chosen not to develop the corresponding theory.

□

As a consequence of Lemma 6.2.6, one immediately obtains the following theorem.

THEOREM 6.2.7. *The effect algebra* $\langle \mathcal{E}, \boxplus, \mathbf{0}, \mathbf{1} \rangle$ *of any effect-state system* $(\mathcal{E}, \mathcal{S})$ *is a* σ*-effect algebra.*

In this framework, the notion of *observable* (or *effect-valued measure*) of an effect-state system can be naturally defined. Let us first introduce the notion of σ-homomorphism.

DEFINITION 6.2.8. σ*-homomorphism*
Let $\mathcal{A} = \langle A, \boxplus, \mathbf{0}, \mathbf{1} \rangle$ and $\mathcal{B} = \langle B, \boxplus, \mathbf{0}, \mathbf{1} \rangle$ be two σ-effect algebras and let $h : A \mapsto B$. Then, h is a σ*-homomorphism* from $\mathcal{A}$ to $\mathcal{B}$ in case:

(i) $h(\mathbf{1}) = \mathbf{1}$;
(ii) h preserves existing countable sums, i.e., if $\boxed{+}\{a_n\}_{n\in\mathbb{N}}$ exists in $\mathcal{A}$, then $\boxed{+}\{h(a_n)\}_{n\in\mathbb{N}}$ exists in $\mathcal{B}$ and
$$h(\boxed{+}\{a_n\}_{n\in\mathbb{N}}) = \boxed{+}\{h(a_n)\}_{n\in\mathbb{N}}.$$

One can easily show that the $[0,1]$-effect algebra $\langle [0,1], \boxplus, \mathbf{0}, \mathbf{1} \rangle$ is a σ-effect algebra. As a consequence, one obtains that any state s is a σ-homomorphism of the σ-effect algebra $\langle \mathcal{E}, \boxplus, \mathbf{0}, \mathbf{1} \rangle$ into the σ-effect algebra $\langle [0,1], \boxplus, 0, 1 \rangle$.

DEFINITION 6.2.9. An *observable* of an effect-state system $(\mathcal{E}, \mathcal{S})$ is a σ-homomorphism M of $\mathcal{B}(\mathbb{R})$ into the σ-effect algebra $\langle \mathcal{E}, \boxplus, \mathbf{0}, \mathbf{1} \rangle$. This is equivalent to saying that the following conditions hold:

(1) $M(\emptyset) = \mathbf{0}$, $M(\mathbb{R}) = \mathbf{1}$;
(2) If Δ and Γ are two disjoint Borel sets, then $M(\Delta) \perp M(\Gamma)$ and $M(\Delta \cup \Gamma) = M(\Delta) \boxplus M(\Gamma)$;
(3) Let $\{\Delta_n\}_{n\in\mathbb{N}}$ be a set of pairwise disjoint Borel sets. Then:
$$M\left(\bigcup\{\Delta_n\}_{n\in\mathbb{N}}\right) = \boxed{+}\{M(\Delta_n)\}_{n\in\mathbb{N}}.$$

Interestingly enough, Axiom U7 permits one to characterize Hilbert effect-state systems in the class of all pre-Hilbert systems.

As we already know, a pre-Hilbert space is a vector space over a division ring (whose elements are the reals or the complexes or the quaternions) equipped with an inner product.[4] Unlike Hilbert spaces, pre-Hilbert spaces are not necessarily metrically complete.

Take now a (complex separable) pre-Hilbert space $\mathcal{K}$. The notions of *effect*, *projection* and *density operator* will be defined as in the Hilbert space case. Let us indicate by $\mathcal{E}(\mathcal{K})$, $\Pi(\mathcal{K})$, $D(\mathcal{K})$ the set of all effects, of all projections and of all density operators of $\mathcal{K}$, respectively.

---

[4]See Definition 1.2.7.

As in $\mathcal{E}(\mathcal{H})$, also in this setting, every density operator ρ will determine a probability measure m_ρ defined on $\mathcal{E}(\mathcal{K})$. It is sufficient to put for any $E \in \mathcal{E}(\mathcal{K})$:

$$m_\rho(E) = \mathrm{Tr}(\rho E).$$

Let $\mathcal{S}(\mathcal{K})$ be the set of all probability measures that are determined by a density operator ρ.

THEOREM 6.2.10. *The pair*

$$(\mathcal{E}(\mathcal{K}), \mathcal{S}(\mathcal{K}))$$

satisfies Axioms U1-U6.

Such concrete systems will be called *pre-Hilbert effect-state systems.*

THEOREM 6.2.11. *A pre-Hilbert effect-state system $(\mathcal{E}(\mathcal{K}), \mathcal{S}(\mathcal{K}))$ satisfies Axiom U7 iff $\mathcal{K}$ is a Hilbert space (i.e., $\mathcal{K}$ is metrically complete).*

PROOF. It follows from Theorem 6.2.4 and from Proposition 3.3 of (Cattaneo and Gudder, 1999). □

The following two theorems describe some important properties of sharp effects in the framework of concrete effect-state systems.

THEOREM 6.2.12. (Dvurečenskij, 1997)
Let $\mathcal{H}$ be a Hilbert space. The following properties hold:

(i) $\forall E \in \mathcal{E}(\mathcal{H})$: $E \in \mathcal{E}_s(\mathcal{H})$ *iff* $E \in \Pi(\mathcal{H})$.
(ii) *Let $\{P_i\}_{i \in I}$ be a set of projections. Then,*
  a) *the meet $\bigwedge_{\Pi(\mathcal{H})} \{P_i\}_{i\in I}$ of $\{P_i\}_{i\in I}$ in $\Pi(\mathcal{H})$ coincides with the meet $\bigwedge_{\mathcal{E}(\mathcal{H})} \{P_i\}_{i\in I}$ of $\{P_i\}_{i\in I}$ in $\mathcal{E}(\mathcal{H})$;*
  b) *the join $\bigvee_{\Pi(\mathcal{H})} \{P_i\}_{i\in I}$ of $\{P_i\}_{i\in I}$ in $\Pi(\mathcal{H})$ coincides with the join $\bigvee_{\mathcal{E}(\mathcal{H})} \{P_i\}_{i\in I}$ of $\{P_i\}_{i\in I}$ in $\mathcal{E}(\mathcal{H})$.*

COROLLARY 6.2.13. *Let $\mathcal{G}(\mathcal{H})$ be a subset of $\mathcal{E}(\mathcal{H})$. Assume that (i) $\Pi(\mathcal{H}) \subseteq \mathcal{G}(\mathcal{H}) \subseteq \mathcal{E}(\mathcal{H})$ and (ii) $(\mathcal{G}(\mathcal{H}), \leq_{\mathcal{G}(\mathcal{H})}, ', \mathbf{0}, \mathbf{1})$ is an orthoposet, where $\leq_{\mathcal{G}(\mathcal{H})}$ is the ordering induced on $\mathcal{G}(\mathcal{H})$ from $\mathcal{E}(\mathcal{H})$. If the existing meets (joins) in $\mathcal{G}(\mathcal{H})$ coincide with their corresponding meets (joins) in $\mathcal{E}(\mathcal{H})$, then $\mathcal{G}(\mathcal{H}) = \Pi(\mathcal{H})$.*

Thus, there is no orthoposet properly including $\Pi(\mathcal{H})$ such that its existing meets (joins) coincide with those in $\mathcal{E}(\mathcal{H})$.

6.3. The sharply dominating principle

Axioms U1-U7 determine a weak axiomatization of unsharp QT. Such a theory can be regarded as an *unsharp version* of Mackey's axioms I-VI. No *normality* principle has, so far, been required. At the same time, any concrete effect structure, based on $\mathcal{E}(\mathcal{H})$, turns out to satisfy some characteristic conditions that are not captured by Axioms U1-U7. Two important properties are the following:

(I) The structure $\langle \Pi(\mathcal{H}), \boxplus_{\Pi(\mathcal{H})}, \mathbf{0}, \mathbf{1} \rangle$, consisting of all sharp elements of $\mathcal{E}(\mathcal{H})$ (where $\boxplus_{\Pi(\mathcal{H})}$ is the restriction of $\boxplus$ to $\Pi(\mathcal{H})$) is an orthoalgebra that is a subalgebra of the larger $\langle \mathcal{E}(\mathcal{H}), \boxplus, \mathbf{0}, \mathbf{1} \rangle$.

(II) Any effect E admits of a *best sharp lower approximation*,[5] which turns out to coincide with its necessity $\nu(E)$. The following conditions hold:
- (i) $\nu(E)$ is sharp (is a projection);
- (ii) $\nu(E) \leq E$;
- (iii) if F is sharp and $F \leq E$, then $F \leq \nu(E)$.

Axiom U8 permits us to also fulfill both properties (I) and (II) in the abstract framework.

We have already defined the abstract notion of *sharp lower approximation*. We have also introduced the map

$$\nu : \mathcal{E} \mapsto \mathcal{E}_s,$$

that associates to any effect E its sharp lower approximation $\nu(E)$. Dually, we can define another map

$$\mu : \mathcal{E} \mapsto \mathcal{E},$$

such that $\forall E \in \mathcal{E}$:

$$\mu(E) = (\nu(E'))'.$$

The element $\mu(E)$ will be called the *sharp upper approximation* of E.

Apparently, the notion of *sharp lower approximation* represents a natural abstraction of the *necessitation* $\nu(E)$ of a concrete effect E.

THEOREM 6.3.1. *Let $(\mathcal{E}, \mathcal{S})$ be an effect-state system.*

1. *The following conditions hold:*
   - (i) $\nu(\mathbf{1}) = \mathbf{1}$;
   - (ii) $\nu(E) \leq E$;
   - (iii) $E \leq F$ *implies* $\nu(E) \leq \nu(F)$;
   - (iv) $E \leq \nu(\mu(E))$;
   - (v) $\nu(\nu(E)) = \nu(E)$;
   - (vi) $\mu(E)' = \nu(E')$ *and* $\mu(E') = \nu(E)'$.
2. *E is sharp iff $E = \nu(E)$.*

PROOF.
(1.i) The effect $\mathbf{1}$ is sharp. Since $\mathbf{1} \leq \mathbf{1}$, we have: $\mathbf{1} \leq \nu(\mathbf{1})$.
(1.ii) It follows from the definition of the map ν.
(1.iii) Suppose $E \leq F$. By (1.ii), $\nu(E) \leq E$. Thus, $\nu(E) \leq F$. Now, $\nu(E)$ is sharp. Thus, by definition of ν, we obtain $\nu(E) \leq \nu(F)$.
(1.iv) First, note that $E \leq \mu(E)$. Since $\mu(E)$ is sharp, we can conclude that $E \leq \nu(\mu(E)))$.
(1.v) By (1.ii), $\nu(\nu(E)) \leq \nu(E)$. Since $\nu(E)$ is sharp, and $\nu(E) \leq \nu(E)$, we have $\nu(E) \leq \nu(\nu(E))$. Thus, $\nu(\nu(E)) = \nu(E)$.
(1.vi) consists of restatements of the definition.

[5] See Section 4.2.

2. By definition, if $E = \nu(E)$, then $E \in \mathcal{E}_s$. Suppose $E \in \mathcal{E}_s$. Since $E \leq E$ and E is sharp, we have: $E \leq \nu(E)$. Thus, by (1.ii), $E = \nu(E)$. □

As a consequence, the map ν turns out to behave as a $\mathbf{S}_5$-like *necessity* operator. Dually, one can show that μ behaves like a *possibility* operator.

As expected, the necessity map ν permits one to define a unary operation $^\sim$ that turns out to behave as a Brouwer complement. Given $(\mathcal{E}, \mathcal{S})$, the Brouwer complement $E^\sim$ of an effect E is defined as follows:

$$E^\sim := \nu(E').$$

In other words, the intuitionistic negation of an effect is the necessity of its fuzzy negation.

THEOREM 6.3.2. *Let $(\mathcal{E}, \mathcal{S})$ be an effect-state system.*

(i) *$(\mathcal{E}, \boxplus, ', ^\sim, \mathbf{0}, \mathbf{1})$ is a BZ-effect algebra (See Definition 5.1.16).*
(ii) *$(\mathcal{E}_s, \boxplus, \mathbf{0}, \mathbf{1})$ is an orthoalgebra that is a subeffect algebra of $(\mathcal{E}, \boxplus, \mathbf{0}, \mathbf{1})$. Furthermore, $\forall E \in \mathcal{E}_s : E^\sim = E'$ and $E = \nu(E) = \mu(E)$.*

PROOF.
(i) See (Cattaneo, 1997).
(ii) We have to prove that

(a) $(\mathcal{E}_s, \boxplus, \mathbf{0}, \mathbf{1})$ is a subeffect algebra of $(\mathcal{E}, \boxplus, \mathbf{0}, \mathbf{1})$;
(b) $\forall E \in \mathcal{E}_s$: $E \perp E$ implies $E = \mathbf{0}$.

(a) We have to prove that $\mathcal{E}_s$ is closed under $\boxplus$. Let $E, F \in \mathcal{E}_s$ and $E \perp F$. We have to show that $E \boxplus F \in \mathcal{E}_s$. Now, $E, F \leq E \boxplus F$. Hence, by Theorem 6.3.1(1iii), $\nu(E), \nu(F) \leq \nu(E \boxplus F) \leq E \boxplus F$. The effects E, F are sharp so that, by Theorem 6.3.1(2.), $E = \nu(E)$ and $F = \nu(F)$. Thus, $E, F \leq \nu(E \boxplus F) \leq E \boxplus F$. By definition of $\leq$, there are three effects G, H, I such that

$$\nu(E \boxplus F) = E \boxplus G;$$
$$\nu(E \boxplus F) = F \boxplus H;$$
$$E \boxplus F = \nu(E \boxplus F) \boxplus I.$$

Thus,

$$E \boxplus F = \nu(E \boxplus F) \boxplus I = (E \boxplus G) \boxplus I = E \boxplus (G \boxplus I)$$

and

$$F \boxplus E = E \boxplus F = \nu(E \boxplus F) \boxplus I = (F \boxplus H) \boxplus I = F \boxplus (H \boxplus I).$$

Hence, by the cancellation law, $F = G \boxplus I$ and $E = H \boxplus I$ so that $I \leq E$ and $I \leq F$. But, by hypothesis, $E \perp F$ so that $I \leq F \leq E'$ as well. Since E is sharp, $I = \mathbf{0}$. Therefore

$$E \boxplus F = \nu(E \boxplus F) \boxplus I = \nu(E \boxplus F).$$

By Theorem 6.3.1(2.), $E \boxplus F$ is sharp.
(b) Suppose $E \in \mathcal{E}_s$ and $E \perp E$. Then, $E \leq E'$ so that $E = E \wedge E' = \mathbf{0}$, because E is sharp. □

LEMMA 6.3.3. *Let $(\mathcal{E}, \mathcal{S})$ be an effect-state system. If $\{E_j\}_{j\in I}$ is a set of effects such that the infimum $\bigwedge\{E_j\}_{j\in I}$ exists in $\mathcal{E}$, then*

$$\nu\left(\bigwedge\{E_j\}_{j\in I}\right) = \bigwedge\{\nu(E_j)\}_{j\in I}.$$

PROOF. By hypothesis, the infimum $\bigwedge\{E_j\}_{j\in I}$ exists in $\mathcal{E}$. We prove that $\nu\left(\bigwedge\{E_j\}_{j\in I}\right)$ is the infimum of the set $\{\nu(E_j)\}_{j\in I}$. We have, $\forall i \in I$, $\bigwedge\{E_j\}_{j\in I} \leq E_i$. Thus, $\nu\left(\bigwedge\{E_j\}_{j\in I}\right) \leq \nu(E_i)$. Suppose that, $\forall j \in I$, $F \leq \nu(E_j)$. Then, $\mu(F) \leq \mu(\nu(E_j))$. Hence,

$$\nu(\mu(E'_j)) = [\mu(\nu(E_j))]' \leq [\mu(F)]'.$$

By Theorem 6.3.1(1.iv), $E'_j \leq \nu(\mu(E'_j))$. Therefore, $E'_j \leq [\mu(F)]'$ so that $\mu(F) \leq E_j$. Thus, $\mu(F) \leq \bigwedge\{E_j\}_{j\in I}$. Accordingly, $\forall j \in I$: $\nu(\mu(F)) \leq \nu\left(\bigwedge\{E_j\}_{j\in I}\right)$. Since $F \leq \nu(\mu(F))$, we obtain that $F \leq \nu\left(\bigwedge\{E_j\}_{j\in I}\right)$. □

Thus, we can conclude that the two properties (I) and (II) (which are satisfied by concrete effects) hold also for abstract effects.

6.4. Abstract unsharp preclusivity spaces

In Chapter 4 we studied the preclusivity spaces $(\mathcal{S}(\mathcal{H}), \perp_{\mathcal{E}(\mathcal{H})})$ that arise in the framework of Hilbert effect-state systems $(\mathcal{E}(\mathcal{H}), \mathcal{S}(\mathcal{H}))$; we proved the following.

1) $\forall E \in \mathcal{E}(\mathcal{H}) : \mathbf{Yes}(E) = \mathbf{Yes}(\nu(E)) = \mathbf{Yes}(E'^{\sim})$.
2) $\forall \rho, \sigma \in D(\mathcal{H}) \left\{m_\rho \perp_{\mathcal{E}(\mathcal{H})} m_\sigma \text{ iff } m_\rho \perp_{\Pi(\mathcal{H})} m_\sigma\right\}$, where: $m_\rho \perp_{\Pi(\mathcal{H})} m_\sigma$ iff $m_\rho \perp m_\sigma$ iff $\exists P \in \Pi(\mathcal{H}) \{m_\rho(P) = 1 \text{ and } m_\sigma(P) = 0\}$.
3) $\forall E \in \mathcal{E}(\mathcal{H}) : \mathbf{Yes}(E) = \mathbf{Yes}(E)^{\perp\perp}$.

We now turn again to the abstract effect-state systems $(\mathcal{E}, \mathcal{S})$. Consider the space $(\mathcal{S}, \perp_{\mathcal{E}})$, where the relation $\perp_{\mathcal{E}}$ (called *unsharp preclusivity*) is defined in the expected way:

$$\forall s, t \in \mathcal{S} \{s \perp_{\mathcal{E}} t \text{ iff } \exists E \in \mathcal{E}[s(E) = 1 \text{ and } t(E) = 0]\}.$$

In other words, two states are *unsharply preclusive* iff they are strongly distinguished by at least one effect, which is certain for the first state and impossible for the second state.

Clearly, $(\mathcal{S}, \perp_{\mathcal{E}})$ is a preclusivity space. Hence, we will have:

$$\mathbf{Yes}(E) \subseteq \mathbf{Yes}(E)^{\perp_{\mathcal{E}}\perp_{\mathcal{E}}},$$

where $^{\perp_{\mathcal{E}}}$ is the preclusive complement determined by the preclusivity relation $\perp_{\mathcal{E}}$.

As in the concrete case, we can define also here the following *sharp preclusivity relation*:

$$s \perp_{\mathcal{E}_s} t \text{ iff } \exists E \in \mathcal{E}_s[s(E) = 1 \text{ and } t(E) = 0].$$

In other words, two states are *sharply preclusive* iff they are strongly distinguished by at least one sharp effect, which is certain for the first state and impossible for the second state.

Clearly:

$$\perp_{\mathcal{E}_s} \subseteq \perp_{\mathcal{E}} .$$

A natural question arises: do abstract effect satisfy the same conditions that hold in the concrete case? In other words:

1) $\forall E \in \mathcal{E}[\mathbf{Yes}(E) = \mathbf{Yes}(\nu(E))]$?
2) $\perp_{\mathcal{E}_s} = \perp_{\mathcal{E}}$?
3) $\forall E \in \mathcal{E}[\mathbf{Yes}(E) = \mathbf{Yes}(E)^{\perp_{\mathcal{E}} \perp_{\mathcal{E}}}]$?

Axiom U9 permits us to give a positive answer to these three questions.

A positive answer to the first question is an immediate consequence Axiom U9(iii).

Let us consider questions 2) and 3).

We know that (by Axiom U9(i)) every state has a *carrier*. Apparently, the carrier of a state s is the smallest effect to which s assigns value 1 (a kind of *unsharp characteristic property* of the physical object described by s). As in the sharp case, the carrier of a state s is denoted by $car(s)$. At the same time (by Axiom U9(ii)) every effect-state system is also sharply rich. It is worthwhile noticing that a sharply rich set of states is generally *rich* only for the sharp part of the effect-system: the *event order* (but not the *effect order!*) is completely determined by the behavior of the states with respect to the certain probability-value.

THEOREM 6.4.1. *Let* $(\mathcal{E}, \mathcal{S})$ *be an effect-state system. For any states* s, t*:*

$$s \perp_{\mathcal{E}} t \ \textit{iff} \ s \perp_{\mathcal{E}_s} t.$$

PROOF. Necessity is trivial.
Suppose $s \perp_{\mathcal{E}} t$. Then, there exists an effect E such that $s(E) = 1$ and $t(E) = 0$. Thus, $s \in \mathbf{Yes}(\nu(E))$ since, by Axiom U9, $(\mathcal{E}, \mathcal{S})$ is sharply normal. Now, $\nu(E) \leq E$ and $t(E) = 0$. Hence, $t(\nu(E)) = 0$. Since $\nu(E)$ is sharp, we can conclude $s \perp_{\mathcal{E}_s} t$. □

Theorem 6.4.1 yields a positive answer to the second question. As a consequence, in the following we will simply write $\perp$ instead of either $\perp_{\mathcal{E}}$ or $\perp_{\mathcal{E}_s}$.

LEMMA 6.4.2. *Let* $(\mathcal{E}, \mathcal{S})$ *be an effect-state system. For any states* $s, t \in \mathcal{S}$*:*

$$s \perp t \ \textit{iff} \ car(s) \perp car(t).$$

PROOF. Suppose $s \perp t$. Then, there exists an effect E such that $s(E) = 1$ and $t(E) = 0$. Now, $car(s) \leq E$, since $s(E) = 1$. Similarly, $car(t) \leq E'$. Hence, $car(s) \perp car(t)$.
Conversely, suppose $car(s) \leq car(t)'$. Now, $s(car(s)) = 1$. Then, $s(car(t)') = 1$. Thus, $s(car(t)) = 0$. Since $t(car(t)) = 1$, we have $s \perp t$. □

LEMMA 6.4.3. *Let $(\mathcal{E}, \mathcal{S})$ be an effect-state system. For any $s \in \mathcal{S}$, the carrier of s is a sharp element.*

PROOF. Let $s \in \mathcal{S}$ and let $\mathcal{E}_1(s) := \{E \in \mathcal{E} : s(E) = 1\}$. Clearly, $car(s) = \bigwedge \mathcal{E}_1(s)$. Let $\nu(\mathcal{E}_1(s)) := \{\nu(E) \in \mathcal{E} : s(E) = 1\}$.

We want to show that $car(s) = \bigwedge \nu(\mathcal{E}_1(s))$. Suppose $E \in \mathcal{E}_1(s)$. Then, $s(E) = 1$. Thus, $s(\nu(E)) = 1$, because $(\mathcal{E}, \mathcal{S})$ is sharply normal, specifically, by Axiom U9 condition (iii). Hence, $car(s) \leq \nu(E)$, for all $E \in \mathcal{E}_1(s)$. Whence, $car(s) \leq \bigwedge \nu(\mathcal{E}_1(s))$. We have, $\bigwedge \nu(\mathcal{E}_1(s)) \leq \nu(E) \leq E$, for all $E \in \mathcal{E}_1(s)$. Consequently, $\bigwedge \nu(\mathcal{E}_1(s)) \leq E$, for all $E \in \mathcal{E}_1(s)$, and $\bigwedge \nu(\mathcal{E}_1(s)) \leq \bigwedge \mathcal{E}_1(s)$. Hence, $car(s) = \bigwedge \mathcal{E}_1(s) = \bigwedge \nu(\mathcal{E}_1(s))$. Consequently, $\nu(car(s)) = \nu(\bigwedge \nu(\mathcal{E}_1(s)))$. By Lemma 6.3.3, $\nu(\bigwedge \nu(\mathcal{E}_1(s))) = \bigwedge \nu(\nu(\mathcal{E}_1(s))) = \bigwedge \nu(\mathcal{E}_1(s))$. Thus, $\nu(car(s)) = \bigwedge \nu(\mathcal{E}_1(s)) = car(s)$. Accordingly, by Theorem 6.3.1(2.), the effect $car(s)$ is sharp. □

THEOREM 6.4.4. *Let $(\mathcal{E}, \mathcal{S})$ be an effect-state system. The structure*

$$(\mathcal{E}_s, \leq, ', \mathbf{0}, \mathbf{1})$$

is a σ-orthocomplete orthomodular poset.

PROOF. (i) Let us first prove that $(\mathcal{E}_s, \leq, ', \mathbf{0}, \mathbf{1})$ is an orthomodular poset. By Theorem 6.3.2 (ii), $(\mathcal{E}_s, \leq, ', \mathbf{0}, \mathbf{1})$ is an orthoalgebra that is a subeffect algebra of $(\mathcal{E}, \boxplus, \mathbf{0}, \mathbf{1})$. In order to prove that $(\mathcal{E}_s, \leq, ', \mathbf{0}, \mathbf{1})$ is an orthomodular poset, by Theorem 5.1.10 and Corollary 5.1.8, it is sufficient to show that $\forall E, F \in \mathcal{E}_s$ such that $E \perp F$, $E \boxplus F = E \vee F$. Suppose $E \perp F$. Clearly, $E, F \leq E \boxplus F$. Suppose $E, F \leq G$, for a given $G \in \mathcal{E}_s$. We want to show that $E \boxplus F \leq G$. Since $E \leq G$, there exists an element H in $\mathcal{E}$ such that $G = E \boxplus H$. We want to show that $F \leq H$. Then, we can conclude $E \boxplus F \leq E \boxplus H = G$. To show $F \leq H$, let us first prove that $\mathbf{Yes}(F) \subseteq \mathbf{Yes}(H)$. Suppose $s(F) = 1$. Then, $s(G) = 1$ and $s(E) = 0$, since $F \leq G$ and $F \leq E'$. Consequently, $1 = s(G) = s(E) + s(H) = s(H)$. Hence, $\mathbf{Yes}(F) \subseteq \mathbf{Yes}(\nu(H))$, because $(\mathcal{E}, \mathcal{S})$ is sharply normal. Since both F and $\nu(H)$ are sharp and $\mathcal{S}$ is sharply rich for $\mathcal{E}$, we can conclude $F \leq \nu(H)$ and therefore $F \leq H$.

(ii) Now we prove that the orthomodular poset $\mathcal{E}_s$ is σ-orthocomplete. Let $E_1, \cdots, E_n$ (where $n \geq 3$) be a finite sequence of pairwise orthogonal sharp effects. One can easily show, by induction, that every E_i is orthogonal to the supremum (= orthogonal sum) of the set $\{E_1, \cdots, E_n\} - \{E_i\}$. Thus, the partial sum $E_1 \boxplus \ldots \boxplus E_n$ can be defined inductively and $E_1 \boxplus \ldots \boxplus E_n = E_1 \vee \ldots \vee E_n$. Let $E_1, E_2, \ldots$ be a sequence of pairwise orthogonal sharp effects. By Theorem 6.2.7, $\mathcal{E}$ is a σ-effect algebra. Thus, there exists an effect E such that

$$E = \bigvee \{E_1 \boxplus \ldots \boxplus E_n\}_{n \in \mathbb{N}} = \bigvee \{E_n\}_{n \in \mathbb{N}}.$$

Now, every E_n $(n \in \mathbb{N})$ is a sharp element. Thus, By Lemma 6.3.3, $\nu(\bigwedge \{E'_n\}_{n \in \mathbb{N}}) = \bigwedge \{\nu(E'_n)\}_{n \in \mathbb{N}} = \bigwedge \{E'_n\}_{n \in \mathbb{N}}$. Therefore $\bigwedge \{E'_n\}_{n \in \mathbb{N}}$ is sharp. Accordingly, also $\bigvee \{E_n\}_{n \in \mathbb{N}}$ is a sharp element. □

Now consider any effect-state system $(\mathcal{E}, \mathcal{S})$. Let $\mathcal{S} \restriction \mathcal{E}_s$ be the set of all states in $\mathcal{S}$ restricted to $\mathcal{E}_s$. Then, by Lemma 6.4.3 and Theorem 6.4.4, the pair $(\mathcal{E}_s, \mathcal{S} \restriction \mathcal{E}_s)$ will be a *normal event-state system.* Thus, we can use Theorem 2.4.9 and Theorem 2.4.10 to obtain, as corollaries, the following results.

THEOREM 6.4.5. *Let $(\mathcal{E}, \mathcal{S})$ be an effect-state system. For any sharp effect E:*

$$E = \bigvee \{car(s) \; : s \in \mathbf{Yes}(E)\} .$$

THEOREM 6.4.6. *Let $(\mathcal{E}, \mathcal{S})$ be an effect-state system. For any sharp effect E:*

$$\mathbf{Yes}(E') = \mathbf{Yes}(E)^{\perp}.$$

On this basis, we can finally prove that for any effect E, $\mathbf{Yes}(E)$ is a closed set of the preclusivity space $(\mathcal{S}, \perp)$.

THEOREM 6.4.7. *Let $(\mathcal{E}, \mathcal{S})$ be an effect-state system. For any effect E:*

$$\mathbf{Yes}(E) = \mathbf{Yes}(E)^{\perp\perp}.$$

PROOF. Let E be an effect. The effect $\nu(E)$ is sharp. Thus, by Theorem 6.4.6, $\mathbf{Yes}(\nu(E)') = \mathbf{Yes}(\nu(E))^{\perp}$. Therefore, $\mathbf{Yes}(\nu(E))^{\perp\perp} = \mathbf{Yes}(\nu(E)')^{\perp}$. Now, $\nu(E)'$ is sharp, because $\nu(E)$ is sharp. Thus, again by Theorem 6.4.6, $\mathbf{Yes}(\nu(E)')^{\perp} = \mathbf{Yes}(\nu(E)'') = \mathbf{Yes}(\nu(E))$. Since $(\mathcal{E}, \mathcal{S})$ is sharply normal, $\mathbf{Yes}(\nu(E)) = \mathbf{Yes}(E)$. Hence, $\mathbf{Yes}(E) = \mathbf{Yes}(E)^{\perp\perp}$. □

By Theorem 6.4.7 we also have a positive answer to the third question.

Finally, let us return to concrete effect-state systems. As anticipated earlier, one can prove that any concrete effect-state system $(\mathcal{E}(\mathcal{H}), \mathcal{S}(\mathcal{H}))$ is an example of an abstract effect state system $(\mathcal{E}, \mathcal{S})$.

THEOREM 6.4.8. *Any Hilbert effect state system $(\mathcal{E}(\mathcal{H}), \mathcal{S}(\mathcal{H}))$ satisfies Axioms U1-U9.*

PROOF. Axioms U1-U7 hold by Theorem 6.2.10 and Theorem 6.2.11. By Theorem 6.2.12, the set of all sharp effects of $\mathcal{E}(\mathcal{H})$ coincides with the set $\Pi(\mathcal{H})$ of all projections of $\mathcal{H}$.
Since $(\mathcal{E}(\mathcal{H}), \mathcal{S}(\mathcal{H}))$ is sharply dominating by Lemma 4.2.9, Axiom U8 holds. We now prove that $(\mathcal{E}(\mathcal{H}), \mathcal{S}(\mathcal{H}))$ is sharply normal. We have to show:

(a) $\forall E, F \in \Pi(\mathcal{H})$: if $\mathbf{Yes}(E) \subseteq \mathbf{Yes}(F)$, then $E \leq F$;
(b) $\forall E \in \mathcal{E}(\mathcal{H})$: $\mathbf{Yes}(E) = \mathbf{Yes}(E'^{\sim})$;
(c) for any state $s \in \mathcal{S}(\mathcal{H})$ the carrier exists.

(a) The proof follows from the fact that the set $\mathcal{S}(\mathcal{H}) \restriction \Pi(\mathcal{H})$ is rich for $\Pi(\mathcal{H})$ (see Theorem 3.3.5).
(b) Since $E'^{\sim} \leq E$ and $\mathbf{Yes}$ is order preserving, it is sufficient to show that $\mathbf{Yes}(E) \subseteq \mathbf{Yes}(E'^{\sim})$, which can be easily proved.

(c) Let $s \in \mathcal{S}(\mathcal{H})$. Then, $s = m_\rho$, where $\rho = \sum_i \lambda_i P_{[\psi_i]}$ is a density operator. We want to show that the carrier of s is $\bigvee_i P_{[\psi_i]}$. By Theorem 3.3.5, $s\left(\bigvee_i P_{[\psi_i]}\right) = 1$ and $\forall E \in \Pi(\mathcal{H})$ such that $s(E) = 1$, $\bigvee_i P_{[\psi_i]} \leq E$. Let E be an effect such that $s(E) = 1$. By (b), $s(E'^{\sim}) = 1$. Since $E'^{\sim}$ is a projection, we can conclude that $\bigvee_i P_{[\psi_i]} \leq E'^{\sim} \leq E$. Hence, $\bigvee_i P_{[\psi_i]} \leq E$. Therefore, $\bigvee_i P_{[\psi_i]}$ is the carrier of s. Consequently, Axiom U9 holds. □

In Chapter 3 we have seen that, after the proof of the Solèr Theorem, the *representation problem* for sharp QT has been solved in a satisfactory way. As far as we know, no similar results have been obtained for unsharp QT: the *unsharp representation problem* seems to be still completely open.

6.5. Sharp and unsharp abstract quantum theory

The abstract event-state systems are not generally models of the axiomatic unsharp QT. One can easily realize that any Gudder event-state system (see Definition 2.3.3) does satisfy Axioms U1-U8. At the same time, Axiom U9 can be violated. Sharp abstract models for the whole axiomatization can be obtained by referring to *normal event-state systems*. However, because much of the quantum logical development has concentrated on the weaker axiom system, it behooves us to understand more deeply some examples of structures that satisfy only Axioms U1-U8. In order to do this, we revisit the notion of a Greechie diagram. As was discussed in Section 1.1, the Greechie diagram of an orthomodular poset $\mathcal{B}$ is a pair $(X, \mathcal{L})$ where X consists of the atoms of $\mathcal{B}$, and $\mathcal{L}$ is the family of all maximal pairwise orthogonal subsets of X. All the Greechie diagrams that we have considered have had the property that, in $(X, \mathcal{L})$,

(h_1) every point is on some line;
(h_2) each line contains at least three points;
(h_3) the intersection of any two distinct lines contains no more than one point.

We call any pair $(X, \mathcal{L})$ a *quasi-geometric hypergraph* in case X is a nonempty set of elements called "points" and $\mathcal{L}$ is a family of subsets of X, called "lines," satisfying (h_1), (h_2) and (h_3). We use the adjective "quasi-geometric" because any 2 points determine at most one line and any two lines determine at most one point.

There are, of course, finite orthomodular posets in which the intersection of two maximal Boolean subalgebras contains more than four elements; these generate a Greechie diagram in which the intersection of two lines contains more than one point.[6] However, the surprising fact is that there are so many orthomodular posets and lattices which are the simplest known structures having some property and in which the intersection of any two lines has at most one point.[7]

---

[6] See (Greechie, 1968; Kalmbach, 1983).

[7] See (Pták and Pulmannová, 1991).

We take this opportunity to present the result which provides the basis for the creation of these examples, the so-called *Loop Lemma.*[8]

This lemma states that a quasi-geometric hypergraph having no triangles is the Greechie diagram of an orthomodular poset; and one having no triangles and no squares is the Greechie diagram of an orthomodular lattice. (A *triangle* is a loop of order 3 and a *square* is a loop of order 4, both of which are defined below.) Those quasi-geometric hypergraphs having triangles represent orthoalgebras that are not orthomodular posets. (Although there are more complicated Greechie diagrams containing triangles, yet representing orthomodular lattices. The interested reader should see the Theorem of Dichtl in (Dichtl, 1984) and (Kalmbach, 1983).)

We first define the notion of a loop in a quasi-geometric hypergraph. Then we give the Loop Lemma.

DEFINITION 6.5.1. *A loop in a quasi-geometric hypergraph*
Let $(X, \mathcal{L})$ be a quasi-geometric hypergraph and let $n \in \mathbb{N}$ with $n \geq 3$. We call a sequence $L_1, L_2, \ldots, L_n$ of lines in $\mathcal{L}$ *a loop of order* n in case

(i) the intersection of any three distinct lines is empty;
(ii) for $0 \leq j < i \leq n-1$, we have

$$L_i \cap L_j \neq \emptyset \text{ iff } i - j \in \{1, n-1\}.$$

LEMMA 6.5.2. The Loop Lemma
Let $(X, \mathcal{L})$ be a quasi-geometric hypergraph. Then, regarded as a Greechie diagram,

(i) *$(X, \mathcal{L})$ represents an orthomodular poset iff there are no loops of order 3;*
(ii) *$(X, \mathcal{L})$ represents an orthomodular lattice iff there are no loops of order 3 or 4.*

This result is essentially proved in (Greechie, 1969). It not difficult to see that any quasi-geometric hypergraph is the Greechie diagram of an orthoalgebra. But we must say how the $\boxplus$ is defined. The lines represent the atoms of Boolean subalgebras and, for x and y in one of these subalgebras (and only then), $x \boxplus y$ is defined only when $x \wedge y = \mathbf{0}$ in that subalgebra, in which case $x \boxplus y = x \vee y$. One must be careful, when $a \in L_1 \cap L_2$, to identify the complement of a in the Boolean algebra generated by L_1 with the complement of a in the Boolean algebra generated by L_2, so that the

[8]See (Greechie, 1969) for the original theorems which are slightly more general than what is presented here, involving possibly nonatomic structures. What we present is the essence of the original theorems, that part which has been repeatedly applied to yield insight into the theory, modified to provide an amenable presentation.

generalized complement is uniquely defined.[9] The interested reader should see (Foulis and Greechie, 2000)

A word of caution: in a finite orthomodular poset or lattice, the lines of the corresponding Greechie diagram are precisely the maximal pairwise orthogonal sets of atoms; but not all such sets of atoms in an orthoalgebra yield lines. Even the Wright triangle, $\mathcal{WT}$ (see Figure 5.1.1), exhibits three elements (the vertices of the triangle) which, although pairwise orthogonal, are not on a line. This is the essence of the lack of orthocoherence.

After our introduction of interval effect algebras in Chapter 5, we defined the notion of a state and of an order determining set of states in an effect algebra. Similarly, we may define the notion of a strongly order determining set of states in an effect algebra. Thus, an effect algebra $\mathcal{A} = \langle A, \boxplus, \mathbf{0}, \mathbf{1}\rangle$ admits a *strongly order determining set of states* in case there exists a set $\mathcal{S}$ of states on $\mathcal{A}$ such that, for any pair $a, b \in A$,

$$a \leq b \text{ iff } \forall s \in \mathcal{S}[s(a) = 1 \text{ implies } s(b) = 1].$$

These notions carry over to orthoalgebras, orthomodular posets and orthomodular lattices.

We now present two examples, the first, called $\mathcal{G}_{52}$ and given in Figure 6.5.1, is an orthomodular poset that admits an order determining set of states but not a strongly order determining set of states. This example is simpler than the one given in Figure 2.4.1 (but it does not have the precise property possessed by $\mathcal{GGM}_{410}$). The second example, called $\mathcal{G}_{58}$ and given in Figure 6.5.2, is of an orthomodular lattice that has the same feature, the state space is order determining but not strongly order determining. The reader is invited to observe that there are no triangles in Figure 6.5.1 and neither triangles nor squares in Figure 6.5.2.

The fact that $\mathcal{G}_{52}$ admits an order determining set of states was proved in (Greechie, 1969) by exhibiting for each pair of nonorthogonal atoms, c and d, a state such that $s(c) + s(d) > 1$. (This was done by exhibiting twenty-seven specific states, fourteen of them dispersion free states.)

To see that $\mathcal{G}_{52}$ does not admit a strongly order determining set of states, note that the Figure 6.5.1 consists of three 3×3-windows amalgamated at the point a and intertwined by two "lines" in the form of semi-circles. Each window has two "diagonals," one containing the point a. Let W_1 denote the North-Western window, W_2 denote the North-Eastern window, and W_3 denote the Southern window. Let $x_{1,i}$ be the center point of the ith window, and let $x_{2,i}$ be the third point on the "diagonal" of the ith window containing a and $x_{1,i}$.

[9] Another way to think of this is to declare the sum of all the elements in each line to exist and equal 1; then, imposing the commutative and associate law infer all possible partial sums; finally impose the cancellation law so that $\boxplus_{b\in L_1-\{a\}} b = \boxplus_{b\in L_2-\{a\}} b$ and the generalized complement of a is clearly uniquely defined.

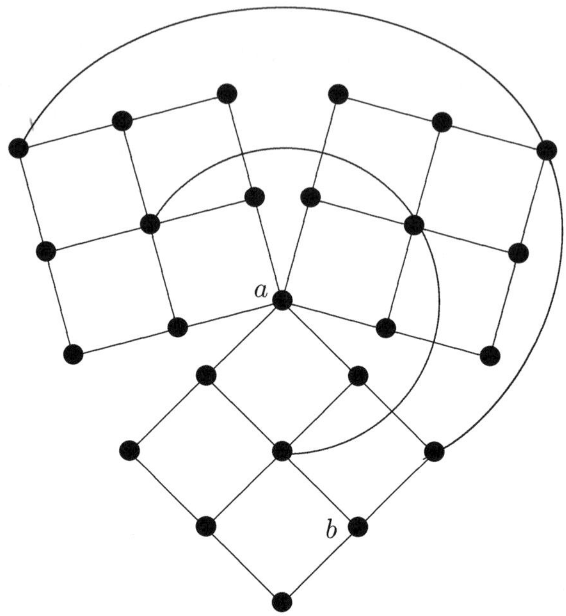

FIGURE 6.5.1. The Greechie diagram of $\mathcal{G}_{52}$

Now let s be any state for which $s(a) = 1$. It is easy to see that $s(x_{1,i}) = s(x_{2,i})$ because of the combinatorics of each window W_i. Moreover, $s(x_{2,1}) + s(x_{2,2}) = 1$ because the third point on the larger semicircle gets mapped to 0 by s. It follows that $s(x_{1,1}) = 1 - s(x_{1,2})$ and therefore,$s(x_{1,3}) = 0$. This forces $s(b) = 1$. (Note that the argument depends on the imperfection in the "near perfect" symmetry of Figure 6.5.1.)

Thus, in $\mathcal{G}_{52}$, $\mathbf{Yes}(b) \subseteq \mathbf{Yes}(a)$ but $b \not\leq a$. Therefore, $\mathcal{G}_{52}$ is not strongly order determining.

The proof that $\mathcal{G}_{58}$ is order determining was given in (Greechie, 1978) by exhibiting 25 specific states.[10]

The proof that $\mathcal{G}_{58}$ is not strongly order determining is again a simple combinatorial argument. But it has a different flavor, reminiscent of the argument used in the discussion of $\mathcal{GGM}_{410}$ (and also that of the orthomodular lattices admitting no states whose creation started the combinatorial approach to understanding quantum structures).[11] We may partition the points X of the Greechie diagram into disjoint sets in two ways, then calculate $\sum_{x \in X} s(x)$ obtaining a restraint on the $s(x)$'s for any state s.

[10] Readers interested in the psychology of mathematical invention may find this paper of interest.

[11] See (Greechie, 1969).

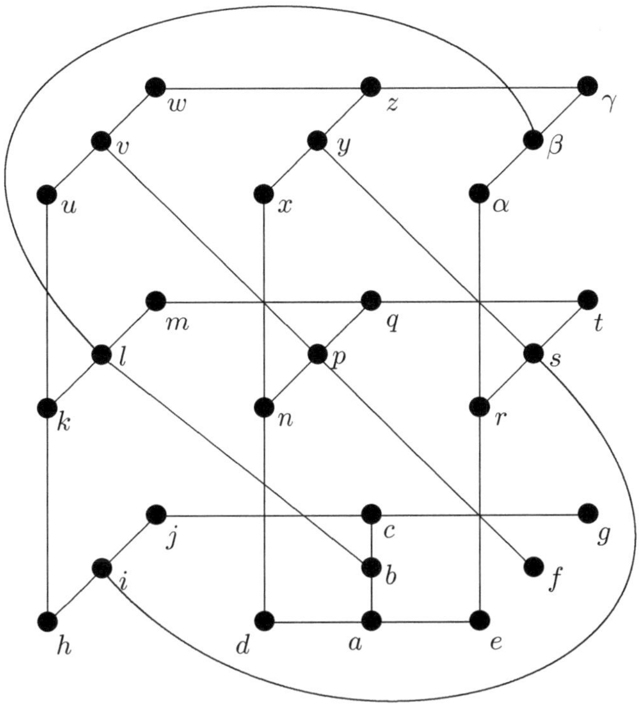

FIGURE 6.5.2. The Greechie diagram of $\mathcal{G}_{58}$

Thus, there is an evident partition of $X - \{b, c, f, g\}$ into eight disjoint lines and another partition of $X - \{a\}$ into nine disjoint lines. After replacing the sum over each line by 1, the following equation emerges:

$$8 + s(b) + s(c) + s(f) + s(g) = 9 + s(a).$$

Therefore $1 + s(a) = s(b) + s(c) + s(f) + s(g)$, so that, if $s(a) = 1$ then $s(g) = 1$ and $s(j) = 0$. It follows that $\mathcal{G}_{58}$ is not strongly order determining, since $\mathbf{Yes}(a) \subseteq \mathbf{Yes}(j')$ but $a \not\leq j'$.

CHAPTER 7

To what extent is quantum ambiguity ambiguous?

A natural question arises in the framework of the unsharp approach to QT: to what extent is "sharp/unsharp" an *unsharp*, *ambiguous* distinction? So far, we have generally called *sharp* any quantum effect (or property) that satisfies the noncontradiction principle. However, different definitions of "sharp effect" can be proposed. One can distinguish two basic kinds of characterizations (Cattaneo, Dalla Chiara and Giuntini, 1999): (i) purely *algebraic definitions*, that only refer to the algebraic structure of the quantum effects; and (ii) *probabilistic definitions*, that also refer to the relationships between *effects* and *states*.

We will first investigate different algebraic characterizations of *quantum sharpness*.

7.1. Algebraic notions of "sharp"

Let us refer to the class of all QMV algebras. Let $\mathcal{M} = \langle M\,,\oplus\,,{}'\,,\mathbf{0}\,,\mathbf{1}\rangle$ be any QMV algebra and let $a \in M$. We introduce six possible definitions of "sharp element."

DEFINITION 7.1.1. *Sharp*

(i) a is *principal* (or *sharp1*) iff $\forall b, c \in M$:
$b, c \leq a$ and $b \leq c'$ imply $b \oplus c \leq a$ (in other words, whenever two orthogonal elements "imply" a given element, then also their Łukasiewicz disjunction "implies" the same element).

(ii) a is *Aristotle-sharp* (or *Aristotelian* or *sharp2*) iff the infimum $a \wedge a'$ exists and is equal to $\mathbf{0}$ (in other words, a satisfies the noncontradiction principle with respect to the standard conjunction $\wedge$).

(iii) a is *Łukasiewicz-sharp* (or *sharp3*) iff $a \Cap a' = \mathbf{0}$ (in other words, a satisfies the noncontradiction principle with respect to the conjunction $\Cap$).

(iv) a is *strongly Łukasiewicz-sharp* (or *sharp4*) iff $a \wedge a' = \mathbf{0}$ and $a \Cap a' = a' \Cap a$ (in other words, a is Aristotelian and the conjunction $\Cap$ commutes for a and its generalized complement a').

(v) a is *Boole-sharp* (or *Boolean* or *sharp5*) iff $a \oplus a = a$ (in other words a is idempotent with respect to the Łukasiewicz disjunction $\oplus$).

(vi) a is *strongly Boolean* (or *sharp6*) iff $\{b \mid b \leq a\}$ is an ideal of $\mathcal{M}$.[1]

[1]An *ideal* of a QMV algebra $\mathcal{M}$ is a nonempty set I of elements of M such that: (i) $\forall a, b \in M$: if $a, b \in I$, then $a \oplus b \in I$; (ii) $\forall a, b \in M$: if $a \in I$ and $b \leq a$, then $b \in I$.

Notice that the notions of *principal* and *Aristotelian* element can be defined also in the case of an effect algebra. In particular, notice that the following theorem holds in the effect algebra setting.

THEOREM 7.1.2. *Let $a \in M$. If a is principal, then a is Aristotelian. But the converse fails.*

PROOF. Let a be a principal element of $\mathcal{M}$. Suppose $b \leq a, a'$. We want to show that $b = \mathbf{0}$. Now, $a \leq a$ and $b \leq a'$. Thus, $a \oplus b \leq a$, since a is principal. Therefore, $a \oplus b = a$. By the cancellation law (Lemma 5.1.4(ii)), there follows $b = \mathbf{0}$. To see that the converse fails, let $\mathcal{WT}$ be the Wright triangle of Example 5.1.11. Note that every element of $\mathcal{WT}$ is *Aristotle-sharp*. It is easy to see that "the vertices of the triangle" a, c, e satisfy $a, c \leq e'$ but $b' = a \oplus c \not\leq e'$ so that the QMV algebra $\mathcal{WT}^{\blacktriangledown}$ contains an element e' that is not principal; and the converse fails. □

We now introduce a particular subset of the set $\mathcal{E}(\mathcal{H})$ of all effects in a Hilbert space $\mathcal{H}$, the set of all *special* effects.[2]

DEFINITION 7.1.3. *Special effect*
An effect E of $\mathcal{E}(\mathcal{H})$ is called *special* iff either E is trivial ($\mathbb{1}$ or $\mathbb{O}$) or E satisfies the following condition: there exist two density operators ρ_1 and ρ_2 such that $\text{Tr}(\rho_1 E) < \frac{1}{2}$ and $\text{Tr}(\rho_2 E) > \frac{1}{2}$.

One can easily show that nontrivial special effects E (different from the null and from the identity projection) are characterized by the following property:

$$E \not\leq E' \text{ and } E' \not\leq E.$$

Let $\mathcal{E}_{sp}(\mathcal{H})$ represent the set of all special effects of $\mathcal{H}$. One immediately realizes that $\mathcal{E}_{sp}(\mathcal{H})$ is closed under the generalized complement $'$ (because $\text{Tr}(\rho E) < \frac{1}{2}$ iff $\text{Tr}(\rho E') > \frac{1}{2}$).

Consider now the structure

$$\langle \mathcal{E}_{sp}(\mathcal{H}), \leq_{\mathcal{E}_{sp}(\mathcal{H})}, ', \mathbf{0}, \mathbf{1} \rangle,$$

where $\leq_{\mathcal{E}_{sp}(\mathcal{H})}$ is the effect order restricted to the set of all special effects.

THEOREM 7.1.4. *The structure*

$$\langle \mathcal{E}_{sp}(\mathcal{H}), \leq_{\mathcal{E}_{sp}(\mathcal{H})}, ', \mathbf{0}, \mathbf{1} \rangle$$

is an orthoposet.

PROOF. The structure $\langle \mathcal{E}_{sp}(\mathcal{H}), \leq_{\mathcal{E}_{sp}(\mathcal{H})}, ', \mathbf{0}, \mathbf{1} \rangle$ is clearly a bounded involution poset. We prove that it is an orthoposet. Let E, F be two special effects such that $F \leq_{\mathcal{E}_{sp}(\mathcal{H})} E$ and $F \leq_{\mathcal{E}_{sp}(\mathcal{H})} E'$. Then, $F \leq_{\mathcal{E}_{sp}(\mathcal{H})} F'$ (by contraposition, double negation, and transitivity). Thus, for any density operator ρ: $\text{Tr}(\rho F) \leq 1/2$. Hence, $F = \mathbf{0}$, because F is a special effect. □

[2]Special effects (first studied by Busch and Garola) are also called *regular effects* or *unsharp properties*. See (Garola, 1980; Busch, 1985).

According to Theorem 7.1.4, special effects do satisfy the noncontradiction principle in the structure $\left\langle \mathcal{E}_{sp}(\mathcal{H}), \leq_{\mathcal{E}_{sp}(\mathcal{H})}, ', \mathbf{0}, \mathbf{1} \right\rangle$. Thus, we will have:

$$E \wedge_{\mathcal{E}_{sp}(\mathcal{H})} E' = \mathbf{0},$$

(where $\wedge_{\mathcal{E}_{sp}(\mathcal{H})}$ is the infimum with respect to $\leq_{\mathcal{E}_{sp}(\mathcal{H})}$). In spite of this, the special effects might violate the same principle in the framework of the larger structure $\langle \mathcal{E}(\mathcal{H}), \leq, ', \mathbf{0}, \mathbf{1} \rangle$ (where $\leq$ is the standard effect order). In fact, two cases are possible:

(i) the infimum $E \wedge E'$ (with respect to $\leq$) exists and is different from $\mathbf{0}$;

(ii) the infimum $E \wedge E'$ does not exist.

As a consequence, the notion of *Aristotelian sharpness* is not significant in the framework of the structure $\left\langle \mathcal{E}_{sp}(\mathcal{H}), \leq_{\mathcal{E}_{sp}(\mathcal{H})}, ', \mathbf{0}, \mathbf{1} \right\rangle$ (where even effects that are "physically unsharp" turn out to satisfy the noncontradiction principle).

As in $\mathcal{E}(\mathcal{H})$, also special effects can be naturally structured as a QMV algebra. As expected, the total sum $\oplus$ will be defined as follows:

$$E \oplus F = \begin{cases} E + F & \text{if } E + F \in \mathcal{E}_{sp}(\mathcal{H}), \\ \mathbb{1} & \text{otherwise.} \end{cases}$$

One can easily show that the structure $\langle \mathcal{E}_{sp}(\mathcal{H}), \oplus, ', \mathbb{0}, \mathbb{1} \rangle$ is a QMV algebra. As in any QMV algebra, the partial order relation $\leq^{QMV}_{\mathcal{E}_{sp}(\mathcal{H})}$ of $\langle \mathcal{E}_{sp}(\mathcal{H}), \oplus, ', \mathbb{0}, \mathbb{1} \rangle$ will be defined as follows:

$$\forall E, F \in \mathcal{E}_{sp}(H) : E \leq^{QMV}_{\mathcal{E}_{sp}(\mathcal{H})} F \text{ iff } E \Cap F = E.$$

Let us now compare the partial order $\leq^{QMV}_{\mathcal{E}_{sp}(\mathcal{H})}$ of $\langle \mathcal{E}_{sp}(\mathcal{H}), \oplus, ', \mathbb{0}, \mathbb{1} \rangle$ with the partial order $\leq^{QMV}_{\mathcal{E}(\mathcal{H})}$ of the bigger structure $\langle \mathcal{E}(\mathcal{H}), \oplus, ', \mathbb{0}, \mathbb{1} \rangle$. As we already know, $\leq^{QMV}_{\mathcal{E}(\mathcal{H})}$ coincides with the *natural* effect order $\leq$. In other words:

$$E \leq^{QMV}_{\mathcal{E}(\mathcal{H})} F \text{ iff } E \leq F \text{ iff for any density operator } \rho : \operatorname{Tr}(\rho E) \leq \operatorname{Tr}(\rho F).$$

One can easily realize that the two order-relations $\leq^{QMV}_{\mathcal{E}_{sp}(\mathcal{H})}$ and $\leq^{QMV}_{\mathcal{E}(\mathcal{H})}$ (the latter being the usual $\leq$) do not coincide for special effects. We only have:

$$\forall E, F \in \mathcal{E}_{sp}(\mathcal{H}) : E \leq^{QMV}_{\mathcal{E}_{sp}(\mathcal{H})} F \text{ implies } E \leq F,$$

but not the other way around. As a counterexample, consider the two special effects aP and P, where P is any nontrivial projection and $a \in [1/2, 1)$. Clearly, $aP \leq P$; however, $aP \not\leq^{QMV}_{\mathcal{E}_{sp}(\mathcal{H})} P$.

Hence, we can conclude that special effects *are not partially ordered* by the natural effect order in the framework of the structure $\langle \mathcal{E}_{sp}(\mathcal{H}), \oplus, ', \mathbb{0}, \mathbb{1} \rangle$.

The following theorem shows that the notions "principal" and "Aristotelian" are equivalent, whenever restricted to the concrete QMV algebra of all effects.

THEOREM 7.1.5. *$\forall E \in \mathcal{E}(\mathcal{H})$: E is a projection iff E is principal iff E is Aristotelian.*

PROOF. From Chapter 5 we know that the partial order determined by the standard QMV algebra $\langle \mathcal{E}(\mathcal{H}), \oplus, ', \mathbf{0}, \mathbf{1} \rangle$ coincides with the partial order determined by the standard effect algebra $\langle \mathcal{E}(\mathcal{H}), \boxplus, \mathbf{0}, \mathbf{1} \rangle$. Thus, by Theorem 6.2.12(i), we can conclude that for any effect $E \in \mathcal{E}(\mathcal{H})$: E is Aristotelian iff E is a projection.
By Theorem 7.1.2 every principal effect is Aristotelian. Hence, to conclude the proof of the theorem it is sufficient to show that any projection is principal. Let P be any projection of $\mathcal{E}(\mathcal{H})$ and let E, F be any two effects such that $E \leq F'$ and $E, F \leq P$. Thus, $E \oplus F = E + F$.
By (Giuntini and Greuling, 1989), if G is any effect and P any projection, then $G \leq P$ iff $G = GP = PG$. Accordingly,

$$E = PE = EP \text{ and } F = PF = FP.$$

Thus,

$$(E \oplus F)P = (E + F)P = P(E \oplus F).$$

Hence, $E \oplus F \leq P$. □

THEOREM 7.1.6. *Let $\mathcal{M} = \langle M, \oplus, ', \mathbf{0}, \mathbf{1} \rangle$ be a QMV algebra. The following conditions are equivalent $\forall a \in M$:*

(i) *a is Łukasiewicz-sharp (sharp3);*
(ii) *a is strongly Łukasiewicz-sharp (sharp4);*
(iii) *a is Boolean (sharp5);*
(iv) *a is strongly Boolean (sharp6).*

PROOF. (i) implies (ii). Suppose a is Łukasiewicz-sharp. Then, $a \Cap a' = \mathbf{0}$. We have to prove that $a \Cap a' = a' \Cap a = \mathbf{0}$ and $a \wedge a' = \mathbf{0}$. By Lemma 5.3.10(ii), $a' \oplus (a \Cap a') = a' \oplus a'$. Hence, $a' \oplus a' = a'$. Therefore, $a' \Cap a = (a' \oplus a') \odot a = a' \odot a = \mathbf{0}$. It remains to show that $a \wedge a' = \mathbf{0}$. Suppose $b \leq a, a'$. By Lemma 5.3.16(iii), $b \Cap a' \leq a \Cap a' = \mathbf{0}$. Thus, $b = b \Cap a' = \mathbf{0}$.
(ii) implies (iii). Suppose a is strongly Łukasiewicz-sharp. Then, $a' \Cap a = a \Cap a'$ and $a \wedge a' = \mathbf{0}$. We have to prove that $a \oplus a = a$. By Lemma 5.3.13(iii), $a \Cap a' \leq a'$ and $a' \Cap a \leq a$. Then, by hypothesis, $a \Cap a' = \mathbf{0}$. By Lemma 5.3.10(i), $a \oplus (a \Cap a') = a \oplus a$. Hence, $a \oplus a = a$.
(iii) implies (iv). Suppose a is Boole-sharp. Then, $a = a \oplus a$. We want to show that the set $I := \{b : b \leq a\}$ is an ideal of $\mathcal{M}$. Clearly if $b \in I$ and $c \leq b$, then $c \in I$. Suppose now $b, c \in I$. Let us show that $b \oplus c \in I$. By hypothesis, $b, c \leq a$. Thus, by Lemma 5.3.16(iv), $b \oplus c \leq a \oplus a$. By hypothesis, $a \oplus a = a$. Hence, $b \oplus c \in I$.

(iv) implies (i). Suppose a is strongly Boolean. Then, the set $I := \{b : b \leq a\}$ is an ideal of $\mathcal{M}$. As a consequence: $a \oplus a \leq a$. By Lemma 5.3.13(ii), $a \leq a \oplus a$. Hence, $a = a \oplus a$. □

THEOREM 7.1.7. *Let $\mathcal{M}$ be a QMV algebra. The following property holds $\forall a \in M$: if a is Łukasiewicz (or strongly Łukasiewicz or Boolean or strongly Boolean), then a is principal.*

PROOF. It follows from the hypothesis and the monotonicity of $\oplus$. □

The following theorem shows that a principal element need not be Boolean (equivalently, Łukasiewicz, strongly Łukasiewicz, strongly Boolean).

THEOREM 7.1.8. *Any nontrivial projection in $\mathcal{E}(\mathcal{H})$ is principal, but not Boolean.*

PROOF. Let P be any nontrivial projection. Then, by Theorem 7.1.5, P is principal. However (by definition of the QMV-sum $\oplus$), $P \oplus P = \mathbb{I} \neq P$. □

THEOREM 7.1.9. *The six notions of "sharp" collapse in the case of MV algebras.*

PROOF. Every MV algebra is a QMV algebra. Thus, by Theorem 7.1.2, Theorem 7.1.6 and Theorem 7.1.7, it is sufficient to show that every Aristotelian element is Łukasiewicz-sharp. In every MV algebra the operation $\Cap$ coincides with the infimum. Accordingly, if $a \wedge a' = \mathbf{0}$, then also $a \Cap a' = \mathbf{0}$. □

Summing up, the implications, first for QMV algebras and then for MV algebras, relating the various algebraic notions of "sharp" are the following.

QMV algebras

$$
\begin{array}{ccccccc}
\text{Łukasiewicz} & \Longleftrightarrow & \text{Strong Łukasiewicz} & \Longleftrightarrow & \text{Boole} & \Longleftrightarrow & \text{Strong Boole} \\
 & & \Downarrow \ \not\Uparrow & & & & \\
 & & \text{Principal} & \begin{array}{c}\Longrightarrow \\ \not\Longleftarrow\end{array} & \text{Aristotle} & &
\end{array}
$$

MV algebras

$$
\begin{array}{ccccc}
\text{Principal} & \Longleftrightarrow & \text{Aristotle} & \Longleftrightarrow & \text{Łukasiewicz} \\
\Updownarrow & & & & \Updownarrow \\
\text{Strong Łukasiewicz} & \Longleftrightarrow & \text{Boole} & \Longleftrightarrow & \text{Strong Boole}
\end{array}
$$

We will now consider a new definition of sharpness that essentially depends on the Brouwer complement $^\sim$. Let us refer to the class of all BZ-posets (which, of course, includes the class of all BZ-effect algebras).

DEFINITION 7.1.10. An element a of a BZ-poset is called *Brouwer-sharp* (or *Brouwerian* or *sharp7*) iff $a = a^{\sim\sim}$ (in other words, a satisfies the double negation principle with respect to the Brouwer complement).

THEOREM 7.1.11. *Brouwer implies Aristotle.*

PROOF. Suppose a is Brouwer-sharp. Then, $a' = a^{\sim\sim\prime} = a^{\sim\sim\sim} = a^{\sim}$. Let b be any element such that $b \leq a, a'$. Since $a' = a^{\sim}$, it follows that $b \leq a \wedge a^{\sim} = \mathbf{0}$. □

Notice that every effect algebra $\mathcal{A} = \langle A\,, \boxplus\,, \mathbf{0}\,, \mathbf{1}\rangle$ can be trivially organized as a BZ-effect algebra. It is sufficient to define the operation $^\sim$ in the following way:

$$a^{\sim} = \begin{cases} \mathbf{1} & \text{if } a = \mathbf{0}, \\ \mathbf{0} & \text{otherwise.} \end{cases}$$

The structure $\mathcal{A} = \langle A\,, \boxplus\,, ^\sim\,, \mathbf{0}\,, \mathbf{1}\rangle$ turns out to be a BZ-effect algebra. In this case, the set of all Brouwerian elements will only consist of $\mathbf{0}$ and $\mathbf{1}$.

THEOREM 7.1.12. *Aristotle does not imply Brouwer.*

PROOF. Let $\mathcal{A}$ be any *orthoalgebra* containing more than two elements. By definition of orthoalgebra (see Definition 5.1.6) every element of $\mathcal{A}$ is Aristotelian. Equip $\mathcal{A}$ with the trivial Brouwer complement. Thus, every nontrivial element of $\mathcal{A}$ is Aristotelian but non-Brouwerian. □

One can easily show that Aristotelian and Brouwerian sharpness coincide for concrete effects.

THEOREM 7.1.13. *Let $E \in \mathcal{E}(\mathcal{H})$: E is Aristotelian iff E is Brouwerian.*

PROOF. Straightforward. □

7.2. Probabilistic definitions of "sharpness"

So far, we have considered purely algebraic definitions of *sharpness*, that only refer to the algebraic structure of quantum effects. We will now investigate possible characterizations of *sharpness* that essentially refer to the interplay between *effects* and *states*. Since the basic role of states is assigning probability-values to the elements of a given effect-structure, we will speak in this connection of *probabilistic definitions of sharpness*.

In Chapter 6 we introduced the abstract notion of *effect-state system* in the framework of an axiomatic approach to unsharp QT. According to this approach, quantum effects are structured as a σ-effect algebra, while states represent σ-homomorphisms from the algebra of all effects into the real effect algebra based on the real interval $[0, 1]$.

Needless to say, one can also refer to abstract effect-state systems $(\mathcal{E}, \mathcal{S})$, where the effect-set $\mathcal{E}$ has the structure of a QMV algebra. In fact, as we already know, concrete effects (belonging to $\mathcal{E}(\mathcal{H})$) give rise to a quasi-linear QMV algebra, where the QMV- sum $\oplus$ represents a total extension of the partial effect-sum $\boxplus$. We have also seen that every abstract effect algebra can be uniquely transformed into a quasi-linear QMV algebra and vice versa (see Theorem 5.4.4).

We call any quasi-linear QMV algebra $\langle M\,, \oplus\,, '\,, \mathbf{0}\,, \mathbf{1}\rangle$, whose associated effect algebra $\langle M\,, \boxplus\,, '\,, \mathbf{0}\,, \mathbf{1}\rangle$ is a σ-effect algebra, a *σ-QMV algebra*. The notion of *state* can also be naturally defined for σ-QMV algebras (in the expected way). As a consequence, we can consider abstract effect-state systems $(\mathcal{E}, \mathcal{S})$, where $\mathcal{E}$ is organized as a σ-QMV algebra.

In this framework, one can introduce the following probabilistic notions of *sharpness*. Let $(\mathcal{E}, \mathcal{S})$ be an effect-state system and let E be an element of $\mathcal{E}$.

DEFINITION 7.2.1.

(i) E is *Tarski-sharp* (or *sharp8*) iff either $E = \mathbf{0}$ or $\exists s \in \mathcal{S}$ such that $s(E) = 1$ (in other words, whenever E is not impossible, E is verified with certainty by at least one state).

(ii) E is *von Neumann-sharp* (or *sharp9*) iff

$$\forall s \in \mathcal{S} : s \in \mathbf{No}(E) \text{ iff } s \perp \mathbf{Yes}(E),$$

(in other words, the *negative domain* of E is maximal).

Von Neumann-sharpness turns out to be stronger than Tarski-sharpness.

THEOREM 7.2.2. *Let $(\mathcal{E}\,, \mathcal{S})$ be an effect-state system. $\forall E \in \mathcal{E}$: if E is von Neumann-sharp, then E is Tarski-sharp.*

PROOF. Suppose $E \neq \mathbf{0}$. Suppose, by contradiction, that $s(E) \neq 1\, \forall s \in \mathcal{S}$. Hence, $\mathbf{Yes}(E) = \emptyset$. Thus, trivially, $s \perp \mathbf{Yes}(E)$ and therefore, $s \in \mathbf{No}(E)$. Thus, $\forall s \in \mathcal{S}$, $s(E) = 0$. Hence, $E = \mathbf{0}$ because $\mathcal{S}$ is order-determining. □

THEOREM 7.2.3. *Consider any concrete effect-state system $(\mathcal{E}(\mathcal{H}), \mathcal{S}(\mathcal{H}))$. There exists an effect E of $\mathcal{E}(\mathcal{H})$ such that E is Tarski-sharp but not von Neumann-sharp.*

PROOF. Let $E := \mathbb{1} - aP$, where $a \in (0, 1/2)$ and P is any nontrivial projection. E is Tarski-sharp, because $\mathbf{Yes}(E) = \mathbf{No}(P) \neq \emptyset$.
We have: $\mathbf{No}(E) = \emptyset$. Thus, to prove that E is not von Neumann-sharp, it is sufficient to show that $\exists s \in \mathcal{S}$ such that $s \in \mathbf{Yes}(E)^{\perp}$.
Let s be any state in $\mathcal{S}$ such that $s(P) = 1$. We want to show that $s \in \mathbf{Yes}(E)^{\perp}$. Since $\mathbf{Yes}(E) = \mathbf{No}(P)$, we have to show that $s \perp \mathbf{No}(P)$, i.e., $\forall t \in \mathcal{S}$ $[t(P) = 0$ implies $s \perp t]$. This holds, because $s(P) = 1$. □

The counterexample described in the proof of Theorem 7.2.3 shows that Tarski-sharpness does not characterize projections in the class of all effects of a Hilbert space. Hence, generally, Tarski-sharpness does not imply Aristotle-sharpness. At the same time, one can easily realize that any projection of a concrete effect-state system $(\mathcal{E}(\mathcal{H}), \mathcal{S}(\mathcal{H}))$ is Tarski-sharp.

What can be said about the notion of *Brouwer sharpness*, in the framework of effect-state systems? Let $(\mathcal{E}, \mathcal{S})$ be an effect-state system. Since $(\mathcal{E}, \mathcal{S})$ is sharply dominating, any effect E admits a *best sharp lower approximation* $\nu(E)$. By Theorem 6.3.2, the structure $\langle \mathcal{E}, \boxplus, ', ^{\sim}, \mathbf{0}, \mathbf{1} \rangle$ is a BZ-effect algebra, where $\forall E \in \mathcal{E}$: $E^{\sim} = \nu(E')$.

LEMMA 7.2.4. *Let $(\mathcal{E}, \mathcal{S})$ be an effect-state system and let $E \in \mathcal{E}$. If E is Aristotelian, then E is Brouwerian.*

PROOF. Suppose E is Aristotelian. Then, by Theorem 6.3.1(2), $E = \nu(E) = E'^{\sim}$. By Theorem 6.3.2(ii), $E^{\sim} = E'$. Therefore, $E^{\sim\sim} = E'^{\sim} = E$. □

As a consequence, we can say that the two notions of *Aristotelian* and *Brouwerian* turn out to coincide in the framework of effect-state systems. It is worthwhile noticing, however, that the *sharply dominating condition* (which is responsible for this collapse) could be also assumed in a purely algebraic context (quite independently of the notion of *state*).[3]

THEOREM 7.2.5. *Let $(\mathcal{E}, \mathcal{S})$ be an effect-state system and let $E \in \mathcal{E}$. E is von Neumann-sharp iff E is Brouwerian.*

PROOF. Let E be any effect in $\mathcal{E}$. By Theorem 6.4.7, $\mathbf{Yes}(E)$ is a closed set of the preclusivity space $(\mathcal{S}, \perp_{\mathcal{E}})$. Thus, $\mathbf{Yes}(E) = \mathbf{Yes}(E)^{\perp\perp}$. Let us consider the set $\mathcal{E}_a$ of all Aristotelian elements of $\mathcal{E}$. By Lemma 6.4.3 and Theorem 6.4.4, the pair $(\mathcal{E}_a, \mathcal{S})$ is a normal event-state system. Consequently, the structure $\langle \mathcal{E}_a, \leq, ', \mathbf{0}, \mathbf{1} \rangle$ is a σ-orthocomplete orthomodular sub-poset of the bounded involution poset of all effects. Now, by Theorem 6.4.1, the preclusivity relations $\perp_{\mathcal{E}}$ and $\perp_{\mathcal{E}_a}$ coincide. Therefore, by Theorem 2.4.12, the restriction of the map $\mathbf{Yes}$ to $\mathcal{E}_a$ is an embedding of the σ-orthocomplete orthomodular poset $\mathcal{E}_a$ into the complete ortholattice of all closed sets of the preclusivity space $(\mathcal{S}, \perp_{\mathcal{E}})$.
We first prove that von Neumann implies Brouwer. Suppose the effect E is von Neumann-sharp, i.e., $\mathbf{Yes}(E)^{\perp} = \mathbf{No}(E) = \mathbf{Yes}(E')$. We have to prove that E is Brouwerian, i.e., $E = E^{\sim\sim}$. Since $E'^{\sim} \leq E$ and $E \leq E^{\sim\sim}$, it is sufficient to prove: $E^{\sim\sim} = E'^{\sim}$. Both $E^{\sim\sim}$ and $E'^{\sim}$ are Brouwerian. Thus, by Theorem 7.1.11, $E^{\sim\sim}$ and $E'^{\sim}$ are Aristotelian. As already noticed, the restriction of the map $\mathbf{Yes}$ to the set $\mathcal{E}_a$ of all Aristotelian elements is injective. Thus, to prove that E is Brouwerian it suffices to show that $\mathbf{Yes}(E'^{\sim}) = \mathbf{Yes}(E^{\sim\sim})$. By hypothesis, $\mathbf{Yes}(E)^{\perp} = \mathbf{No}(E) = \mathbf{Yes}(E')$. Since $\mathbf{Yes}(E)$ is a closed set, we obtain $\mathbf{Yes}(E) = \mathbf{Yes}(E')^{\perp}$. Now, the effect-state system $(\mathcal{E}, \mathcal{S})$ is sharply normal (by Axiom U9). Thus, $\mathbf{Yes}(E) = \mathbf{Yes}(E(\nu(E))) = \mathbf{Yes}(E'^{\sim})$ and $\mathbf{Yes}(E')^{\perp} = \mathbf{Yes}(\nu(E'))^{\perp} =$

[3]See (Gudder, 1998).

$\mathbf{Yes}(E''^{\sim})^{\perp} = \mathbf{Yes}(E^{\sim})^{\perp}$. Since $E^{\sim}$ is Aristotelian (by Theorem 7.1.11), we obtain by Theorem 6.4.6, $\mathbf{Yes}(E^{\sim})^{\perp} = \mathbf{Yes}(E^{\sim\prime}) = \mathbf{Yes}(E^{\sim\sim})$. Consequently, $\mathbf{Yes}(E'^{\sim}) = \mathbf{Yes}(E^{\sim\sim})$.

We now prove that Brouwer implies von Neumann. Suppose E is Brouwerian. Then, $E = E^{\sim\sim}$ and $E^{\sim} = E'$. Thus, $\mathbf{Yes}(E)^{\perp} = \mathbf{Yes}(E^{\sim\sim})^{\perp}$. Since $E^{\sim\sim}$ is Aristotelian (by Theorem 7.1.11), we obtain by Theorem 6.4.6: $\mathbf{Yes}(E^{\sim\sim})^{\perp} = \mathbf{Yes}(E^{\sim\sim\prime}) = \mathbf{Yes}(E^{\sim}) = \mathbf{Yes}(E') = \mathbf{No}(E)$. Consequently, $\mathbf{Yes}(E)^{\perp} = \mathbf{No}(E)$. □

Summing up, the implications, first for abstract effect-state systems and then for concrete effect-state systems, relating the various algebraic and probabilistic notions of "sharp" are the following.

Abstract effect-state systems $(\mathcal{E}, \mathcal{S})$

$$
\begin{array}{ccc}
\text{Łukasiewicz} & \Longleftrightarrow & \text{Strong Boole} \\
\Updownarrow & & \Updownarrow \\
\text{Strong Łukasiewicz} & \Longleftrightarrow & \text{Boole} \\
 & \Downarrow \ \not\Uparrow & \\
 & \text{Principal} & \\
 & \Downarrow \ \Uparrow ? & \\
\text{Aristotle} & \Longleftrightarrow & \text{Brouwer} \\
 & \Updownarrow & \\
 & \text{von Neumann} & \\
 & \Downarrow \ \not\Uparrow & \\
 & \text{Tarski} &
\end{array}
$$

Concrete effect-state systems $(\mathcal{E}(\mathcal{H}), \mathcal{S}(\mathcal{H}))$

$$
\begin{array}{ccccc}
\text{Łukasiewicz} & \Longleftrightarrow & \text{Strong Boole} & & \\
\Updownarrow & & \Updownarrow & & \\
\text{Strong Łukasiewicz} & \Longleftrightarrow & \text{Boole} & & \\
 & \Downarrow \ \not\Uparrow & & & \\
\text{Principal} & \Longleftrightarrow & \text{Aristotle} & & \\
\Updownarrow & & \Updownarrow & & \\
\text{Brouwer} & \Longleftrightarrow & \text{von Neumann} & \Longleftrightarrow & \text{Projection} \\
 & \Downarrow \ \not\Uparrow & & & \\
 & \text{Tarski} & & &
\end{array}
$$

PART II Quantum Logics as Logic

Introduction

In Part I we have investigated a number of structures that describe indeterministic and ambiguous features. From an intuitive point of view, in classical contexts, the origin of all uncertainties may be regarded as depending on human limitations. In the quantum world, however, both indeterminism and ambiguity, seems to play an essential and unavoidable role.

Interestingly enough, the origin of uncertainties cannot simply be described as a mere *interpretation-question*. We will see that some formal properties of our abstract structures (mainly connected with possible failures of the distributivity relations) seem to be strongly connected with the problem: to what extent do indeterminism and ambiguity represent "essential features" of the concrete situations that are described by our theories?

In the quantum logical literature, some of the abstract structures we have analyzed in Part I are currently termed *logics*. In particular, by a (sharp) *quantum logic* one often means an example of a σ-orthocomplete orthomodular poset. As already noticed, such a terminology sounds somewhat misleading for logicians. We will now study how different categories of the quantum structures we have investigated give rise to different forms of *logic*, in the sense of the standard logical tradition.

Henceforth we will use the term logic in its most traditional sense: as a theory for a *consequence-relation* that may hold between well-formed formulas of a given language.

In principle, the languages we are referring to are not necessarily *sharp*: as happens in the case of natural languages, fuzzy borders for the sets of well-formed expressions might be admitted. One could also consider *fuzzy inference rules*: hence, even the consequence-relation would turn out to be, to a certain extent, an ambiguous concept.[4] However, we will not follow here such an "extreme fuzzy choice." For the sake of simplicity, we will restrict our attention to traditional formal languages with *sharp borders*. Also the different consequence-relations we are going to study will represent sharp notions. At the same time, we will admit the possibility of *ambiguous meanings* for the well-formed expressions. In this sense, we will study different forms of *fuzzy quantum semantics*.

[4]Fuzzy inference rules have been studied for instance in (Hájek, 1998).

As is well known, many "important" logics can be semantically characterized by means of two privileged kinds of semantics: an *algebraic semantics* and a *possible-world semantics* (also called *Kripkean semantics*).

From an intuitive point of view, the algebraic semantic approach can be described as founded on the following idea: *interpreting* a language essentially means associating to any sentence an abstract meaning that corresponds to an element of a given abstract structure. At the same time, the logical constants are interpreted as particular operations of the same structure. Hence, generally, an *algebraic realization* for a logic **L** has the following form

$$\mathcal{A} = \langle \mathcal{B} \, , \, v \rangle \, ,$$

where $\mathcal{B}$ is an element of given category of algebraic structures, while v is a *valuation-function* that transforms sentences into elements of $\mathcal{B}$, preserving the logical form.

On the other hand, in the Kripkean semantics one assumes that *interpreting* a language essentially means associating to any sentence α the set of the *possible worlds* or *situations* where α holds. This set, which represents the *extensional meaning* of α, is called the *proposition* associated to α (or simply *the proposition of* α). Hence, generally, a Kripkean realization for a logic **L** has the following form:

$$\mathcal{K} = \left\langle I \, , \, \overrightarrow{\mathcal{R}_i} \, , \, \overrightarrow{o_j} \, , \, Pr \, , \, V \right\rangle ,$$

where

(i) I is a nonempty set of possible worlds possibly correlated by relations in the sequence $\overrightarrow{\mathcal{R}_i}$ and operations in the sequence $\overrightarrow{o_j}$. The subsystem $\left\langle I \, , \, \overrightarrow{\mathcal{R}_i} \, , \, \overrightarrow{o_j} \right\rangle$ is also called the *frame* of $\mathcal{K}$. In most cases, we have only one binary relation $\mathcal{R}$, called the *accessibility* relation.

(ii) Pr is a set of sets of possible worlds, representing possible propositions of sentences. Any proposition and the total set of propositions Pr must satisfy convenient closure conditions that depend on the logic **L**.

(iii) V transforms sentences into propositions preserving the logical form.

In many "happy" situations, the algebraic realizations (of a certain type) can be canonically transformed into Kripkean realizations (of a corresponding type), and vice versa. When this is the case, one can conclude that one is dealing with an algebraic semantics and with a Kripkean semantics that *characterize the same logic*.

It is expedient to recall the algebraic semantic characterization and the Kripkean semantic characterization for two important examples of logic: classical logic (**CL**) and intuitionistic logic (**IL**); this provides an interesting standard of comparison for the quantum logics. At the *sentential level* the language of both logics consists of a countable set of atomic sentences and of the following logical connectives: $\neg$ (not), $\curlywedge$ (and), $\curlyvee$ (or), $\rightarrow$ (if ...

then). The notion of *sentence* (or *formula*) of the language is defined in the expected way. We will use the following metavariables: $\mathbf{p}, \mathbf{q}, \mathbf{r}, \ldots$ for atomic sentences and α, β, $\gamma, \ldots$ for sentences.[5] We indicate the basic metalogical constants as follows: not, and, or, $\curvearrowright$ (if...then), iff (if and only if), $\forall$ (for all), $\exists$ (for at least one).

As is well known, in the algebraic semantics, **CL** is characterized by the class of all algebraic realizations based on Boolean algebras. Hence, any classical algebraic realization has the following form:

$$\mathcal{A} = \langle \mathcal{B}\,, v \rangle\,,$$

where $\mathcal{B} = \langle B\,, \wedge\,, \vee\,, '\,, \mathbf{0}\,, \mathbf{1} \rangle$ is a Boolean algebra and v is a *valuation-function* that associates to any sentence α of the language an element in B, satisfying the following conditions:

(i) $v(\neg\beta) = v(\beta)'$;
(ii) $v(\beta \curlywedge \gamma) = v(\beta) \wedge v(\gamma)$;
(iii) $v(\beta \curlyvee \gamma) = v(\beta) \vee v(\gamma)$;
(iv) $v(\beta \to \gamma) = v(\beta) \rightsquigarrow v(\gamma) := v(\beta)' \vee v(\gamma)$.

On this basis, one can give the usual definitions for the notions of *truth*, *logical truth*, *consequence* and *logical consequence*.

DEFINITION. *Truth and logical truth*
A sentence α is *true* in a realization $\mathcal{A} = \langle \mathcal{B}\,, v \rangle$ (abbreviated $\models_{\mathcal{A}} \alpha$) iff $v(\alpha) = \mathbf{1}$;
α is a *logical truth* of **CL** ($\models \alpha$) iff for any algebraic realization $\mathcal{A} = \langle \mathcal{B}\,, v \rangle$, $\models_{\mathcal{A}} \alpha$.

When $\models_{\mathcal{A}} \alpha$, we will also say that $\mathcal{A}$ is a *model* of α; $\mathcal{A}$ will be called a *model* of a set of formulas T ($\models_{\mathcal{A}} T$) iff $\mathcal{A}$ is a model of any $\beta \in T$.

DEFINITION. *Consequence in a realization and logical consequence*
Let T be a set of sentences and let $\mathcal{A} = \langle \mathcal{B}\,, v \rangle$ be a realization. A sentence α is a *consequence in* $\mathcal{A}$ of T ($T \models_{\mathcal{A}} \alpha$) iff for any element a of B:
if, for any $\beta \in T$, $a \leq v(\beta)$, then $a \leq v(\alpha)$.
A sentence α is a *classical logical consequence* of T ($T \models \alpha$) iff $T \models_{\mathcal{A}} \alpha$ for any algebraic realization $\mathcal{A}$.

An algebraic semantic characterization for intuitionistic logic can be obtained by referring to the category of all *Heyting algebras* (instead of Boolean algebras).

DEFINITION. *Heyting algebra*
A *Heyting algebra* is a structure

$$\mathcal{B} = \langle B\,, \wedge\,, \vee\,, \rightsquigarrow\,, \mathbf{0}\,, \mathbf{1} \rangle\,,$$

where

[5] A metavariable is a variable of the metalanguage ranging over expressions of the object-language.

(i) $\langle B\,,\wedge\,,\vee\,,\mathbf{0}\,,\mathbf{1}\rangle$ is a bounded lattice;

(ii) $\rightsquigarrow$ is a *residual operation* (representing the *intuitionistic conditional*) that satisfies the condition:

$$\forall a,b \in B[a \rightsquigarrow b = \max\{c\,:\,a \wedge c \leq b\}].$$

On this basis, a unary operation $'$ (which represents the *intuitionistic negation*) can be defined:

$$\forall a \in B[a' := a \rightsquigarrow \mathbf{0}].$$

One can prove that any Heyting algebra is a distributive lattice. The notion of *intuitionistic algebraic realization* and all the other semantic concepts can be then defined as in classical logic, *mutatis mutandis*.

Let us now turn to Kripkean semantics. Classical logic is characterized by the class of all Kripkean realizations

$$\mathcal{K} = \langle I\,,\mathcal{R}\,,Pr\,,V\rangle\,,$$

where:

- $\mathcal{R}$ is the total relation defined on I (hence every world is accessible to any other world);
- Pr coincides with the power set $\mathcal{P}(I)$ of I;
- V is a function that associates to any sentence α a proposition in Pr, satisfying the following conditions:
  $V(\neg\beta) = V(\beta)^c$;
  $V(\beta \curlywedge \gamma) = V(\beta) \cap V(\gamma)$;
  $V(\beta \curlyvee \gamma) = V(\beta) \cup V(\gamma)$;
  $V(\beta \rightarrow \gamma) = V(\beta)^c \cup V(\gamma)$.

Instead of $i \in V(\alpha)$, we will also write $i \models \alpha$ (or, $i \models_{\mathcal{K}} \alpha$, in case of possible confusion) and we will read: "α is true in the world i". If T is a set of sentences, $i \models T$ will mean $i \models \beta$ for any $\beta \in T$.

On this basis, we define the notions of *truth*, *logical truth*, *consequence*, *logical consequence* in the expected way.

DEFINITION. *Truth and logical truth*
A sentence α is *true* in a realization $\mathcal{K} = \langle I\,,\mathcal{R}\,,\mathrm{Pr}\,,V\rangle$ (abbreviated $\models_{\mathcal{K}} \alpha$) iff $V(\alpha) = I$;
α is a *logical truth* of **CL** ($\models_{CL} \alpha$) iff $\models_{\mathcal{K}} \alpha$ for any realization $\mathcal{K}$.

When $\models_{\mathcal{K}} \alpha$, we will also say that $\mathcal{K}$ is a *model* of α. Similarly in the case of a set of sentences T.

DEFINITION. *Consequence in a realization and logical consequence*
Let T be a set of sentences and let $\mathcal{K}$ be a realization. A sentence α is a *consequence in* $\mathcal{K}$ of T ($T \models_{\mathcal{K}} \alpha$) iff for any world i of $\mathcal{K}$, $i \models T \curvearrowright i \models \alpha$.
A sentence α is a *logical consequence* of T ($T \models \alpha$) iff for any realization $\mathcal{K}$: $T \models_{\mathcal{K}} \alpha$. When no confusion is possible we will simply write $T \models \alpha$.

As we have seen, the accessibility relation does not play any role for classical Kripkean realizations (since $\mathcal{R}$ coincides with the total relation and the definition of V does not refer to $\mathcal{R}$). As a consequence, the classical possible world semantics turns out to be nothing but a rephrasing of the usual model theory for **CL**.

On the other hand accessibility relations have an essential role for **IL**. Every intuitionistic Kripkean realization has the following form:

$$\mathcal{K} = \langle I \,, \mathcal{R} \,, Pr \,, V \rangle \,,$$

where:

- $\mathcal{R}$ is a reflexive and transitive relation defined on I;
- Pr is the set of all *possible propositions*; where a *proposition* is a set X of worlds that is $\mathcal{R}$-closed. In other words:
  $X \in Pr$ iff $X \subseteq I$ and $\forall i, j \in I[i \in X$ and $i\mathcal{R}j \;\curvearrowright\; j \in X]$;
- V is a function that associates to any sentence α a proposition in Pr, satisfying the following conditions:
  $V(\neg\beta) = \{i \in I : \forall j[i\mathcal{R}j \;\curvearrowright\; j \notin V(\beta)]\}$;
  $V(\beta \curlywedge \gamma) = V(\beta) \cap V(\gamma)$;
  $V(\beta \curlyvee \gamma) = V(\beta) \cup V(\gamma)$;
  $V(\beta \to \gamma) = \{i \in I : \forall j[(i\mathcal{R}j$ and $j \in V(\beta)) \;\curvearrowright\; j \in V(\gamma)]\}$.

The notions of *truth*, *logical truth*, *consequence* and *logical consequence* are then defined as in the case of **CL**.

According to a canonical intuitive interpretation, the possible worlds of intuitionistic Kripkean realizations can be regarded as *states of knowledge in progress*. When j is accessible to i, the state of knowledge corresponding to j is more informative with respect to the state of knowledge represented by i. In this framework, knowledge is *conservative*, because the following condition holds:

$$i \models \alpha \text{ and } i\mathcal{R}j \;\curvearrowright\; j \models \alpha.$$

This is the reason why **IL** is often described as a form of *epistemic logic.*

As we will see in the next chapters, the Kripkean characterization of quantum logics will be based on a quite different intuitive idea.

Both in the case of **CL** and of **IL** one can prove that the algebraic semantics and the Kripkean semantics characterize the same logic. In other words:

- α is a logical truth according to the algebraic semantics iff α is a logical truth according to the Kripkean semantics.
- α is a logical consequence of T according to the algebraic semantics iff α is a logical consequence of T according to the Kripkean semantics.

In the following chapters we will investigate different forms of *sharp and unsharp quantum logics* that can be characterized by different forms of algebraic and Kripkean semantics. One is dealing with a class of logics that have a common feature: they admit an algebraic semantic characterization that can be naturally exemplified in the framework of abstract and concrete quantum structures, centered around orthomodular lattices, orthoalgebras, effect algebras, QMV algebras and BZ-lattices. Furthermore, some of these logics admit a Kripkean semantic characterization that can be exemplified in the framework of event-state systems and of effect-state systems.

CHAPTER 8

Sharp quantum logics

We will first study two interesting examples of logics that represent a natural logical abstraction from the class of all Hilbert lattices. These are represented respectively by *orthomodular quantum logic* (**OQL**) and by the weaker *orthologic* (**OL**), which for a long time has been also termed *minimal quantum logic*. In fact, the name "minimal quantum logic" appears today quite inappropriate for two reasons. First, a number of, in a sense, weaker forms of quantum logic have recently attracted much attention; and second, the models for the "minimal quantum logic" do not provide for the possibility of an adequate modeling of the generalized probabilities that are induced by states of QT. However these probabilities do not usually play a fundamental role in the logical developments that follow. And the "minimal quantum logic" provides a "floor" for the other logics, so we include it. In the following we will use **QL** as an abbreviation for either **OL** or **OQL**.

The language of **QL** consists of a set of atomic sentences and of two primitive connectives: $\neg$ (not), $\curlywedge$ (and). The notion of *sentence* (or *formula*) of the language is defined in the expected way. The connective disjunction ($\curlyvee$) is supposed defined via de Morgan's law:

$$\alpha \curlyvee \beta := \neg(\neg\alpha \curlywedge \neg\beta) .$$

The problem concerning the possibility of a well behaved conditional connective will be discussed in Section 8.3.

8.1. Algebraic and Kripkean semantics for sharp quantum logics

Because of its historical origin, the most natural characterization of **QL** can be carried out in the framework of an algebraic semantics.

Orthologic **OL** is the logic that is semantically characterized by the class of all algebraic realizations based on ortholattices. Orthomodular quantum logic **OQL** is semantically characterized by the class of all algebraic realizations based on orthomodular lattices. We will first study the logic **OL**.

The notion of *algebraic realization* for **OL** is defined in a fashion similar to the case of **CL** and **IL** (*mutatis mutandis*).

DEFINITION 8.1.1. *Algebraic realization for* **OL**
An *algebraic realization* for **OL** is a pair $\mathcal{A} = \langle \mathcal{B}, v \rangle$, consisting of an ortholattice $\mathcal{B} = \langle B, \leq, {}', \mathbf{0}, \mathbf{1} \rangle$ and a *valuation-function* v that associates to any sentence α of the language an element in B, satisfying the following conditions:

(i) $v(\neg\beta) = v(\beta)'$;
(ii) $v(\beta \curlywedge \gamma) = v(\beta) \wedge v(\gamma)$.

The notions of *truth*, *logical truth*, *consequence* and *logical consequence* are then defined in the expected way.

DEFINITION 8.1.2. *Truth and logical truth*
A sentence α is *true* in a realization $\mathcal{A} = \langle \mathcal{B}, v \rangle$ (abbreviated $\models_{\mathcal{A}} \alpha$) iff $v(\alpha) = \mathbf{1}$;
α is a *logical truth* of **OL** ($\models_{\overline{\mathbf{OL}}} \alpha$) iff $\models_{\mathcal{A}} \alpha$ for any algebraic realization $\mathcal{A} = \langle \mathcal{B}, v \rangle$.

When $\models_{\mathcal{A}} \alpha$, we will also say that $\mathcal{A}$ is a *model* of α; $\mathcal{A}$ will be called a *model* of a set of sentences T ($\models_{\mathcal{A}} T$) iff $\mathcal{A}$ is a model of any $\beta \in T$.

DEFINITION 8.1.3. Let T be a set of sentences and let $\mathcal{A} = \langle \mathcal{B}, v \rangle$ be a realization. A sentence α is a *consequence in* $\mathcal{A}$ of T ($T \models_{\mathcal{A}} \alpha$) iff for any element a of B: if for any $\beta \in T$, $a \leq v(\beta)$, then $a \leq v(\alpha)$.
A sentence α is a *logical consequence* of T ($T \models_{\overline{\mathbf{OL}}} \alpha$) iff $T \models_{\mathcal{A}} \alpha$ for any algebraic realization $\mathcal{A}$.

Instead of $\{\alpha\} \models_{\overline{\mathbf{OL}}} \beta$ we will write $\alpha \models_{\overline{\mathbf{OL}}} \beta$. If T is finite and equal to $\{\alpha_1, \ldots, \alpha_n\}$, we will obviously have: $T \models_{\overline{\mathbf{OL}}} \alpha$ iff $v(\alpha_1) \wedge \cdots \wedge v(\alpha_n) \leq v(\alpha)$. One can easily check that $\models_{\overline{\mathbf{OL}}} \alpha$ iff $T \models_{\overline{\mathbf{OL}}} \alpha$ for any T.

OL can be equivalently characterized by means of a Kripkean semantics, which was first proposed by Dishkant (1972).

Unlike classical and intuitionistic logic, the Kripkean realizations that turn out to be adequate for **OL** are based on frames equipped with a single accessibility relation, which is reflexive and symmetric. From an intuitive point of view, one can easily understand the reason why semantic models with a reflexive and symmetric accessibility relation may be physically significant. In fact, physical theories are not generally concerned with *possible evolutions of states of knowledge* with respect to a constant world (as happens in the case of **IL**), but rather with *sets of physical situations* that may be *similar*, where *states of knowledge* must single out some *invariants*. And similarity relations are reflexive and symmetric, but generally not transitive.

In Part I (Section 2.4) we have investigated the notion of *similarity space*: a pair consisting of a set of objects (representing the *universe*) and a *similarity relation*. We have seen how this notion plays an important role both for abstract and for concrete event-state systems (and effect-state systems). We will now use similarity spaces for the construction of Kripkean realizations of **QL**.

In semantic contexts, similarity spaces $(I, \mathcal{R})$ (where I represents a set of *possible worlds*, while $\mathcal{R}$ represents an *accessibility relation* that is reflexive and symmetric) are often called *orthoframes*.

Given an orthoframe, we will use $i, j, k, \ldots$ as variables ranging over the set of worlds. Sometimes we write $i \not\perp j$ for $i\mathcal{R}j$.

As we already know, any similarity space has a "twin space" that is a preclusivity space. The preclusivity relation, corresponding to the accessibility relation $\not\perp$ will be denoted by $\perp$. Hence, we will have:

$$i \perp j \text{ iff not } i \not\perp j.$$

Whenever $i \perp j$ we will say that j is *inaccessible* or *orthogonal* to i.

We have already learned that any preclusivity space $(I, \perp)$ permits one to define a *preclusive complement* $^{\perp}$ on the power set $\mathcal{P}(I)$ of I:

$$\forall X \subseteq I[X^{\perp} := \{i \in I : i \perp X\}].$$

The following conditions hold:

- the map $^{\perp\perp} : \mathcal{P}(I) \mapsto \mathcal{P}(I)$ is a closure operator;
- the structure $\langle \mathbb{C}(\mathcal{P}(I)), \subseteq, {}^{\perp}, \emptyset, I \rangle$ based on the set of all closed subsets of I is an ortholattice. Hence, in particular, $^{\perp}$ is an orthocomplement;
- X is a closed subset of I iff $\forall i[i \in X \text{ iff } \forall j \not\perp i \exists k \not\perp j(k \in X)]$.

In the framework of our semantic application, the closed subsets of I will be also called *propositions* of the orthoframe $(I, \not\perp)$.

LEMMA 8.1.4. *Let $(I, \not\perp)$ be an orthoframe and X a set of worlds of I. Then,*

$$X \textit{ is a proposition of } (I, \not\perp) \textit{ iff } \forall i\, [i \notin X \curvearrowright \exists j(i \not\perp j \textit{ and } j \perp X)].$$

LEMMA 8.1.5. *Let $(I, \mathcal{R})$ be an orthoframe.*

(i) *I and $\emptyset$ are propositions;*
(ii) *If X is any set of worlds, then $X^{\perp}$ is a proposition;*
(iii) *If C is a family of propositions, then $\bigcap C$ is a proposition.*

DEFINITION 8.1.6. *Kripkean realization for* **OL**
A *Kripkean realization* for **OL** is a system $\mathcal{K} = \langle I, \mathcal{R}, Pr, V \rangle$, where:

(i) $(I, \mathcal{R})$ is an orthoframe and Pr is a set of propositions of the frame that contains $\emptyset, I$ and is closed under the orthocomplement $^{\perp}$ and set-theoretic intersection $\cap$;
(ii) V is a function that associates to any sentence α a proposition in Pr, satisfying the following conditions:
$V(\neg\beta) = V(\beta)^{\perp}$;
$V(\beta \curlywedge \gamma) = V(\beta) \cap V(\gamma)$.

As in **CL** and in **IL**, instead of $i \in V(\alpha)$, we will also write $i \models \alpha$ (or, $i \models_{\mathcal{K}} \alpha$, in case of possible confusion) and we will read: "α is true in the world i". If T is a set of sentences, $i \models T$ will mean $i \models \beta$ for any $\beta \in T$.

THEOREM 8.1.7. *For any Kripkean realization $\mathcal{K}$ and any sentence α:*

$$i \models \alpha \textit{ iff } \forall j \not\perp i \exists k \not\perp j\, (k \models \alpha).$$

PROOF. By definition of a Kripkean realization and by Lemma 8.1.5, $V(\alpha)$ is a closed set of worlds, for any α. Hence, by Lemma 2.4.4 we have:

$$i \in V(\alpha) \text{ iff } \forall j \not\perp i \, \exists k \not\perp j \, [k \in V(\alpha)].$$

□

LEMMA 8.1.8. *In any Kripkean realization $\mathcal{K}$:*

(i) $i \models \neg\beta$ *iff* $\forall j \not\perp i \, [j \not\models \beta]$;
(ii) $i \models \beta \curlywedge \gamma$ *iff* $i \models \beta$ *and* $i \models \gamma$.

On this basis, the notions of *truth*, *logical truth*, *consequence*, *logical consequence* are defined in the expected way.

DEFINITION 8.1.9. *Truth and logical truth*
A sentence α is *true* in a realization $\mathcal{K} = \langle I\,, \mathcal{R}\,, Pr\,, V\rangle$ (abbreviated $\models_{\mathcal{K}} \alpha$) iff $V(\alpha) = I$;
α is a *logical truth* of **OL** ($\models_{\mathbf{OL}} \alpha$) iff $\models_{\mathcal{K}} \alpha$ for any realization $\mathcal{K}$.

When $\models_{\mathcal{K}} \alpha$, we will also say that $\mathcal{K}$ is a *model* of α. Similarly in the case of a set of sentences T.

DEFINITION 8.1.10. *Consequence in a realization and logical consequence*
Let T be a set of sentences and let $\mathcal{K}$ be a realization. A sentence α is a *consequence in* $\mathcal{K}$ of T ($T \models_{\mathcal{K}} \alpha$) iff for any world i of $\mathcal{K}$, $i \models T \curvearrowright i \models \alpha$.
A sentence α is a *logical consequence* of T ($T \models_{\mathbf{OL}} \alpha$) iff for any realization $\mathcal{K}$, $T \models_{\mathcal{K}} \alpha$. When no confusion is possible we will simply write $T \models \alpha$.

Now we will prove that the algebraic and the Kripkean semantics for **OL** characterize the same logic. Let us abbreviate the metalogical expressions "α is a logical truth of **OL** according to the algebraic semantics", "α is a logical consequence in **OL** of T according to the algebraic semantics", "α is a logical truth of **OL** according to the Kripkean semantics", "α is a logical consequence in **OL** of T according to the Kripkean semantics", by $\overset{\text{A}}{\models}_{\mathbf{OL}} \alpha\,, T \overset{\text{A}}{\models}_{\mathbf{OL}} \alpha\,, \overset{\text{K}}{\models}_{\mathbf{OL}} \alpha\,, T \overset{\text{K}}{\models}_{\mathbf{OL}} \alpha$, respectively.

THEOREM 8.1.11. $\overset{A}{\models}_{\mathbf{OL}} \alpha$ *iff* $\overset{K}{\models}_{\mathbf{OL}} \alpha$ *for any sentence* α.

The theorem is an immediate corollary of the following lemma:

LEMMA 8.1.12.

(1) *For any algebraic realization $\mathcal{A}$ there exists a Kripkean realization $\mathcal{K}^{\mathcal{A}}$ such that, for any α, $\models_{\mathcal{A}} \alpha$ iff $\models_{\mathcal{K}^{\mathcal{A}}} \alpha$.*
(2) *For any Kripkean realization $\mathcal{K}$ there exists an algebraic realization $\mathcal{A}^{\mathcal{K}}$ such that, for any α, $\models_{\mathcal{K}} \alpha$ iff $\models_{\mathcal{A}^{\mathcal{K}}} \alpha$.*

Sketch of the proof.
(1) The basic intuitive idea of the proof is the following: any algebraic realization can be canonically transformed into a Kripkean realization by identifying the set of worlds with the set of all nonzero elements of the

algebra, the accessibility relation with the nonorthogonality relation in the algebra, and finally the set of propositions with the set of all *principal quasi-ideals* (i.e., all $X - \{\mathbf{0}\}$ where X is a principal ideal). More precisely, given $\mathcal{A} = \langle \mathcal{B}, v \rangle$, the Kripkean realization $\mathcal{K}^{\mathcal{A}} = \langle I, \mathcal{R}, Pr, V \rangle$ is defined as follows:

$$\begin{aligned} &I = \{b \in B : b \neq \mathbf{0}\}; \\ &i\mathcal{R}j \text{ iff } i \not\leq j'; \\ &Pr = \{\{b \in B : b \neq \mathbf{0} \text{ and } b \leq a\} : a \in B\}; \\ &V(\mathbf{p}) = \{b \in I : b \leq v(\mathbf{p})\}. \end{aligned}$$

One can easily check that $\mathcal{K}^{\mathcal{A}}$ is a Kripkean realization. Furthermore, we will have, for any α: $V(\alpha) = \{b \in B : b \neq \mathbf{0} \text{ and } b \leq v(\alpha)\}$. Consequently, $\models_{\mathcal{A}} \alpha$ iff $\models_{\mathcal{K}^{\mathcal{A}}} \alpha$.

(2) Any Kripkean realization $\mathcal{K} = \langle I, \mathcal{R}, Pr, V \rangle$ can be canonically transformed into an algebraic realization $\mathcal{A}^{\mathcal{K}} = \langle \mathcal{B}, v \rangle$ by putting:

$$\begin{aligned} &B = Pr; \\ &\forall a, b \in B[a \leq b \text{ iff } a \subseteq b]; \\ &a' = \{i \in I : i \perp a\}; \\ &\mathbf{1} = I; \ \ \mathbf{0} = \emptyset; \\ &v(\mathbf{p}) = V(\mathbf{p}). \end{aligned}$$

It turns out that $\mathcal{B}$ is an ortholattice. Furthermore, for any α, $v(\alpha) = V(\alpha)$. Consequently: $\models_{\mathcal{K}} \alpha$ iff $\models_{\mathcal{A}^{\mathcal{K}}} \alpha$. □

THEOREM 8.1.13. $T \overset{A}{\underset{\mathbf{OL}}{\models}} \alpha$ *iff* $T \overset{K}{\underset{\mathbf{OL}}{\models}} \alpha$.

PROOF. In order to prove the left to right implication, suppose by contradiction: $T \overset{\mathrm{A}}{\underset{\mathbf{OL}}{\models}} \alpha$ and $T \overset{\mathrm{K}}{\underset{\mathbf{OL}}{\not\models}} \alpha$. Hence, there exists a Kripkean realization $\mathcal{K} = \langle I, \mathcal{R}, Pr, V \rangle$ and a world i of $\mathcal{K}$ such that $i \models T$ and $i \not\models \alpha$. One can easily see that $\mathcal{K}$ can be transformed into $\mathcal{K}^{\circ} = \langle I, \mathcal{R}, Pr^{\circ}, V \rangle$ where Pr° is the smallest subset of the power set of I, that includes Pr and is closed under arbitrary intersection. Owing to Lemma 8.1.5(iii), $\mathcal{K}^{\circ}$ is a Kripkean realization for **OL** and, for any β, $V(\beta)$ turns out to be the same proposition in $\mathcal{K}$ and in $\mathcal{K}^{\circ}$. Consequently, also in $\mathcal{K}^{\circ}$, we will have: $i \models T$ and $i \not\models \alpha$. Let us now consider $\mathcal{A}^{\mathcal{K}^{\circ}}$. The algebra $\mathcal{B}$ of $\mathcal{A}^{\mathcal{K}^{\circ}}$ is complete, because Π° is closed under arbitrary intersection. Hence, $\bigcap \{V(\beta) : \beta \in T\}$ is an element of B. Since $i \models \beta$ for any $\beta \in T$, we will have: $i \in \bigcap \{V(\beta) : \beta \in T\}$. Thus, there is an element of B, which is less or equal than $v(\beta)(= V(\beta))$ for any $\beta \in T$, but is not less or equal than $v(\alpha)(= V(\alpha))$, because $i \notin V(\alpha)$. This contradicts the hypothesis $T \overset{\mathrm{A}}{\underset{\mathbf{OL}}{\models}} \alpha$.
The right to left implication is trivial. □

Let us now turn to a semantic characterization of **OQL**. As we already know, an orthomodular lattice is an ortholattice $\mathcal{B} = \langle B, \leq, ', \mathbf{0}, \mathbf{1} \rangle$ such that for any $a, b \in B$,

$$a \leq b \ \curvearrowright \ b = a \vee (b \wedge a').$$

Orthomodularity clearly represents a weak form of distributivity.

LEMMA 8.1.14. *Let $\mathcal{B}$ be an ortholattice. The following conditions are equivalent:*

(i) $\mathcal{B}$ *is orthomodular;*
(ii) $a \wedge (a' \vee (a \wedge b)) \leq b$;
(iii) $\forall a, b \in B[a \leq b$ *iff* $a \wedge (a \wedge b)' = \mathbf{0}]$;
(iv) $\forall a, b \in B[a \leq b$ *and* $b \wedge a' = \mathbf{0} \curvearrowright a = b]$;

The property considered in (iii) represents a significant weakening of the Boolean condition:

$$a \leq b \text{ iff } a \wedge b' = \mathbf{0}.$$

DEFINITION 8.1.15. *Algebraic realization for* **OQL**
An *algebraic realization* for **OQL** is an algebraic realization $\mathcal{A} = \langle \mathcal{B}, v \rangle$ for **OL**, where $\mathcal{B}$ is an orthomodular lattice.

The definitions of truth, logical truth and logical consequence in **OQL** are analogous to the corresponding definitions of **OL**.

As happens in the case of **OL**, also **OQL** can be characterized by means of a Kripkean semantics.

DEFINITION 8.1.16. *Kripkean realization for* **OQL**
A *Kripkean realization* for **OQL** is a Kripkean realization $\mathcal{K} = \langle I, \mathcal{R}, Pr, V \rangle$ for **OL**, where the set of propositions Pr satisfies the *orthomodular property*: $X \not\subseteq Y \curvearrowright X \cap (X \cap Y)' \neq \emptyset$.

The definitions of truth, logical truth and logical consequence in **OQL** are analogous to the corresponding definitions of **OL**. Also in the case of **OQL** one can show:

THEOREM 8.1.17. $\models^{A}_{\mathbf{OQL}} \alpha$ *iff* $\models^{K}_{\mathbf{OQL}} \alpha$.

The theorem is an immediate corollary of Theorem 8.1.11 and of the following lemma:

LEMMA 8.1.18.

(1) *If* $\mathcal{A}$ *is orthomodular then* $\mathcal{K}^{\mathcal{A}}$ *is orthomodular;*
(2) *If* $\mathcal{K}$ *is orthomodular then* $\mathcal{A}^{\mathcal{K}}$ *is orthomodular.*

PROOF. (1) We have to prove $X \not\subseteq Y \curvearrowright X \cap (X \cap Y)' \neq \emptyset$ for any propositions X, Y of $\mathcal{K}^{\mathcal{A}}$. Suppose $X \not\subseteq Y$. By definition of proposition in $\mathcal{K}^{\mathcal{A}}$:

$$X = \{b : b \neq \mathbf{0} \text{ and } b \leq x\} \text{ for a given } x;$$
$$Y = \{b : b \neq \mathbf{0} \text{ and } b \leq y\} \text{ for a given } y.$$

Consequently, $x \not\leq y$, and by Lemma 8.1.14: $x \wedge (x \wedge y)' \neq \mathbf{0}$, because $\mathcal{A}$ is orthomodular. Hence, $x \wedge (x \wedge y)'$ is a world in $\mathcal{K}^{\mathcal{A}}$. In order to prove $X \cap (X \cap Y)' \neq \emptyset$, it is sufficient to prove $x \wedge (x \wedge y)' \in X \cap (X \cap Y)'$. Trivially:

$x \wedge (x \wedge y)' \in X$. Furthermore, $x \wedge (x \wedge y)' \in (X \cap Y)'$, because $(x \wedge y)'$ is the generator of the quasi-ideal $(X \cap Y)'$. Consequently, $x \wedge (x \wedge y)' \in X \cap (X \cap Y)'$.

(2) Let $\mathcal{K}$ be orthomodular. Then, for any $X, Y \in Pr$:

$$X \not\subseteq Y \curvearrowright X \cap (X \cap Y)' \neq \emptyset.$$

The converse of this implication is easily verified. Hence, by Lemma 8.1.14, the algebra $\mathcal{B}$ of $\mathcal{A}^{\mathcal{K}}$ is orthomodular. □

As to the concept of logical consequence, the proof we have given for **OL** (Theorem 8.1.13) cannot be automatically extended to the case of **OQL**. The critical point is represented by the transformation of $\mathcal{K}$ into $\mathcal{K}^{\circ}$ whose set of propositions is closed under arbitrary intersections: $\mathcal{K}^{\circ}$ is trivially an **OL**-realization; at the same time, it is not granted that $\mathcal{K}^{\circ}$ preserves the orthomodular property. One can easily prove:

THEOREM 8.1.19. $T \overset{K}{\underset{\mathbf{OQL}}{\models}} \alpha \curvearrowright T \overset{A}{\underset{\mathbf{OQL}}{\models}} \alpha.$

The inverse relation has been proved by Minari (1987):

THEOREM 8.1.20. $T \overset{A}{\underset{\mathbf{OQL}}{\models}} \alpha \curvearrowright T \overset{K}{\underset{\mathbf{OQL}}{\models}} \alpha.$

Are there any significant structural relations between $\mathcal{A}$ and $\mathcal{A}^{\mathcal{K}^{\mathcal{A}}}$ and between $\mathcal{K}$ and $\mathcal{K}^{\mathcal{A}^{\mathcal{K}}}$? The question admits a very strong answer in the case of $\mathcal{A}$ and $\mathcal{A}^{\mathcal{K}^{\mathcal{A}}}$.

THEOREM 8.1.21. $\mathcal{A} = \langle \mathcal{B}, v \rangle$ *and* $\mathcal{A}^{\mathcal{K}^{\mathcal{A}}} = \langle \mathcal{B}^*, v^* \rangle$ *are isomorphic realizations.*

Sketch of the proof. Let us define a map $h : B \mapsto B^*$ such that:

$$h(a) = \{b : b \neq \mathbf{0} \text{ and } b \leq a\} \text{ for any } a \in B.$$

One can easily check that: (1) h is an isomorphism (from $\mathcal{B}$ onto $\mathcal{B}^*$); (2) $v^*(\mathbf{p}) = h(v(\mathbf{p}))$ for any atomic sentence $\mathbf{p}$. □

At the same time, in the case of $\mathcal{K}$ and $\mathcal{K}^{\mathcal{A}^K}$, there is no natural correspondence between I and Pr. As a consequence, one can prove only the weaker relation:

THEOREM 8.1.22.
Given $\mathcal{K} = \langle I, \mathcal{R}, Pr, V \rangle$ *and* $\mathcal{K}^{\mathcal{A}^K} = \langle I^*, \mathcal{R}^*, Pr^*, V^* \rangle$,
the following holds:

$$V^*(\alpha) = \{X \in Pr : X \subseteq V(\alpha)\}, \textit{ for any } \alpha.$$

In the class of all Kripkean realizations for **QL**, the realizations $\mathcal{K}^{\mathcal{A}}$ (which have been obtained by canonical transformation of an algebraic realization $\mathcal{A}$) present some interesting properties that are summarized by the following theorem.

THEOREM 8.1.23. *In any* $\mathcal{K}^{\mathcal{A}} = \langle I\, , \mathcal{R}\, , Pr\, , V\rangle$ *there is a one-to-one correspondence* h *between the set of worlds* I *and the set of propositions* $Pr - \{\emptyset\}$ *such that:*

(1) $i \in h(i)$;
(2) $i \not\perp j$ *iff* $h(i) \not\subseteq h(j)'$;
(3) $\forall X \in Pr$: $i \in X$ *iff* $\forall k \in h(i)(k \in X)$.

Sketch of the proof. Let $h(i)$ be the quasi-ideal generated by i. □

Theorem 8.1.23 suggests the isolation, in the class of all $\mathcal{K}$, of an interesting subclass of Kripkean realizations that we will call *algebraically adequate*.

DEFINITION 8.1.24. *Algebraically adequate*
A Kripkean realization $\mathcal{K}$ is *algebraically adequate* iff it satisfies the conditions of Theorem 8.1.23.

When restricting to the class of all algebraically adequate Kripkean realizations one can prove the following.

THEOREM 8.1.25. $\mathcal{K} = \langle I\, , \mathcal{R}\, , Pr\, , V\rangle$ *and* $\mathcal{K}^{\mathcal{A}^{K}} = \langle I^*\, , \mathcal{R}^*\, , Pr^*\, , V^*\rangle$ *are isomorphic realizations; i.e., there exists a bijective function* h *from* I *onto* I^* *such that:*

(1) $i\mathcal{R}j$ *iff* $h(i)\mathcal{R}^*h(j)$, *for any* $i, j \in I$;
(2) $Pr^* = \{h(X) : X \in Pr\}$, *where* $h(X) := \{h(i) : i \in X\}$;
(3) $V^*(p) = h(V(\mathbf{p}))$, *for any atomic formula* $\mathbf{p}$.

One can easily show that the class of all algebraically adequate Kripkean realizations determines the same concept of logical consequence that is determined by the larger class of all possible realizations.

8.2. Algebraic and Kripkean realizations of Hilbert event-state systems

Both the algebraic and the Kripkean realizations of **OQL** admit of a quite natural interpretation in the framework of the Hilbert event-state systems (see Section 3.3).

Consider a quantum system $\mathfrak{S}$ with associated Hilbert space $\mathcal{H}$. Let $(\Pi(\mathcal{H}), \mathcal{S}(\mathcal{H}))$ be the (sharp) event-state system based on $\mathcal{H}$. As we already know, $\Pi(\mathcal{H})$ (the set of all projections of $\mathcal{H}$) represents the set of all possible sharp events that may occur to system $\mathfrak{S}$, while $\mathcal{S}(\mathcal{H})$ (the set of all probability measures m_ρ determined by a density operator ρ of $\mathcal{H}$) represents the set of all pure and mixed states of $\mathfrak{S}$. Consider now a sentential language $\mathcal{L}^{\mathfrak{S}}$ for $\mathfrak{S}$, whose intended interpretation refers to possible questions of the event-state system $(\Pi(\mathcal{H}), \mathcal{S}(\mathcal{H}))$. In other words, every atomic sentence $\mathbf{p}$ represents a question (M, Δ) asserting that the value of the observable M lies in the Borel set Δ. Consider now the event $M(\Delta)$ associated to our question (M, Δ) and the set $\mathbf{Yes}(M(\Delta))$, consisting of all the states that assign probability-value 1 to the event $M(\Delta)$. As we already know, $\mathbf{Yes}(M(\Delta))$

is a closed subset of $\mathcal{S}(\mathcal{H})$. On this basis, we can define both the algebraic and the Kripkean realizations for $\mathfrak{S}$.

The algebraic realization for $\mathfrak{S}$

Consider the following algebraic realization

$$\mathcal{A}^{\mathfrak{S}} = \langle \mathcal{B}, v \rangle, \text{ where :}$$

- $\mathcal{B}$ is the orthomodular lattice $\langle \Pi(\mathcal{H}), \leq, ', \mathbf{0}, \mathbf{1} \rangle$, based on the set of all projections of the Hilbert space $\mathcal{H}$ associated to $\mathfrak{S}$;
- for any atomic sentence $\mathbf{p}$,

  $$v(\mathbf{p}) = M(\Delta),$$

  where $M(\Delta)$ is the projection corresponding to the question (M, Δ), expressed by $\mathbf{p}$.
  In other words, the *meaning* of an atomic sentence is the projection that corresponds to the question represented by our sentence.

By definition, $\mathcal{A}^{\mathfrak{S}}$ is an algebraic realization.

The Kripkean realization for $\mathfrak{S}$

Consider the following Kripkean realization

$$\mathcal{K}^{\mathfrak{S}} = \langle I, \mathcal{R}, Pr, V \rangle, \text{ where:}$$

- I is the set $\mathcal{S}(\mathcal{H})$ of the states of $\mathfrak{S}$;
- $\mathcal{R}$ is the similarity relation that is defined on $\mathcal{S}(\mathcal{H})$. In other words:

  $$i\mathcal{R}j \text{ iff not } \exists E \in \Pi(\mathcal{H})[i(E) = 1 \text{ and } j(E) = 0];$$

- $Pr = \mathbb{C}(\mathcal{P}(\mathcal{S}))$ (= the set of all closed subsets of $\mathcal{S}(\mathcal{H})$);
- for any atomic sentence $\mathbf{p}$,

  $$V(\mathbf{p}) = \mathbf{Yes}(M(\Delta)),$$

  where $M(\Delta)$ is the projection corresponding to the question (M, Δ), expressed by $\mathbf{p}$.

One immediately realizes that $\mathcal{K}^{\mathfrak{S}}$ is a Kripkean realization. For:

- $\mathcal{R}$ is a similarity relation (reflexive and symmetric);
- Pr is a set of propositions, because every element X of $\mathbb{C}(\mathcal{P}(\mathcal{S}))$ is a closed set such that $X = X^{\perp\perp}$.
  Furthermore, Pr contains $\emptyset$ and I, and is closed under the operations $^{\perp}$ and $\cap$;
- for any $\mathbf{p}$, $V(\mathbf{p}) \in Pr$.

Interestingly enough, the accessibility relation turns out to have the following physical meaning: $i\mathcal{R}j$ iff j is a state into which i can be transformed after the performance of a physical measurement that concerns an observable of the system (by application of von Neumann-Lüders axiom).

By the isomorphism between the lattice of all closed subsets of $\mathcal{S}(\mathcal{H})$ and the lattice of all projections, one can also say that the *meaning* of any sentence α of the language concerning $\mathfrak{S}$ corresponds to a sharp event.

As expected, such a perfect symmetry between propositions and events will break down in the case of unsharp quantum logics.

8.3. The implication problem in quantum logic

Unlike most weak logics, **QL** gives rise to a critical "implication-problem." All conditional connectives that can be reasonably introduced in **QL** are, to a certain extent, anomalous; for, they do not share most of the characteristic properties that are satisfied by the *positive conditionals* (which are governed by a logic that is at least as strong as *positive logic*).[1] In the past, simply the failure of a positive-like conditional sometimes led to the conclusion that **QL** cannot be a "real" logic. In spite of these difficulties, these days one cannot help recognizing that **QL** admits a set of different implicational connectives, even if none of them has a *positive* behavior. Let us first propose a general semantic condition for a logical connective to be classified as an implication-connective.

DEFINITION 8.3.1. *Implication-connective*
In any semantics, a binary connective $\overset{*}{\rightarrow}$ is called an *implication-connective*[2] iff it satisfies at least the two following conditions:

(1) $\alpha \overset{*}{\rightarrow} \alpha$ is always true (*identity*);
(2) if α is true and $\alpha \overset{*}{\rightarrow} \beta$ is true then β is true (*modus ponens*).

In the particular case of **QL**, one can easily obtain:

LEMMA 8.3.2. *A sufficient condition for a connective $\overset{*}{\rightarrow}$ to be an implication-connective is:*

(i) *in the algebraic semantics: for any realization $\mathcal{A} = \langle \mathcal{A}, v \rangle$,*
$\models_{\mathcal{A}} \alpha \overset{*}{\rightarrow} \beta$ *iff* $v(\alpha) \leq v(\beta)$;
(ii) *in the Kripkean semantics: for any realization $\mathcal{K} = \langle I, \mathcal{R}, Pr, V \rangle$,*
$\models_{\mathcal{K}} \alpha \overset{*}{\rightarrow} \beta$ *iff* $V(\alpha) \subseteq V(\beta)$.

In **QL** it seems reasonable to assume the sufficient condition of Lemma 8.3.2 as a minimal condition for a connective to be an implication-connective.

Suppose we have defined two different implication-connectives in the algebraic and in the Kripkean semantics. When shall we admit that they represent the "same logical connective"? A reasonable answer to this question is represented by the following convention:

---

[1] *Positive logic* can be characterized as the *positive fragment* of intuitionistic logic: the sublogic of **IL** that only concerns the *positive connectives* (conjunction, disjunction, conditional).

[2] Implication-connectives are also called *conditionals* or *material conditionals*.

DEFINITION 8.3.3. Let $\overset{\mathrm{A}}{*}$ be a binary connective defined in the algebraic semantics and $\overset{\mathrm{K}}{*}$ a binary connective defined in the Kripkean semantics; $\overset{\mathrm{A}}{*}$ and $\overset{\mathrm{K}}{*}$ represent the *same logical connective* iff the following conditions are satisfied:

(1) given any $\mathcal{A} = \langle \mathcal{B}, v \rangle$ and given the corresponding $\mathcal{K}^{\mathcal{A}} = \langle I\,, \mathcal{R}\,, Pr\,, V \rangle$, $V(\alpha \overset{\mathrm{K}}{*} \beta)$ is the quasi-ideal generated by $v(\alpha \overset{\mathrm{A}}{*} \beta)$;

(2) given any $\mathcal{K} = \langle I\,, \mathcal{R}\,, Pr\,, V \rangle$ and given the corresponding $\mathcal{A}^{\mathcal{K}} = \langle \mathcal{B}\,, v \rangle$, $v(\alpha \overset{\mathrm{A}}{*} \beta) = V(\alpha \overset{\mathrm{K}}{*} \beta)$.

We will now consider different possible semantic characterizations of an implication-connective in **QL**. Unlike classical logic, a material conditional defined by the so called *Philo-law*, $\alpha \to \beta := \neg\alpha \curlyvee \beta$, does not give rise to an implication-connective **QL**. For, there are algebraic realizations $\mathcal{A} = \langle \mathcal{B}, v \rangle$ such that $v(\neg\alpha \curlyvee \beta) = \mathbf{1}$, while $v(\alpha) \not\leq v(\beta)$. Furthermore, unlike the case of Boolean algebras and of Heyting algebras, ortholattices and orthomodular lattices are not, generally, *pseudocomplemented* lattices: in other words, given $a, b \in B$, the maximum c such that $a \wedge c \leq b$ does not necessarily exist in B. In fact, one can prove that any pseudocomplemented lattice is distributive.[3]

8.4. Five polynomial conditionals

We will first consider the case of *polynomial conditionals*, that can be defined in terms of the connectives $\curlywedge$, $\curlyvee$ and $\neg$. In the algebraic semantics, the minimal requirement of Lemma 8.3.2 restricts the choice only to five possible candidates.[4] This result follows from the fact that in the orthomodular lattice freely generated by two elements there are only five polynomial binary operations $\circ$ satisfying the condition $a \leq b$ iff $a \circ b = \mathbf{1}$. These are our five candidates:

(i) $v(\alpha \to_1 \beta) = v(\alpha)' \vee (v(\alpha) \wedge v(\beta))$.

(ii) $v(\alpha \to_2 \beta) = v(\beta) \vee (v(\alpha)' \wedge v(\beta)')$.

(iii) $v(\alpha \to_3 \beta) = (v(\alpha)' \wedge v(\beta)) \vee (v(\alpha) \wedge v(\beta)) \vee (v(\alpha)' \wedge v(\beta)')$.

(iv) $v(\alpha \to_4 \beta) = (v(\alpha)' \wedge v(\beta)) \vee (v(\alpha) \wedge v(\beta)) \vee ((v(\alpha)' \vee v(\beta)) \wedge v(\beta)')$.

(v) $v(\alpha \to_5 \beta) = (v(\alpha)' \wedge v(\beta)) \vee (v(\alpha)' \wedge v(\beta)') \vee (v(\alpha) \wedge (v(\alpha)' \vee v(\beta)))$.

The corresponding five implication-connectives in the Kripkean semantics can be easily obtained.

THEOREM 8.4.1. *The polynomial conditionals* $\to_i$ $(1 \leq i \leq 5)$ *are implication-connectives in* **OQL**; *but they are not implication-connectives in* **OL**.

---

[3]See, for instance, (Birkhoff, 1967).

[4]See (Kalmbach, 1983).

PROOF. Since $\to_i$ represent the same connective in both semantics, it will be sufficient to refer to the algebraic semantics.

As an example, let us prove the theorem for $i = 1$ (the other cases are similar). First, we have to prove $v(\alpha) \leq v(\beta)$ iff $\mathbf{1} = v(\alpha \to_1 \beta) = v(\alpha)' \vee (v(\alpha) \wedge v(\beta))$, which is equivalent to $v(\alpha) \leq v(\beta)$ iff $v(\alpha) \wedge (v(\alpha) \wedge v(\beta))' = \mathbf{0}$. From Lemma 8.1.14, we know that the latter condition holds for any pair of elements of B iff $\mathcal{B}$ is orthomodular. This proves, in one stroke, that $\to_1$ is an implication-connective in **OQL**, but cannot be an implication-connective in **OL**.

□

Interestingly enough, each polynomial conditional $\to_i$ represents a good weakening of the classical material conditional.

In Section 1.1 we have defined the *compatibility relation* that may hold between two elements of an orthomodular lattice.

Two elements a, b of an orthomodular lattice $\mathcal{B}$ are *compatible* iff

$$a = (a \wedge b') \vee (a \wedge b).$$

From an intuitive point of view, such a relation represents a kind of "Boolean mutual behavior." For, two elements a, b of an orthomodular lattice are compatible iff the subalgebra of $\mathcal{B}$ generated by $\{a, b\}$ is Boolean.

THEOREM 8.4.2. *For any algebraic realization $\mathcal{A} = \langle \mathcal{B}, v \rangle$, for any i $(1 \leq i \leq 5)$ and for any α, β:*

$$v(\alpha \to_i \beta) = v(\alpha)' \vee v(\beta) \text{ iff } v(\alpha) \text{ and } v(\beta) \text{ are compatible.}$$

Boolean algebras are pseudocomplemented lattices. Therefore they satisfy the following condition for any a, b, c:

$$c \wedge a \leq b \text{ iff } c \leq a \rightsquigarrow b,$$

where: $a \rightsquigarrow b := a' \vee b$.

An orthomodular lattice $\mathcal{B}$ is a Boolean algebra iff for any algebraic realization $\mathcal{A} = \langle \mathcal{B}, v \rangle$, any i $(1 \leq i \leq 5)$ and any α, β the following *import-export* condition is satisfied:

$$v(\gamma) \wedge v(\alpha) \leq v(\beta) \text{ iff } v(\gamma) \leq v(\alpha \to_i \beta).$$

In order to single out a unique polynomial conditional, various weakenings of the import-export condition have been proposed. For instance the following condition (which we will call *weak import-export*):

$$v(\gamma) \wedge v(\alpha) \leq v(\beta) \text{ iff } v(\gamma) \leq v(\alpha) \to_i v(\beta),$$

provided that $v(\alpha)$ and $v(\beta)$ are compatible.

One can prove (Hardegree, 1975; Mittelstaedt, 1972) that a polynomial conditional $\to_i$ satisfies the weak import-export condition iff $i = 1$. As a consequence, we can conclude that $\to_1$ represents, in a sense, the best possible approximation for a material conditional in quantum logic. This connective (often called *Sasaki hook*) was originally proposed by Mittelstaedt (1972) and Finch (1970), and was further investigated by Hardegree (1976) and other authors. In the following, we will usually write $\rightarrow$ instead of $\rightarrow_1$ and we will ignore the other four polynomial conditionals.

Some important *positive logical truths* that are violated by our quantum logical conditional are the following:[5]

$$\alpha \rightarrow (\beta \rightarrow \alpha);$$
$$(\alpha \rightarrow (\beta \rightarrow \gamma)) \rightarrow ((\alpha \rightarrow \beta) \rightarrow (\alpha \rightarrow \gamma));$$
$$(\alpha \rightarrow \beta) \rightarrow ((\beta \rightarrow \gamma) \rightarrow (\alpha \rightarrow \gamma));$$
$$(\alpha \curlywedge \beta \rightarrow \gamma) \rightarrow (\alpha \rightarrow (\beta \rightarrow \gamma));$$
$$(\alpha \rightarrow (\beta \rightarrow \gamma)) \rightarrow (\beta \rightarrow (\alpha \rightarrow \gamma)).$$

8.5. The quantum logical conditional as a counterfactual conditional

The somewhat anomalous behavior of the quantum logical conditional $\rightarrow$ has suggested that one is dealing with a kind of *counterfactual conditional* (Hardegree, 1975). Such a conjecture seems to be confirmed by some important physical examples. Let us consider again the Kripkean realizations that are associated to a quantum system $\mathfrak{S}$ (Section 8.2). For the sake of simplicity, we restrict our attention to the case of pure states. As a consequence, the Kripkean realizations we shall consider will have the following form:

$$\mathcal{K}^{\mathfrak{S}} = \langle I\,, \mathcal{R}\,, Pr\,, V\rangle\,, \text{ where :}$$

- I is the set of all pure states of $\mathfrak{S}$;
- $\mathcal{R}$ is the nonorthogonality relation defined on I;
- Pr is the set of all *pure propositions*[6] of the event-state system $(\Pi(\mathcal{H}), \mathcal{S}(\mathcal{H}))$. In other words: $Z \in Pr$ iff Z is a closed set of pure states (i.e., such that $Z = Z^{\perp\perp}$);
- $V(\mathbf{p})$ is the pure proposition consisting of all pure states that assign probability-value 1 to the question expressed by $\mathbf{p}$.

As Hardegree has shown, in such a case the conditional $\rightarrow$ turns out to receive a quite natural counterfactual interpretation (in the sense of Stalnaker (1981)). More precisely, one can define, for any sentence α of the language $\mathcal{L}^{\mathfrak{S}}$, a partial *Stalnaker-function* f_α in the following way:

$$f_\alpha : \mathrm{Dom}(f_\alpha) \mapsto I,$$

where:

$$\mathrm{Dom}(f_\alpha) := \{i \in I \,:\, i \not\perp V(\alpha)\}\,.$$

In other words, f_α is defined exactly for all the pure states that are not orthogonal to the proposition of α.

If $i \in \mathrm{Dom}(f_\alpha)$, then:

$$f_\alpha(i) := P^{V(\alpha)}i,$$

[5]We follow the standard bracket-convention according to which $\curlywedge$ and $\curlyvee$ are more binding than $\rightarrow$.

[6]The notion of *pure proposition* has been introduced in Section 4.4.

where $P^{V(\alpha)}$ is the projection that is uniquely associated with the *pure proposition* $V(\alpha)$. The following condition holds:

$$i \models \alpha \to \beta \text{ iff either } \forall j \not\perp i (j \not\models \alpha) \text{ or } f_\alpha(i) \models \beta.$$

From an intuitive point of view, one can say that $f_\alpha(i)$ represents the "pure state nearest" to i, that verifies α, where "nearest" is here defined in terms of the metric of the Hilbert space $\mathcal{H}$. By definition and in virtue of von Neumann- Lüders axiom (*the collapse of the wave-function*), $f_\alpha(i)$ turns out to have the following physical meaning: it represents the transformation of state i after the performance of a measurement concerning the physical event expressed by α, provided the result was positive. As a consequence, one obtains: $\alpha \to \beta$ is true in a state i iff either α is impossible for i or the state into which i has been transformed after a positive α-test, verifies β.

Another interesting characteristic of our connective $\to$, is a *weak non-monotonic* behavior. In fact, in the algebraic semantics the inequality

$$v(\alpha \to \gamma) \leq v(\alpha \curlywedge \beta \to \gamma)$$

can be violated (a counterexample can be easily obtained in the orthomodular lattice based on all subspaces of $\mathbb{R}^3$). As a consequence:

$$\alpha \to \gamma \not\models \alpha \curlywedge \beta \to \gamma.$$

8.6. Implication-connectives

Of course, polynomial conditionals are not the only significant examples of implication-connectives in **QL**. In the framework of a Kripkean semantic approach, it seems quite natural to introduce a conditional connective, that represents a kind of *strict implication*. Given a Kripkean realization $\mathcal{K} = \langle I , \mathcal{R} , Pr , V \rangle$ one would like to require:

$$i \models \alpha \multimap \beta \text{ iff } \forall j \not\perp i \, (j \models \alpha \curvearrowright j \models \beta).$$

However such a condition does not automatically represent a correct semantic definition, because it is not granted that $V(\alpha \multimap \beta)$ is an element of Pr. In order to overcome this difficulty, let us first define a new operation in the power set of an orthoframe $\langle I , \mathcal{R} \rangle$.

DEFINITION 8.6.1. *Strict-implication operation* ($\boxed{\multimap}$)
Given an orthoframe $\langle I , \mathcal{R} \rangle$ and $X, Y \subseteq I$:

$$X \boxed{\multimap} Y := \{ i : \forall j \, (i \not\perp j \text{ and } j \in X \curvearrowright j \in Y) \}.$$

If X and Y are sets of worlds in the orthoframe, then $X \boxed{\multimap} Y$ turns out to be a proposition of the frame.

When the set Pr of $\mathcal{K}$ is closed under $\boxed{\multimap}$, we will say that $\mathcal{K}$ is a *realization for a strict-implication language*.

DEFINITION 8.6.2. *Strict implication* ($\multimap$)
If $\mathcal{K} = \langle I , \mathcal{R} , Pr , V \rangle$ is a realization for a strict-implication language, then

$$V(\alpha \multimap \beta) := V(\alpha) \boxed{\multimap} V(\beta).$$

One can easily check that $\multimap$ is an implication-connective, that behaves as a strict implication. For, we have:

$$i \models \alpha \multimap \beta \text{ iff } \forall j \not\perp i \, (j \models \alpha \curvearrowright j \models \beta).$$

Another interesting implication that can be defined in **QL** is represented by an *entailment-connective*.

DEFINITION 8.6.3. *Entailment* ($\twoheadrightarrow$)
Given $\mathcal{K} = \langle I\,, \mathcal{R}\,, Pr\,, V\rangle$,

$$V(\alpha \twoheadrightarrow \beta) := \begin{cases} I & \text{if } V(\alpha) \subseteq V(\beta), \\ \emptyset & \text{otherwise.} \end{cases}$$

Since $I, \emptyset \in Pr$, V is well defined. One can trivially check that $\twoheadrightarrow$ is an implication-connective. Interestingly enough, our strict implication and our entailment represent implication-connectives also for **OL**.

The general relations between $\rightarrow$, $\multimap$ and $\twoheadrightarrow$ are described by the following theorem:

THEOREM 8.6.4.

(i) *For any realization $\mathcal{K}$ for a strict-implication language of* **OL**:

$$\models_{\mathcal{K}} (\alpha \twoheadrightarrow \beta) \twoheadrightarrow (\alpha \multimap \beta).$$

(ii) *For any realization $\mathcal{K}$ for a strict-implication language of* **OQL**:

$$\models_{\mathcal{K}} (\alpha \twoheadrightarrow \beta) \twoheadrightarrow (\alpha \rightarrow \beta); \quad \models_{\mathcal{K}} (\alpha \multimap \beta) \twoheadrightarrow (\alpha \rightarrow \beta).$$

But the inverse relations do not generally hold!

Are the connectives $\multimap$ and $\twoheadrightarrow$ definable also in the algebraic semantics? The possibility of defining $\twoheadrightarrow$ is straightforward.

DEFINITION 8.6.5. *Entailment in the algebraic semantics*
Given $\mathcal{A} = \langle \mathcal{B}, v\rangle$,

$$v(\alpha \twoheadrightarrow \beta) := \begin{cases} \mathbf{1} & \text{if } v(\alpha) \leq v(\beta), \\ \mathbf{0} & \text{otherwise.} \end{cases}$$

One can easily check that $\twoheadrightarrow$ represents the same connective in the two semantics. As for $\multimap$, given $\mathcal{A} = \langle \mathcal{B}, v\rangle$, one would like to require:

$$v(\alpha \multimap \beta) =$$

$$\bigvee \left\{ b \in B \,:\, b \neq \mathbf{0} \text{ and } \forall c [c \neq \mathbf{0} \text{ and } b \not\leq c' \text{ and } c \leq v(\alpha) \curvearrowright c \leq v(\beta)] \right\}.$$

However such a definition presupposes the algebraic completeness of $\mathcal{B}$. Furthermore, we can prove that $\multimap$ represents the same connective in the two semantics only if we restrict our consideration to the class of all algebraically adequate Kripkean realizations.

CHAPTER 9

Metalogical properties and anomalies of quantum logic

9.1. The failure of the Lindenbaum property

Some metalogical distinctions that are not interesting in the case of a number of familiar logics weaker than classical logic turn out to be significant for **QL** (and for nondistributive logics in general).

We have already defined (both in the algebraic and in the Kripkean semantics) the concepts of *model* and of *logical consequence*. Now we will introduce, in both semantics, the notions of *quasi-model*, *weak consequence* and *quasi-consequence*. Let T be any set of sentences.

DEFINITION 9.1.1. *Quasi-model*

Algebraic semantics	*Kripkean semantics*
A realization $\mathcal{A} = \langle \mathcal{B}, v \rangle$ is a *quasi-model* of T iff $\exists a[a \in B$ and $a \neq \mathbf{0}$ and $\forall \beta \in T(a \leq v(\beta))]$.	A realization $\mathcal{K} = \langle I, \mathcal{R}, Pr, V \rangle$ is a *quasi-model* of T iff $\exists i[i \in I$ and $i \models T]$.

The following definitions can be expressed in both semantics.

DEFINITION 9.1.2. *Realizability* and *verifiability*
T is *realizable* (Real T) iff it has a quasi-model; T is *verifiable* (Verif T) iff it has a model.

DEFINITION 9.1.3. *Weak consequence*
A sentence α is a *weak consequence* of T ($T \models\!\!\!\equiv \alpha$) iff any model of T is a model of α.

DEFINITION 9.1.4. *Quasi-consequence.*
A sentence α is a *quasi-consequence* of T ($T \mid\!\approx \alpha$) iff any quasi-model of T is a quasi-model of α.

One can easily check that the algebraic notions of verifiability, realizability, weak consequence and quasi-consequence turn out to coincide with the corresponding Kripkean notions. In other words, T is Kripke-realizable iff T is algebraically realizable. Similarly for the other concepts.

In both semantics one can trivially prove the following lemmas.

LEMMA 9.1.5. Verif $T \curvearrowright$ Real T.

LEMMA 9.1.6. Real T *iff* $T \not\mid\!\approx \beta \curlywedge \neg\beta$ *for any contradiction* $\beta \curlywedge \neg\beta$.

Hence, one can also say that T is realizable iff T is semantically noncontradictory.

LEMMA 9.1.7. $T \models \alpha \curvearrowright T \models\!\!\!\equiv \alpha$; $T \models \alpha \curvearrowright T \models_{\approx} \alpha$.

LEMMA 9.1.8. $\alpha \models\!\!\!\equiv \beta$ *iff* $\neg\beta \models_{\approx} \neg\alpha$.

DEFINITION 9.1.9. We say that a set of sentences T is *complete*, written $\text{Compl}\, T$, in case $\forall\alpha\, [\alpha \in T \text{ or } \neg\alpha \in T]$.

Most familiar logics, that are stronger than positive logic, turn out to satisfy the following metalogical properties, that we will call *Herbrand-Tarski*, *verifiability* and *Lindenbaum*, respectively.

- *Herbrand-Tarski*

  $T \models \alpha$ iff $T \models\!\!\!\equiv \alpha$ iff $T \models_{\approx} \alpha$
- *Verifiability*

  $\text{Verif}\, T$ iff $\text{Real}\, T$
- *Lindenbaum*

  $\text{Real}\, T \curvearrowright \exists T^* [T \subseteq T^* \text{ and } \text{Real}\, T^* \text{ and } \text{Compl}\, T^*]$.

The Herbrand-Tarski property represents a semantic version of the deduction theorem. The Lindenbaum property asserts that any semantically noncontradictory set of sentences admits a semantically noncontradictory complete extension. In the algebraic semantics, canonical proofs of these properties essentially use some versions of Stone's Theorem, according to which any *proper filter* F in an algebra $\mathcal{B}$ can be extended to a *proper complete* filter F^* (i.e., satisfying $\forall a[a \in F^* \text{ or } a' \in F^*]$).[1] However, Stone's Theorem does not generally hold for nondistributive orthomodular lattices! In the case of ortholattices, one can still prove that every proper filter can be extended to a *maximal filter* (one that does not admit any extension that is a proper filter). However, unlike Boolean algebras, maximal filters need not be complete.

A counterexample to the Herbrand-Tarski property in **OL** can be obtained using the "non-valid" part of the distributive law. We know that (owing to the failure of distributivity in ortholattices):

$$\alpha \curlywedge (\beta \curlyvee \gamma) \not\models (\alpha \curlywedge \beta) \curlyvee (\alpha \curlywedge \gamma).$$

At the same time

$$\alpha \curlywedge (\beta \curlyvee \gamma) \models\!\!\!\equiv (\alpha \curlywedge \beta) \curlyvee (\alpha \curlywedge \gamma),$$

since one can easily calculate that for any realization $\mathcal{A} = \langle \mathcal{B}, v \rangle$ the hypothesis $v(\alpha \curlywedge (\beta \curlyvee \gamma)) = \mathbf{1}$, $v((\alpha \curlywedge \beta) \curlyvee (\alpha \curlywedge \gamma)) \neq \mathbf{1}$ leads to a contradiction.

In **OQL** a counterexample in two variables can be obtained by using the failure of the contraposition law for $\rightarrow$. One has: $\alpha \rightarrow \beta \not\models \neg\beta \rightarrow \neg\alpha$. At the same time $\alpha \rightarrow \beta \models\!\!\!\equiv \neg\beta \rightarrow \neg\alpha$; since for any realization $\mathcal{A} = \langle \mathcal{B}, v \rangle$ the

[1] See Definitions 1.1.19 and 1.1.21. For Stone's Theorem see, for instance (Bell and Slomson, 1969).

hypothesis $v(\alpha \to \beta) = \mathbf{1}$ implies $v(\alpha) \leq v(\beta)$ by Lemma 8.1.14(iii), and therefore $v(\neg\beta \to \neg\alpha) = v(\beta) \vee (v(\alpha)' \wedge v(\beta)') = v(\beta) \vee v(\beta)' = \mathbf{1}$.

A counterexample to the verifiability-property is given by the negation of the *a fortiori* principle for the quantum logical conditional $\to$:

$$\gamma := \neg(\alpha \to (\beta \to \alpha)) = \neg(\neg\alpha \curlyvee (\alpha \curlywedge (\neg\beta \curlyvee (\alpha \curlywedge \beta)))).$$

This γ has an algebraic quasi-model. For instance the realization $\mathcal{A} = \langle \mathcal{B}, v \rangle$, where $\mathcal{B}$ is the orthomodular lattice determined by all subspaces of the plane (as shown in Figure 9.1.1). The following condition holds: $v(\gamma) = v(\alpha) \neq \mathbf{0}$. But one can easily check that γ cannot have any model, since the hypothesis that $v(\gamma) = \mathbf{1}$ leads to a contradiction in any algebraic realization of **QL**.

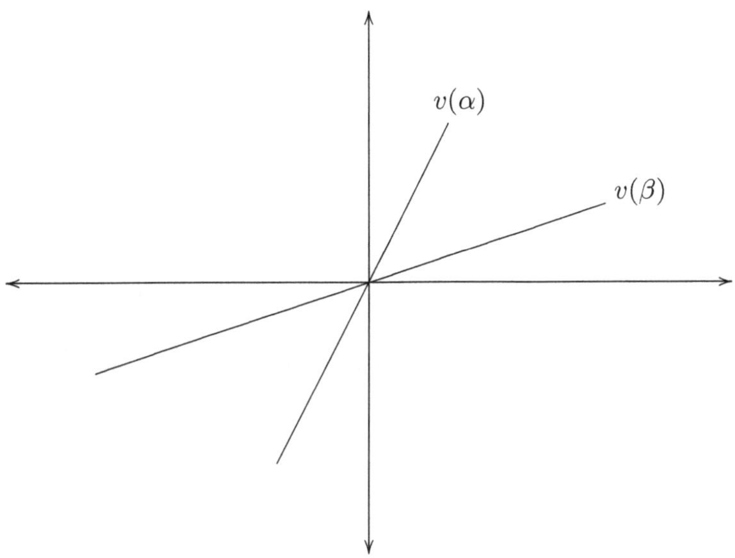

FIGURE 9.1.1. Quasi-model for γ in $\mathbb{R}^2$

The same γ also represents a counterexample to the Lindenbaum-property. Let us first prove the following lemma.

LEMMA 9.1.10. *If T is realizable and $T \subseteq T^*$, where T^* is realizable and complete, then T is verifiable.*

Sketch of the proof. Let us define a realization $\mathcal{A} = \langle \mathcal{B}, v \rangle$ such that

(i) $B = \{\mathbf{0}, \mathbf{1}\}$;

(ii) $v(\alpha) = \begin{cases} \mathbf{1} & \text{if } T^* \models \alpha, \\ \mathbf{0} & \text{otherwise.} \end{cases}$

Since T^* is realizable and complete, $\mathcal{A}$ is a realization that is trivially a model of T. □

Now, one can easily show that γ violates Lindenbaum. Suppose, by contradiction, that γ has a realizable and complete extension. Then, by Lemma 9.1.10, γ must have a model, and which we have shown to be impossible.

The failure of the metalogical properties we have considered represents, in a sense, a relevant "anomaly" of quantum logics. Just these anomalies suggest the following observation: the distinction between *epistemic logics* (characterized by Kripkean models where the accessibility relation is at least reflexive and transitive) and *similarity logics* (characterized by Kripkean models where the accessibility relation is at least reflexive and symmetric) seems to represent a possibly significant dividing line in the class of all logics that are weaker than classical logic.

9.2. A modal interpretation of sharp quantum logics

QL admits a modal interpretation, which is formally very similar to the modal interpretation of intuitionistic logic.[2] Any modal interpretation of a given nonclassical logic is quite interesting from an intuitive point of view, since it permits us to associate a classical meaning to a given system of nonclassical logical constants. As is well known, intuitionistic logic can be translated into the modal system **S4**.[3] The modal basis that turns out to be adequate for **OL** is instead the logic **B**, given below. Such a result is perhaps not surprising, since both the **B**-realizations and the **OL**-realizations are characterized by frames where the accessibility relation is reflexive and symmetric.

Suppose we have a modal language $\mathcal{L}^{\mathbf{B}}$ whose alphabet contains the same atomic sentences as **QL** and the following primitive logical constants: the classical connectives $\neg_c$ (*not*), $\curlywedge_c$ (*and*) and the modal operator L (*necessarily*). At the same time, the connectives $\curlyvee_c$ (*or*), $\rightarrow_c$ (*if ... then*), $\leftrightarrow_c$ (*if and only if*), and the modal operator M (*possibly*) are supposed defined from these in the standard way.

The modal logic **B** is semantically characterized by a class of Kripkean realizations that we will call **B**-*realizations.*

DEFINITION 9.2.1. **B**-*realization*
A **B**-*realization* is a system $\mathcal{M} = \langle I\,,\,\mathcal{R}\,,\,Pr\,,\,V\rangle$ where:

(i) $\langle I\,,\,\mathcal{R}\rangle$ is an orthoframe;
(ii) Pr is a subset of the power set of I satisfying the following conditions:
  (iia) $I, \emptyset \in Pr$;
  (iib) Pr is closed under the set-theoretic relative complement ^{c}, the set-theoretic intersection $\cap$ and the modal operation $\Box$, which is defined as follows:
  for any $X \subseteq I$, $\Box X := \{i : \forall j\,[i\mathcal{R}j \curvearrowright j \in X]\}$;

[2]See (Goldblatt, 1974; Dalla Chiara, 1981).
[3]See, for instance, (Fitting, 1969).

(iii) V associates to any sentence α of $L^{\mathbf{B}}$ a proposition in Pr satisfying the conditions: $V(\neg_c \beta) = V(\beta)^c$; $V(\beta \curlywedge \gamma) = V(\beta) \cap V(\gamma)$; $V(L\beta) = \Box V(\beta)$.

Instead of $i \in V(\alpha)$, we will write $i \models \alpha$. The definitions of truth, logical truth, and logical consequence for **B** are analogous to the corresponding definitions in the Kripkean semantics for **QL**.

Let us now define a translation τ of the language of **QL** into the language $\mathcal{L}^{\mathbf{B}}$.

DEFINITION 9.2.2. *Modal translation of* **OL**

- $\tau(\mathbf{p}) = LM\mathbf{p}$;
- $\tau(\neg\beta) = L\neg_c\tau(\beta)$;
- $\tau(\beta \curlywedge \gamma) = \tau(\beta) \curlywedge_c \tau(\gamma)$.

In other words, τ translates any atomic sentence as the necessity of the possibility of the same sentence; furthermore, the quantum logical negation is interpreted as the necessity of the classical negation, while the quantum logical conjunction is interpreted as the classical conjunction. We will indicate the set $\{\tau(\beta) : \beta \in T\}$ by $\tau(T)$.

THEOREM 9.2.3. *For any α and T of* **OL**: $T \models_{\mathbf{OL}} \alpha$ *iff* $\tau(T) \models_{\mathbf{B}} \tau(\alpha)$.

Theorem 9.2.3 is an immediate corollary of the following Lemmas 9.2.4 and 9.2.5.

LEMMA 9.2.4. *Any* **OL**-*realization* $\mathcal{K} = \langle I, \mathcal{R}, Pr, V\rangle$ *can be transformed into a* **B**-*realization* $\mathcal{M}^{\mathcal{K}} = \langle I^*, \mathcal{R}^*, Pr^*, V^*\rangle$ *such that:*

$$I^* = I;\ \mathcal{R}^* = \mathcal{R};$$

$$\forall i\,[i \models_{\mathcal{K}} \alpha \text{ iff } i \models_{\mathcal{M}^{\mathcal{K}}} \tau(\alpha)].$$

Sketch of the proof. Take Pr^* as the smallest subset of the power set of I that contains $V(\mathbf{p})$ for any atomic sentence $\mathbf{p}$ and that is closed under $I, \emptyset, {}^c, \cap$ and $\Box$. Also, take $V^*(\mathbf{p})$ equal to $V(\mathbf{p})$. □

LEMMA 9.2.5. *Any* **B**-*realization* $\mathcal{M} = \langle I, \mathcal{R}, Pr, V\rangle$ *can be transformed into a* **OL**-*realization* $\mathcal{K}^{\mathcal{M}} = \langle I^*, \mathcal{R}^*, Pr^*, V^*\rangle$ *such that:*

$$I^* = I;\ \mathcal{R}^* = \mathcal{R};$$

$$\forall i\,[i \models_{\mathcal{K}^{\mathcal{M}}} \alpha \text{ iff } i \models_{\mathcal{M}} \tau(\alpha)].$$

Sketch of the proof. Take Pr^* as the smallest subset of the power set of I that contains $V(LM\mathbf{p})$ for any atomic formula $\mathbf{p}$ and that is closed under $I, \emptyset, {}^\perp, \cap$ (where, for any set X of worlds, $X^\perp := \{j : \text{not}\, i\mathcal{R}j\}$). Furthermore, take $V^*(\mathbf{p})$ equal to $V(LM\mathbf{p})$. The set $V^*(\mathbf{p})$ is a proposition in the orthoframe $\langle I^*, \mathcal{R}^*\rangle$, owing to the **B**-logical truth: $LM\alpha \leftrightarrow_c LMLM\alpha$. □

The translation of **OL** into **B** is technically very useful, since it permits us to transfer to **OL** some important metalogical properties such as *decidability* and the *finite-model property*.

LEMMA 9.2.6. (Goldblatt, 1974)

(i) **OL** *is decidable. In other words, the set of the logical truths is* recursive *(for any sentence α, one can decide, in a finite number of steps, whether α is a logical truth).*
(ii) **OL** *has the finite model property. In other words, whenever α is not a logical truth, there is a finite realization $\mathcal{K}$ (with a finite set of worlds) where α is not true.*

Does **OQL** also admit a modal interpretation? The question has a somewhat trivial answer. It is sufficient to apply the technique used for **OL** by referring to a convenient modal system $\mathbf{B^o}$ (stronger than **B**), which is founded on a modal version of the orthomodular principle. Semantically $\mathbf{B^o}$ can be characterized by a particular class of realizations. In order to determine this class, let us first define the concept of *quantum proposition* in a **B**-realization.

DEFINITION 9.2.7. *Quantum proposition*
Given a **B**-realization $\mathcal{M} = \langle I, \mathcal{R}, Pr, V\rangle$ the set Pr_Q of all *quantum propositions* of $\mathcal{M}$ is the smallest subset of the power set of I which contains $V(LMp)$ for any atomic **p** and is closed under $^\perp$ and $\cap$.

LEMMA 9.2.8. *In any* **B**-*realization* $\mathcal{M} = \langle I, \mathcal{R}, Pr, V\rangle$, $Pr_Q \subseteq Pr$.

Sketch of the proof. The only nontrivial point of the proof concerns the closure of Pr under $^\perp$. This holds since one can prove: $\forall X \in Pr\,[X^\perp = \Box X^c]$. □

LEMMA 9.2.9. *Given* $\mathcal{M} = \langle I, \mathcal{R}, Pr, V\rangle$ *and* $\mathcal{K}^{\mathcal{M}} = \langle I, \mathcal{R}, Pr^*, V^*\rangle$, $Pr_Q = Pr^*$.

LEMMA 9.2.10. *Given* $\mathcal{K} = \langle I, \mathcal{R}, Pr, V\rangle$ *and* $\mathcal{M}^{\mathcal{K}} = \langle I, \mathcal{R}, Pr^*, V^*\rangle$, $Pr \supseteq Pr^*_Q$.

DEFINITION 9.2.11. $\mathbf{B^o}$-*realization*
A $\mathbf{B^o}$-*realization* is a **B**-realization $\langle I, \mathcal{R}, Pr, V\rangle$ where the quantum propositions satisfy the orthomodular property:

$$\forall X, Y \in Pr_Q[X \not\subseteq Y \;\curvearrowright\; X \cap (X \cap Y)' \neq \emptyset].$$

We will also call the $\mathbf{B^o}$-realizations *orthomodular realizations*.

THEOREM 9.2.12. *For any T and α of* **OQL**: $T \models_{\mathbf{OL}} \alpha$ *iff* $\tau(T) \models_{\mathbf{B^o}} \tau(\alpha)$.

The theorem is an immediate corollary of Lemmas 9.2.4, 9.2.5 and the following lemma:

LEMMA 9.2.13.

(i) *If $\mathcal{K}$ is orthomodular then $\mathcal{M}^{\mathcal{K}}$ is orthomodular.*
(ii) *If $\mathcal{M}$ is orthomodular then $\mathcal{K}^{\mathcal{M}}$ is orthomodular.*

Unfortunately, the modal interpretation of **OQL** is not particularly interesting from a logical point of view. Unlike the **OL**-case, $\mathbf{B}^{\mathbf{o}}$ does not correspond to a familiar modal system with well behaved metalogical properties. A characteristic logical truth of this logic is a modal version of orthomodularity:

$$\alpha \curlywedge_c \neg_c \beta \rightarrow_c M\,(\alpha \curlywedge_c L \neg_c (\alpha \curlywedge_c \beta)),$$

where α, β are modal translations of formulas of **OQL** into the language $\mathcal{L}^{\mathbf{B}}$.

CHAPTER 10

An axiomatization of OL and OQL

10.1. The calculi for OL and OQL

QL is an axiomatizable logic. Many axiomatizations are known: both in the Hilbert-Bernays style and in the Gentzen-style (*natural deduction* and *sequent-calculi*).[1]

We will present here a **QL**-calculus (in the natural deduction style) which is a slight modification of a calculus proposed by Goldblatt (1974). The principal advantage of this axiomatization is the fact that it is formally very close to the algebraic definition of ortholattice; also it is independent of any idea of quantum logical implication.

Our calculus (which has no axioms) is determined as a set of *rules*. Let $T_1, \ldots, T_n$ be finite or infinite (possibly empty) sets of sentences. Any rule has the form

$$\frac{T_1 \models \alpha_1, \ldots, T_n \models \alpha_n}{T \models \alpha}$$

(if α_1 has been inferred from $T_1, \ldots,$ α_n has been inferred from T_n, then α can be inferred from T). We will call any expression of the form $T \models \alpha$ a *configuration*. The configurations $T_1 \models \alpha_1, \ldots, T_n \models \alpha_n$ represent the *premisses* of the rule, while $T \models \alpha$ is the *conclusion*. As a limit case, we may have a rule in which the set of premisses is empty; in such a case we will speak of an *improper rule*. Instead of $\frac{\emptyset}{T \models \alpha}$ we will write $T \models \alpha$; instead of $\emptyset \models \alpha$, we will write $\models \alpha$.

Rules of **OL**

(OL1) $\quad T \cup \{\alpha\} \models \alpha$ (identity)

(OL2) $\quad \dfrac{T \models \alpha,\ T^* \cup \{\alpha\} \models \beta}{T \cup T^* \models \beta}$ (transitivity)

(OL3) $\quad T \cup \{\alpha \curlywedge \beta\} \models \alpha$ ($\curlywedge$-elimination)

(OL4) $\quad T \cup \{\alpha \curlywedge \beta\} \models \beta$ ($\curlywedge$-elimination)

[1] For an axiomatization of **QL** in the Hilbert-Bernays style see, for instance, (Hardegree, 1976). Sequent calculi for different forms of quantum logic have been investigated by Nishimura (1980) and by Battilotti and Sambin (1999). See also (Battilotti and Faggian, 2002).

$$\text{(OL5)} \qquad \frac{T \vdash \alpha,\ T \vdash \beta}{T \vdash \alpha \curlywedge \beta} \qquad (\curlywedge\text{-introduction})$$

$$\text{(OL6)} \qquad \frac{T \cup \{\alpha, \beta\} \vdash \gamma}{T \cup \{\alpha \curlywedge \beta\} \vdash \gamma} \qquad (\curlywedge\text{-introduction})$$

$$\text{(OL7)} \qquad \frac{\{\alpha\} \vdash \beta,\ \{\alpha\} \vdash \neg\beta}{\neg\alpha} \qquad \text{(absurdity)}$$

$$\text{(OL8)} \qquad T \cup \{\alpha\} \vdash \neg\neg\alpha \qquad \text{(weak double negation)}$$

$$\text{(OL9)} \qquad T \cup \{\neg\neg\alpha\} \vdash \alpha \qquad \text{(strong double negation)}$$

$$\text{(OL10)} \qquad T \cup \{\alpha \curlywedge \neg\alpha\} \vdash \beta \qquad \text{(Duns Scotus)}$$

$$\text{(OL11)} \qquad \frac{\{\alpha\} \vdash \beta}{\{\neg\beta\} \vdash \neg\alpha} \qquad \text{(contraposition)}$$

DEFINITION 10.1.1. *Derivation*
A *derivation* of **OL** is a finite sequence of configurations $T \vdash \alpha$, where any element of the sequence is either the conclusion of an improper rule or the conclusion of a proper rule whose premisses are previous elements of the sequence.

DEFINITION 10.1.2. *Derivability*
A sentence α is *derivable* from T $(T \vdash_{\mathbf{OL}} \alpha)$ iff there is a derivation such that the configuration $T \vdash \alpha$ is the last element of the derivation.

Instead of $\{\alpha\} \vdash_{\mathbf{OL}} \beta$ we will write $\alpha \vdash_{\mathbf{OL}} \beta$. When no confusion is possible, we will write $T \vdash \alpha$ instead of $T \vdash_{\mathbf{OL}} \alpha$.

DEFINITION 10.1.3. *Logical theorem*
A sentence α is a *logical theorem* of **OL** $(\vdash_{\mathbf{OL}} \alpha)$ iff $\emptyset \vdash_{\mathbf{OL}} \alpha$.

One can easily prove the following syntactical lemmas.

LEMMA 10.1.4. $\alpha_1, \ldots, \alpha_n \vdash \alpha$ *iff* $\alpha_1 \curlywedge \cdots \curlywedge \alpha_n \vdash \alpha$.

LEMMA 10.1.5. Syntactical compactness
$T \vdash \alpha$ *iff* $\exists T^* \subseteq T$ $[T^*$ *is finite and* $T^* \vdash \alpha]$.

LEMMA 10.1.6.
$T \vdash \alpha$ *iff* $\exists \alpha_1, \ldots, \alpha_n [\alpha_1 \in T$ *and... and* $\alpha_n \in T$ *and* $\alpha_1 \curlywedge \cdots \curlywedge \alpha_n \vdash \alpha]$.

DEFINITION 10.1.7. *Consistency*
T is an *inconsistent* set of sentences iff $\exists \alpha\, (T \vdash \alpha \curlywedge \neg\alpha)$; T is *consistent*, otherwise.

DEFINITION 10.1.8. *Deductive closure*
The *deductive closure* $\overline{T}$ of a set of sentences T is the set $\{\alpha : T \vdash \alpha\}$. T is called *deductively closed* iff $T = \overline{T}$.

DEFINITION 10.1.9. *Syntactical compatibility*
Two sets of sentences T_1 and T_2 are called *syntactically compatible* iff

$$\forall \alpha \, [T_1 \vdash \alpha \curvearrowright T_2 \nvdash \neg\alpha].$$

It is easy to see that this is a symmetric relation; if this relation holds, we sometimes say that T_1 is *syntactically compatible with* T_2. The following theorem represents a kind of "weak Lindenbaum Theorem."

THEOREM 10.1.10. Weak Lindenbaum Theorem.
If $T \nvdash \neg\alpha$, then there exists a set of sentences T^* such that T^* is compatible with T and $T^* \vdash \alpha$.

PROOF. Suppose $T \nvdash \neg\alpha$. Take $T^* = \{\alpha\}$. Trivially: $T^* \vdash \alpha$. Let us prove the compatibility between T and T^* . Suppose, by contradiction, T and T^* incompatible. Then, for a certain β, $T^* \vdash \beta$ and $T \vdash \neg\beta$. Hence, by definition of T^*, $\alpha \vdash \beta$ and by contraposition, $\neg\beta \vdash \neg\alpha$. Consequently, because $T \vdash \neg\beta$, one obtains by transitivity, $T \vdash \neg\alpha$, contradicting our hypothesis. □

10.2. The soundness and completeness theorems

We will now prove a soundness and a completeness theorem with respect to the Kripkean semantics.

THEOREM 10.2.1. Soundness theorem.

$$T \vdash \alpha \curvearrowright T \models \alpha.$$

PROOF. Straightforward. □

As happens in the case of many logics characterized by a Kripkean or by an algebraic semantics, the completeness theorem is proved by applying a standard technique which refers to a *canonical model*.[2]

THEOREM 10.2.2. Completeness theorem

$$T \models \alpha \curvearrowright T \vdash \alpha.$$

PROOF. It is sufficient to construct a *canonical model* $\mathcal{K} = \langle I \, , \, \mathcal{R} \, , \, Pr \, , \, V \rangle$ such that:

$$T \vdash \alpha \text{ iff } T \models_{\mathcal{K}} \alpha.$$

As a consequence, we will immediately obtain:

$$T \nvdash \alpha \curvearrowright T \not\models_{\mathcal{K}} \alpha \curvearrowright T \not\models \alpha.$$

---

[2]A *canonical model* of a logic **L** is a special kind of model whose characteristic properties are the following: (a) the model entirely consists of *linguistic objects*; (b) the notion of *truth* is defined in terms of *syntactical notions*. In the case of Kripkean semantics, the possible worlds of a canonical model are generally identified with particular sets of sentences; while a world i *verifies* a sentence α iff α is *derivable* from i. Notice that a logic may have different canonical models that need not be isomorphic.

Definition of a canonical model for **OL**

(i) I is the set of all consistent and deductively closed sets of sentences;
(ii) $\mathcal{R}$ is the compatibility relation between sets of sentences;
(iii) Pr is the set of all propositions in the frame $\langle I, \mathcal{R}\rangle$;
(iv) $V(\mathbf{p}) = \{i \in I : \mathbf{p} \in i\}$.

In order to recognize that $\mathcal{K}$ is an **OL**-realization, it is sufficient to prove that: (a) $\mathcal{R}$ is reflexive and symmetric; (b) $V(\mathbf{p})$ is a proposition in the frame $\langle I, \mathcal{R}\rangle$.
The proof of (a) is immediate (reflexivity follows from the consistency of any i, and symmetry follows from the weak double negation rule).

In order to prove (b), it is sufficient to show (by Lemma 8.1.4): $i \notin V(\mathbf{p}) \curvearrowright \exists j \not\perp i\,[j \perp V(\mathbf{p})]$. Let $i \notin V(\mathbf{p})$. Then, by definition of $V(\mathbf{p})$, $\mathbf{p} \notin i$; and, since i is deductively closed, $i \not\vdash p$. Consequently, by the weak Lindenbaum Theorem (and by the strong double negation rule), for a certain j: $j \not\perp i$ and $\neg\mathbf{p} \in j$. Hence, $j \perp V(\mathbf{p})$.

LEMMA 10.2.3. Lemma of the canonical model

For any α *and any* $i \in I$, $i \models \alpha$ *iff* $\alpha \in i$.

Sketch of the proof. By induction on the length of α. The case $\alpha = \mathbf{p}$ holds by definition of $V(\mathbf{p})$. The case $\alpha = \neg\beta$ can be proved by using Lemma 8.1.8(i) and the weak Lindenbaum Theorem. The case $\alpha = \beta \curlywedge \gamma$ can be proved using the $\curlywedge$-introduction and the $\curlywedge$-elimination rules. □

Finally we can show that $T \vdash \alpha$ iff $T \models_{\mathcal{K}} \alpha$. Since the left to right implication is a consequence of the soundness theorem, it is sufficient to prove: $T \not\vdash \alpha \curvearrowright T \not\models_{\mathcal{K}} \alpha$. Let $T \not\vdash \alpha$; then, by Duns Scotus, T is consistent. Take $i := \overline{T}$. Then, $i \in I$ and $T \subseteq i$. As a consequence, by the lemma of the canonical model, $i \models T$. At the same time $i \not\models \alpha$. For, should $i \models \alpha$ be the case, we would obtain $\alpha \in i$ and by definition of i, $T \vdash \alpha$, contradicting our hypothesis. □

An axiomatization of **OQL** *can be obtained by adding to the* **OL**-*calculus the following rule*:

$$\text{(OQL)} \qquad \alpha \curlywedge \neg(\alpha \curlywedge \neg(\alpha \curlywedge \beta)) \vdash \beta. \qquad \text{(orthomodularity)}$$

All the syntactical definitions we have considered for **OL** can be extended to **OQL**. Also Lemmas 10.1.4, 10.1.5, 10.1.6 and the weak Lindenbaum Theorem can be proved exactly in the same way. Since **OQL** admits a material conditional, we are able to prove a *deduction theorem* for **OQL**.

THEOREM 10.2.4. $\alpha \vdash_{\mathbf{OQL}} \beta$ *iff* $\vdash_{\mathbf{OQL}} \alpha \to \beta$.

It is worthwhile noticing that the deduction theorem is not in contrast with the failure in **QL** of the semantic property we have called Herbrand-Tarski ($T \models \alpha$ iff $T \models\!\!\!\equiv \alpha$ iff $T \models\!\!\!\approx \alpha$). Unlike the case of other logics, in **OQL** the syntactical relation $\vdash$ does not correspond to the weak consequence relation $\models$!

The soundness theorem can be easily proved, from the following, which holds in any orthomodular realization $\mathcal{K}$:

$$\alpha \curlywedge \neg(\alpha \curlywedge \neg(\alpha \curlywedge \beta)) \models_{\mathcal{K}} \beta.$$

As for the completeness theorem, we need a slight modification of the proof we have given for **OL**. In fact, should we try to construct a canonical model $\mathcal{K}$ by taking Pr as the set of all possible propositions of the frame, we would not be able to prove the orthomodularity of $\mathcal{K}$. In order to obtain an orthomodular canonical model $\mathcal{K} = \{I, \mathcal{R}, Pr, V\}$, it is sufficient to define Pr as the set of all propositions X of $\mathcal{K}$ such that $X = V(\alpha)$ for some α. One immediately recognizes that $V(\mathbf{p}) \in Pr$ and that Pr is closed under $^\perp$ and $\cap$. Hence, $\mathcal{K}$ is an **OL**-realization. Also, for this $\mathcal{K}$, one can easily show that $i \models \alpha$ iff $\alpha \in i$. In order to prove the orthomodularity of $\mathcal{K}$, one has to prove for any propositions $X, Y \in Pr$, $X \not\subseteq Y \curvearrowright X \cap (X \cap Y)^\perp \neq \emptyset$; which is equivalent, by Lemma 8.1.14, to $X \cap (X \cap (X \cap Y)^\perp)^\perp \subseteq Y$. By construction of Pr, $X = V(\alpha)$ and $Y = V(\beta)$ for some α, β. By the orthomodular rule we have $\alpha \curlywedge \neg(\alpha \curlywedge \neg(\alpha \curlywedge \beta)) \vdash \beta$. Consequently, for any $i \in I$,

$$i \models \alpha \curlywedge \neg(\alpha \curlywedge \neg(\alpha \curlywedge \beta)) \curvearrowright i \models \beta.$$

Hence, $V(\alpha) \cap (V(\alpha) \cap (V(\alpha) \cap V(\beta))^\perp)^\perp \subseteq V(\beta)$.

Of course, a canonical model of **OL** could also be constructed by taking Pr as the set of all propositions that are "meanings" of formulas. Nevertheless, in this case, we would lose the following important information: the canonical model, given earlier, of **OL** gives rise to an algebraically complete realization (closed under arbitrary intersection).[3]

---

[3]Even **OL** admits two canonical models that are not isomorphic!

CHAPTER 11

The metalogical intractability of orthomodularity

As we have seen, the proposition-ortholattice in a Kripkean realization $\mathcal{K} = \langle I, \mathcal{R}, Pr, V \rangle$ does not generally coincide with the (algebraically) *complete* ortholattice of *all* propositions of the orthoframe $\langle I, \mathcal{R} \rangle$. When Pr is the set of all propositions, $\mathcal{K}$ will be called *standard*. Thus, a *standard orthomodular Kripkean realization* is a standard realization, where Pr is orthomodular. In the case of **OL**, every nonstandard Kripkean realization can be naturally extended to a standard one (see the proof of Theorem 8.1.13). In particular, Pr can always be embedded into the complete ortholattice of all propositions of the orthoframe at issue. Moreover, as we have learned from the completeness proof, the canonical model of **OL** is standard. In the case of **OQL**, however, there are various reasons that make significant the distinction between standard and nonstandard realizations:

(i) Orthomodularity is not elementary (Goldblatt, 1984). In other words, there is no way to express the orthomodular property of the ortholattice Pr in an orthoframe $\langle I, \mathcal{R} \rangle$ as an elementary (first-order) property.

(ii) It is not known whether every orthomodular lattice is embeddable into a complete orthomodular lattice.

(iii) It is an open question whether **OQL** is characterized by the class of all standard orthomodular Kripkean realizations.

(iv) It is not known whether **OQL** admits a *standard canonical model.* If we try to construct a canonical realization for **OQL** by taking Pr as the set of all possible propositions as in the **OL**-case (call such a realization a *pseudo canonical realization*), do we obtain an **OQL**-realization, satisfying the orthomodular property? In other words, is the pseudo canonical realization a model of **OQL**?

In order to prove that **OQL** is characterized by the class of all standard Kripkean realizations it would be sufficient to show that some canonical model belongs to this class. Were orthomodularity elementary, then, by a general result proved by Fine, this problem would amount to showing the following statement: there is an elementary condition (or a set thereof) implying the orthomodularity of the standard pseudo canonical realization. Result (i), however, makes this path definitively unpracticable.

Notice that a positive solution to problem (iv) would automatically provide a proof of the full equivalence between the algebraic and the Kripkean consequence relation ($T \models^{\mathrm{A}}_{\mathbf{OQL}} \alpha$ iff $T \models^{\mathrm{K}}_{\mathbf{OQL}} \alpha$). If **OQL** is characterizable by a standard canonical model, then we would be able to apply the same argument used in the case of **OL**, the ortholattice Pr of the canonical model being orthomodular. For similar reasons, a positive solution to problem (ii) would provide a direct proof of the same result. For, the orthomodular lattice Pr of the (not necessarily standard) canonical model of **OQL** would be embeddable into a complete orthomodular lattice.

We will now present Goldblatt's result proving that orthomodularity is not elementary. Furthermore, we will show how orthomodularity leaves defeated one of the most powerful embedding techniques: the Mac Neille completion method.

11.1. Orthomodularity is not elementary

Let us consider a first-order language $\mathcal{L}^2$ with a single predicate denoting a binary relation $\mathcal{R}$. Any frame $\langle I\,,\,\mathcal{R}\rangle$ (where I is a nonempty set and $\mathcal{R}$ is a binary relation) will represent a classical (first-order) realization of $\mathcal{L}^2$. Following a standard notation, we will indicate by $\alpha(x_1,\ldots,x_n)$ a formula of $\mathcal{L}^2$, where the variables $x_1,\ldots,x_n$ *occur free*.[1] Furthermore, $\langle I\,,\,\mathcal{R}\rangle \models \alpha[i_1,\ldots i_n]$ will be an abbreviation for the following semantic relation: the individuals $i_1,\ldots i_n$, belonging to the universe I, *satisfy* the formula $\alpha(x_1,\ldots,x_n)$ in the realization $\langle I\,,\,\mathcal{R}\rangle$.

DEFINITION 11.1.1. *Elementary property and elementary class*
Let Γ be a class of frames.

(i) A property P that may hold for the elements of Γ is called *first-order* (or *elementary*) iff there exists a sentence η of $\mathcal{L}^2$ such that for any $\langle I\,,\,\mathcal{R}\rangle \in \Gamma$:

$$\langle I\,,\,\mathcal{R}\rangle \models \eta \text{ iff } \langle I\,,\,\mathcal{R}\rangle \text{ has the property } P.$$

(ii) Γ is said to be an *elementary class* iff the property of being in Γ is an elementary property of Γ.

Thus, Γ is an elementary class iff there is a sentence η of $\mathcal{L}^2$ such that

$$\Gamma = \{\langle I\,,\,\mathcal{R}\rangle : \langle I\,,\,\mathcal{R}\rangle \models \eta\}\,.$$

DEFINITION 11.1.2. *Elementary substructure*
Let $\langle I_1\,,\,\mathcal{R}_1\rangle$, $\langle I_2\,,\,\mathcal{R}_2\rangle$ be two frames.

(i) $\langle I_1\,,\,\mathcal{R}_1\rangle$ is a *substructure* of $\langle I_2\,,\,\mathcal{R}_2\rangle$ iff the following conditions are satisfied:
 (i) $I_1 \subseteq I_2$;
 (ii) $\mathcal{R}_1 = \mathcal{R}_2 \cap (I_1 \times I_1)$;

(ii) $\langle I_1\,,\,\mathcal{R}_1\rangle$ is an *elementary substructure* of $\langle I_2\,,\,\mathcal{R}_2\rangle$ iff the following conditions hold:

[1] A variable x is *free* in a formula α iff x occurs in α, and there exists at least one occurrence of x in α that is not bound by a quantifier.

(iia) $\langle I_1, \mathcal{R}_1\rangle$ is a substructure of $\langle I_2, \mathcal{R}_2\rangle$;
(iib) For any formula $\alpha(x_1, \ldots, x_n)$ of $\mathcal{L}^2$ and any $i_1, \ldots, i_n$ of I_1:

$$\langle I_1, \mathcal{R}_1\rangle \models \alpha[i_1, \ldots i_n] \text{ iff } \langle I_2, \mathcal{R}_2\rangle \models \alpha[i_1, \ldots i_n].$$

In other words, the elements of the "smaller" structure satisfy exactly the same $\mathcal{L}^2$-formulas in both structures. The following theorem provides an useful criterion to check whether a substructure is an elementary substructure.[2]

THEOREM 11.1.3. *Let $\langle I_1, \mathcal{R}_1\rangle$ be a substructure of $\langle I_2, \mathcal{R}_2\rangle$. Then, $\langle I_1, \mathcal{R}_1\rangle$ is an elementary substructure of $\langle I_2, \mathcal{R}_2\rangle$ iff the following condition holds. Suppose we have a formula $\alpha(x_1, \cdots, x_n, y)$ of $\mathcal{L}^2$, whose free variables are precisely $x_1, \cdots, x_n, y$, and let $i_1, \cdots, i_n$ be elements of I_1 such that for some $j \in I_2$, $\langle I_2, \mathcal{R}_2\rangle \models \alpha[i_1, \cdots, i_n, j]$. Then, there exists some $i \in I_1$ such that $\langle I_2, \mathcal{R}_2\rangle \models \alpha[i_1, \cdots, i_n, i]$.*

Let us now consider a *pre-Hilbert space* $\mathcal{H}$ and let $\mathcal{H}^* := \{\psi \in \mathcal{H} : \psi \neq \underline{0}\}$, where $\underline{0}$ is the null vector.[3] The pair

$$\langle \mathcal{H}^*, \not\perp\rangle$$

is an orthoframe, where $\forall\psi, \phi \in \mathcal{H}^*$: $\psi \not\perp \phi$ iff the inner product of ψ and ϕ is different from the null vector $\underline{0}$ (i.e., $\langle\psi|\phi\rangle \neq \underline{0}$). Let $Pr(\mathcal{H})$ be the ortholattice of all propositions of $\langle \mathcal{H}^*, \not\perp\rangle$, which turns out to be isomorphic to the ortholattice $\mathcal{C}(\mathcal{H})$ of all (not necessarily closed) subspaces of $\mathcal{H}$ (a proposition is simply a subspace minus the null vector). The following deep theorem, due to Amemiya and Halperin,[4] permits us to characterize the class of all Hilbert spaces in the larger class of all pre-Hilbert spaces, by means of the orthomodular property.

THEOREM 11.1.4. Amemiya-Halperin Theorem
Let $\mathcal{H}$ be a pre-Hilbert space. Then, $\mathcal{C}(\mathcal{H})$ is orthomodular iff $\mathcal{H}$ is a Hilbert space.

In other words, $\mathcal{C}(\mathcal{H})$ is orthomodular iff $\mathcal{H}$ is metrically complete.

As is well known, the property of "being metrically complete" is not elementary. Thus, it is a reasonable conjecture that the orthomodular property also is not elementary. The key-lemma in Goldblatt's proof is the following:

LEMMA 11.1.5. *Let Y be an infinite dimensional (not necessarily closed) subspace of a separable Hilbert space $\mathcal{H}$. If α is any formula of $\mathcal{L}^2$ and $\psi_1, \cdots, \psi_n$ are vectors of Y such that for some $\phi \in \mathcal{H}$,*

$$\langle \mathcal{H}^*, \not\perp\rangle \models \alpha[\psi_1, \cdots, \psi_n, \phi],$$

then there is a vector $\psi \in Y$ such that

$$\langle \mathcal{H}^*, \not\perp\rangle \models \alpha[\psi_1, \cdots, \psi_n, \psi].$$

---

[2]See, for instance, (Bell-Slomson 1969).

[3]The notion of *pre-Hilbert space* has been defined in Definition 1.2.7.

[4]See (Varadarajan, 1985).

As a consequence one obtains:

THEOREM 11.1.6. *The orthomodular property is not elementary.*

PROOF. Let $\mathcal{H}$ be any *metrically incomplete* pre-Hilbert space. Let $\overline{\mathcal{H}}$ be its metric completion, regarded as a Hilbert space. Thus, $\mathcal{H}$ is an infinite dimensional subspace of the Hilbert space $\overline{\mathcal{H}}$. By Lemma 11.1.5 and by Theorem 11.1.3, $\langle \mathcal{H}^*, \not\perp \rangle$ is an elementary substructure of $\left\langle \overline{\mathcal{H}}^*, \not\perp \right\rangle$. But, by the Amemiya-Halperin's Theorem, $\mathcal{C}(\mathcal{H})$ is not orthomodular, because $\mathcal{H}$ is metrically incomplete. However, $\mathcal{C}(\overline{\mathcal{H}})$ is orthomodular. As a consequence, orthomodularity cannot be expressed as an elementary property. □

11.2. The embeddability problem

As we have seen in Chapter 8, the class of all propositions of an orthoframe is a complete ortholattice. Conversely, the representation theorem for ortholattices states that every ortholattice $\mathcal{B} = \langle B, \leq, ', \mathbf{0}, \mathbf{1} \rangle$ is embeddable into the complete ortholattice of all propositions of the orthoframe $\langle B^+, \not\perp \rangle$, where $B^+ := B - \{\mathbf{0}\}$ and $\forall a, b \in B\ [a \not\perp b \text{ iff } a \not\leq b']$. The embedding is given by the map

$$h : a \rightarrowtail (a\,],$$

where $(a\,] = \{b \in B^+ : b \leq a\}$ is the quasi-ideal generated by a. One can prove the following theorem.

THEOREM 11.2.1. *Let* $\mathcal{B} = \langle B, \leq, ', \mathbf{0}, \mathbf{1} \rangle$ *be an ortholattice.* $\forall X \subseteq B$, X *is a proposition of* $\langle B^+, \not\perp \rangle$ *iff* $X = l(u(X))$, *where:*

$$u(Y) := \left\{b \in B^+ : \forall a \in Y, a \leq b\right\} \text{ and } l(Y) := \left\{b \in B^+ : \forall a \in Y, b \leq a\right\}.$$

Accordingly, the complete ortholattice of all propositions of the orthoframe $\langle B^+, \not\perp \rangle$ is isomorphic to the *Mac Neille completion* of $\mathcal{B}$ (see Section 4.3).

At the same time, orthomodularity (as with distributivity and modularity) is not preserved by the Mac Neille completion, as the following example shows (Kalmbach, 1983).

Let $\mathcal{C}^0_{(2)}(\mathbb{R})$ be the class of all continuous complex-valued functions f on $\mathbb{R}$ such that

$$\int_{-\infty}^{+\infty} |\, f(x)\, |^2\, dx < \infty.$$

Let us define the following bilinear form $(.,.) : \mathcal{C}^0_{(2)}(\mathbb{R}) \times \mathcal{C}^0_{(2)}(\mathbb{R}) \to \mathbb{C}$ (representing an inner product):

$$(f, g) = \int_{-\infty}^{+\infty} f^*(x) g(x) dx,$$

where $f^*(x)$ is the complex conjugate of $f(x)$. It turns out that $\mathcal{C}^0_{(2)}(\mathbb{R})$, equipped with this inner product gives rise to a metrically incomplete infinite dimensional pre-Hilbert space. Thus, by the Amemiya-Halperin Theorem

(Theorem 11.1.4) , the algebraically complete ortholattice $\mathcal{C}(\mathcal{C}^0_{(2)}(\mathbb{R}))$ of all subspaces of $\mathcal{C}^0_{(2)}(\mathbb{R})$ cannot be orthomodular. Now consider the sublattice $\mathcal{FI}$ of $\mathcal{C}(\mathcal{C}^0_{(2)}(\mathbb{R}))$, consisting of all finite or cofinite dimensional subspaces. It is not hard to see that $\mathcal{FI}$ is orthomodular. One can prove that $\mathcal{C}(\mathcal{C}^0_{(2)}(\mathbb{R}))$ is *sup-dense* in $\mathcal{FI}$; in other words, any $X \in \mathcal{C}(\mathcal{C}^0_{(2)}(\mathbb{R}))$ is the *sup* of a set of elements of $\mathcal{FI}$. Thus, by a theorem proved by MacLaren (Kalmbach, 1983), the Mac Neille completion of $\mathcal{C}(\mathcal{C}^0_{(2)}(\mathbb{R}))$ is isomorphic to the Mac Neille completion of $\mathcal{FI}$. Since $\mathcal{C}(\mathcal{C}^0_{(2)}(\mathbb{R}))$ is algebraically complete, the Mac Neille completion of $\mathcal{C}(\mathcal{C}^0_{(2)}(\mathbb{R}))$ is isomorphic to $\mathcal{C}(\mathcal{C}^0_{(2)}(\mathbb{R}))$ itself. As a consequence, $\mathcal{FI}$ is orthomodular, while its Mac Neille completion is not.

11.3. Hilbert quantum logic and the orthomodular law

As we have seen, the prototypical models of **OQL** that are interesting from the physical point of view are based on the class $\mathbb{H}$ of all Hilbert lattices, whose support is the set $\mathcal{C}(\mathcal{H})$ of all closed subspaces of a Hilbert space $\mathcal{H}$. Let us call *Hilbert quantum logic* (**HQL**) the logic that is semantically characterized by $\mathbb{H}$. A question naturally arises: do **OQL** and **HQL** represent one and the same logic? This question has a negative answer (Greechie, 1981): there is an ortholattice-theoretical equation, the so-called *orthoarguesian law*[5] that holds in $\mathbb{H}$, but fails in a particular orthomodular lattice. As a consequence, **OQL** does not represent a faithful logical abstraction from its quantum theoretical origin.

DEFINITION 11.3.1. Let Γ be a class of orthomodular lattices. We say that **OQL** is *characterized* by Γ iff for any T and any α the following condition is satisfied:

$$T \models_{\mathbf{OQL}} \alpha \text{ iff for any } \mathcal{B} \in \Gamma \text{ and any } \mathcal{A} = \langle \mathcal{B}, v \rangle \,,\; T \models_{\mathcal{A}} \alpha.$$

In order to formulate the orthoarguesian law in an equational way, let us first introduce the notion of *Sasaki projection*.

DEFINITION 11.3.2. *The Sasaki projection*

[5]The basic idea underlying this law is to assume that, in the triangles referred to in the usual presentation of a Desargues configuration in projective geometry (See (Bennett, 1995)), the corresponding points are orthogonal. This allows the statement of Desargues Theorem to be expressed as an equation in an ortholattice, the orthoarguesian law. One can then ask in which structures it holds. This unpublished idea of the late distinguished mathematician Alan Day was communicated orally to one of the authors. Day was the first to observe that it held in Hilbert space. See also (Kalmbach, 1983).

Let $\mathcal{B}$ be an orthomodular lattice and let a, b be any two elements of B. The *Sasaki projection* of a onto b, denoted by $a \Cap b$, is defined as follows:[6]

$$a \Cap b := (a \vee b') \wedge b.$$

It is easy to see that two elements a, b of an orthomodular lattice are compatible $(a = (a \wedge b') \vee (a \wedge b))$ iff $a \Cap b = a \wedge b$. Consequently, in any Boolean lattice, $\Cap$ coincides with $\wedge$.

DEFINITION 11.3.3. *The orthoarguesian law*[7]

$$a \leq b \vee \{(a \Cap b') \wedge [(a \Cap c') \vee ((b \vee c) \wedge ((a \Cap b') \vee (a \Cap c')))]\} \qquad \text{(OAL)}$$

It is an elementary exercise to show that (OAL) holds in $\mathbb{H}$. In the following theorem, we show that (OAL) fails in the orthomodular lattice $\mathcal{G}_{30}$ depicted in the carefully labeled Greechie diagram in Figure 11.3.1.

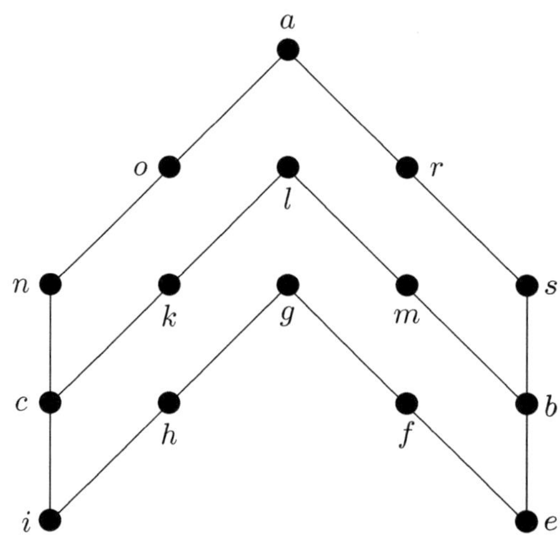

FIGURE 11.3.1. The Greechie diagram of $\mathcal{G}_{30}$

THEOREM 11.3.4. (OAL) fails in the orthomodular lattice $\mathcal{G}_{30}$.

---

[6]If we represent the orthomodular lattice $\mathcal{B}$ as a QMV algebra (where $\oplus$ coincides with $\vee$), then the Sasaki projection turns out to coincide with the QMV-operation $\Cap$. Needless to say, the Sasaki projection $((a \vee b') \wedge b)$ and the Sasaki hook $(a' \vee (a \wedge b))$ are different operations!

[7]Note that this law can be expressed as an equation, since any expression of the form $x \leq y$ can be written in the form $x = x \wedge y$.

PROOF. The following holds: $a \Cap b' = (a \vee b) \wedge b' = s' \wedge b' = e, a \Cap c' = (a \vee c) \wedge c' = n' \wedge c' = i$ and $b \vee c = l'$. Thus,

$$\begin{aligned} b \vee \{(a \Cap b') \wedge [(a \Cap c') \vee ((b \vee c) \wedge ((a \Cap b') \vee (a \Cap c')))]\} \\ &= b \vee \{e \wedge [i \vee (l' \wedge (e \vee i))]\} \\ &= b \vee \{e \wedge [i \vee (l' \wedge g')]\} \\ &= b \vee (e \wedge (i \vee \mathbf{0})) \\ &= b \vee (e \wedge i) \\ &= b \\ &\not\geq a. \end{aligned}$$

□

Hence, there are two sentences α and β (whose valuations in a convenient realization represent two sides of (OAL), respectively) such that $\alpha \not\models_{\mathbf{OQL}} \beta$. At the same time, for any $\mathcal{C}(\mathcal{H}) \in \mathbb{H}$ and for any realization $\mathcal{A} = \langle \mathcal{C}(\mathcal{H}), v \rangle$, we will have: $\alpha \models_{\mathcal{A}} \beta$.

As a consequence, we have the following result.

COROLLARY 11.3.5. **OQL** *is not characterized by* $\mathbb{H}$

Accordingly, **HQL** is definitely stronger than **OQL**. We are faced with the problem of finding a calculus, if any exists, that is sound and complete with respect to $\mathbb{H}$. The main question is whether the class of all sentences valid in $\mathbb{H}$ is recursively enumerable. In order to solve this problem, it would be sufficient (but not necessary) to show that a canonical model of **HQL** is isomorphic to the subdirect product of a class of Hilbert lattices. So far, very little is known about this question.

CHAPTER 12

First-order quantum logics and quantum set theories

12.1. First-order semantics

The most significant logical and metalogical peculiarities of **QL** arise at the sentential level. At the same time the extension of sentential **QL** to a first-order logic seems to be quite natural. As in the case of sentential **QL**, we will characterize first-order **QL** both by means of an algebraic and a Kripkean semantics.

Suppose we have a standard first-order language with predicates P^n_m and individual constants a_m. For the sake of simplicity, we will not assume functional symbols. The primitive logical constants are the connectives $\neg$, $\curlywedge$ and the universal quantifier $\forall$. The concepts of *term*, *formula* and *sentence* are defined in the usual way (sentences are formulas where no free variables occur). We will use $x, y, z, x_1, \cdots, x_n, \cdots$ as metavariables ranging over the individual variables, and $t, t_1, t_2, \cdots$ as metavariables ranging over terms. The existential quantifier $\exists$ is defined by a generalized de Morgan law:

$$\exists x\alpha := \neg\forall x\neg\alpha.$$

DEFINITION 12.1.1. *Algebraic realization for first-order* **OL**
An *algebraic realization* for (first-order) **OL** is a system $\mathcal{A} = \langle \mathcal{B}_{\mathcal{C}}\,, D\,, v\,\rangle$ where:

(i) $\mathcal{B}_{\mathcal{C}} = \langle B\,, \leq\,, {}'\,, \mathbf{0}\,, \mathbf{1}\rangle$ is an ortholattice closed under the infimum $\bigwedge$ and the supremum $\bigvee$ for any $F \in \mathcal{C}$, where $\mathcal{C}$ is a particular family of subsets of B.
(ii) D is a nonempty set (disjoint from B) called the *domain* of $\mathcal{A}$.
(iii) v is the *valuation-function* satisfying the following conditions:
- for any constant a_m: $v(a_m) \in D$; for any predicate P^n_m, $v(P^n_m)$ is an *n-ary attribute* in $\mathcal{A}$, i.e., a function that associates to any n-tuple $\langle \mathbf{d}_1, \cdots, \mathbf{d}_n\rangle$ of elements of D an element of B;
- for any *interpretation* σ of the variables in the domain D (i.e., for any function from the set of all variables into D) the pair $\langle v\,, \sigma\rangle$ (abbreviated v^σ and called a *generalized valuation*) associates to any term an element in D and to any formula an element of B, according to the conditions:

$$v^\sigma(a_m) = v(a_m)$$
$$v^\sigma(x) = \sigma(x)$$
$$v^\sigma(P^n_m t_1, \cdots, t_n) = v(P^n_m)(v^\sigma(t_1), \cdots, v^\sigma(t_n))$$
$$v^\sigma(\neg\beta) = v^\sigma(\beta)'$$
$$v^\sigma(\beta \curlywedge \gamma) = v^\sigma(\beta) \wedge v^\sigma(\gamma)$$
$$v^\sigma(\forall x \beta) = \bigwedge \{v^{\sigma[x/\mathbf{d}]}(\beta) : \mathbf{d} \in D\} \text{ , where } \{v^{\sigma[x/\mathbf{d}]}(\beta) : \mathbf{d} \in D\} \in C$$

($\sigma^{[x/\mathbf{d}]}$ is the interpretation that associates to x the individual $\mathbf{d}$ and differs from σ at most in the value attributed to x).

DEFINITION 12.1.2. *Truth and logical truth*
A formula α is *true* in $\mathcal{A} = \langle \mathcal{B}_C, D, v \rangle$ (abbreviated $\models_\mathcal{A} \alpha$) iff, for any interpretation of the variables σ, $v^\sigma(\alpha) = \mathbf{1}$;
α is a *logical truth* of **OL** ($\models_{\mathbf{OL}} \alpha$) iff, for any $\mathcal{A}$, $\models_\mathcal{A} \alpha$

DEFINITION 12.1.3. *Consequence in a realization and logical consequence*
Let $\mathcal{A} = \langle \mathcal{B}_C, D, v \rangle$ be a realization. A formula α is a *consequence* of T in $\mathcal{A}$ (abbreviated $T \models_\mathcal{A} \alpha$) iff for any element a of B and any interpretation σ: if, for any $\beta \in T$, $a \leq v^\sigma(\beta)$, then $a \leq v^\sigma(\alpha)$;
α is a logical consequence of T ($T \models_{\mathbf{OL}} \alpha$) iff, for any realization $\mathcal{A}$, $T \models_\mathcal{A} \alpha$.

DEFINITION 12.1.4. *Kripkean realization for (first-order)* **OL**
A Kripkean realization for (first-order) **OL** is a system $\mathcal{K} = \langle I, \mathcal{R}, Pr_C, U, V \rangle$, where:

(i) $\langle I, \mathcal{R}, Pr_C \rangle$ satisfies the same conditions as in the sentential case; furthermore Pr_C is closed under the intersection $\bigcap$ for any $F \subseteq Pr$ such that $F \in C$ (where C is a particular family of subsets of Pr_C);
(ii) U, called the *domain* of $\mathcal{K}$, is a nonempty set, disjoint from the set of worlds I. The elements of U are *individual concepts* $\mathbf{u}$ such that for any world i: $\mathbf{u}(i)$ is an *individual* (called the *reference* of $\mathbf{u}$ in the world i). An individual concept $\mathbf{u}$ is called *rigid* iff, for any pairs of worlds i, j, $\mathbf{u}(i) = \mathbf{u}(j)$. The set $U_i = \{\mathbf{u}(i) : \mathbf{u} \in U\}$ represents the *domain of individuals in the world* i. Whenever $U_i = U_j$ for all i,j we will say that the realization $\mathcal{K}$ has a *constant domain*.
(iii) V associates a meaning to any individual constant a_m and to any predicate P^n_m according to the following conditions:
$V(a_m)$ is an individual concept in U.
$V(P^n_m)$ is a *predicate-concept*, i.e., a function that associates to any n-tuple of individual concepts $\langle \mathbf{u}_1, \cdots, \mathbf{u}_n \rangle$ a proposition in Pr_C;
(iv) for any interpretation of the variables σ in the domain U, the pair $\langle V, \sigma \rangle$ (abbreviated V^σ and called *valuation*) associates

to any term t an individual concept in U and to any formula a proposition in $Pr_{\mathcal{C}}$ according to the conditions:

$$V^{\sigma}(x) = \sigma(x)$$
$$V^{\sigma}(a_m) = V(a_m)$$
$$V^{\sigma}(P^n_m t_1, \cdots, t_n) = V(P^n_m)(V^{\sigma}(t_1), \cdots, V^{\sigma}(t_n))$$
$$V^{\sigma}(\neg\beta) = V^{\sigma}(\beta)'$$
$$V^{\sigma}(\beta \curlywedge \gamma) = V^{\sigma}(\beta) \cap V^{\sigma}(\gamma)$$
$$V^{\sigma}(\forall x\beta) = \bigcap \left\{ V^{\sigma[x/\mathbf{u}]}(\beta) \,:\, \mathbf{u} \in U \right\}, \text{ where } \left\{ V^{\sigma[x/\mathbf{u}]}(\beta) \,:\, \mathbf{u} \in U \right\} \in C.$$

For any world i and any interpretation σ of the variables, the triple $\langle V, i, \sigma \rangle$ (abbreviated V_i^{σ}) will be called a *world-valuation.*

DEFINITION 12.1.5. *Satisfaction*
$V_i^{\sigma} \models \alpha$ (V_i^{σ} *satisfies* α) iff $i \in V^{\sigma}(\alpha)$.

DEFINITION 12.1.6. *Verification*
$V_i \models \alpha$ (*i verifies* α) iff $V_i^{\sigma} \models \alpha$ for any σ.

DEFINITION 12.1.7. *Truth and logical truth*
$\models_{\mathcal{K}} \alpha$ (α is *true* in $\mathcal{K}$) iff, for any i: $i \models \alpha$;
$\models_{\mathbf{OL}} \alpha$ (α is a *logical truth* of **OL**) iff, for any $\mathcal{K}$: $\models_{\mathcal{K}} \alpha$.

DEFINITION 12.1.8. *Consequence in a realization and logical consequence*
$T \models_{\mathcal{K}} \alpha$ iff, for any i of $\mathcal{K}$ and any σ: $V_i^{\sigma} \models T \;\curvearrowright\; V_i^{\sigma} \models \alpha$;
$T \models_{\mathbf{OL}} \alpha$ iff, for any realization $\mathcal{K}$: $T \models_{\mathcal{K}} \alpha$.

The algebraic and the Kripkean characterization for first-order **OQL** can be obtained, in a similar way, by requiring that any realization be orthomodular.

In both semantics for first-order **QL** one can prove a coincidence lemma.

LEMMA 12.1.9. *Given* $\mathcal{A} = \langle \mathcal{B}_{\mathcal{C}}, D, v \rangle$ *and* $\mathcal{K} = \langle I, \mathcal{R}, Pr_{\mathcal{C}}, U, V \rangle$:

(1) *If* σ *and* σ^* *coincide in the values attributed to the variables occurring in a term t, then* $v^{\sigma}(t) = v^{\sigma^*}(t)$; $V^{\sigma}(t) = V^{\sigma^*}(t)$.
(2) *If* σ *and* σ^* *coincide in the values attributed to the free variables occurring in a formula* α, *then* $v^{\sigma}(\alpha) = v^{\sigma^*}(\alpha)$; $V^{\sigma}(\alpha) = V^{\sigma^*}(\alpha)$.

One can easily prove the following lemma, as in the sentential case.

LEMMA 12.1.10.

(1) *For any algebraic realization* $\mathcal{A}$ *there exists a Kripkean realization* $\mathcal{K}^{\mathcal{A}}$ *such that for any* α: $\models_{\mathcal{A}} \alpha$ *iff* $\models_{\mathcal{K}^{\mathcal{A}}} \alpha$. *Furthermore, if* $\mathcal{A}$ *is orthomodular then* $\mathcal{K}^{\mathcal{A}}$ *is orthomodular.*
(2) *For any Kripkean realization* $\mathcal{K}$, *there exists an algebraic realization* $\mathcal{A}^{\mathcal{K}}$ *such that for any for any* α: $\models_{\mathcal{K}} \alpha$ *iff* $\models_{\mathcal{A}^{\mathcal{K}}} \alpha$. *Furthermore, if* $\mathcal{K}$ *is orthomodular then* $\mathcal{A}^{\mathcal{K}}$ *is orthomodular.*

An axiomatization of first-order **OL** (**OQL**) can be obtained by adding to the rules of **OL** (**OQL**)-sentential calculus the following new rules:

(PR1) $T \cup \{\forall x \alpha\} \vdash \alpha(x/t)$, where $\alpha(x/t)$ indicates a legitimate substitution.[1]

(PR2) $\dfrac{T \vdash \alpha}{T \vdash \forall x \alpha}$ (provided x is not free in T).

All the basic syntactical notions are defined as in the sentential case. One can prove that any consistent set of sentences T admits of a consistent *inductive* extension T^*, such that $T^* \vdash \forall x \alpha(x)$ whenever for any constant term t, $T^* \vdash \alpha(t)$. The "weak Lindenbaum Theorem" can be strengthened as follows: if $T \nvdash \neg\alpha$ then there exists a consistent and inductive T^* such that:

$$T \text{ is syntactically compatible with } T^* \text{ and } T^* \vdash \alpha.$$

(where: T is syntactically compatible with T^* iff there is no formula α such that $T \vdash \alpha$ and $T^* \vdash \neg\alpha$).

One can prove a soundness and a completeness theorem for this calculus with respect to the Kripkean semantics.

THEOREM 12.1.11. Soundness

$$T \vdash \alpha \curvearrowright T \models \alpha.$$

PROOF. Straightforward. □

THEOREM 12.1.12. Completeness

$$T \models \alpha \curvearrowright T \vdash \alpha.$$

Sketch of the proof. As in the sentential case, it is sufficient to construct a canonical model $\mathcal{K} = \langle I, \mathcal{R}, Pr_{\mathcal{C}}, U, V \rangle$ such that $T \vdash \alpha$ iff $T \models_{\mathcal{K}} \alpha$.

Definition of a canonical model for first-order **QL**

(i) I is the set of all consistent, deductively closed and inductive sets of sentences expressed in a common language $\mathcal{L}^K$, which is an extension of the original language.

(ii) $\mathcal{R}$ is determined as in the sentential case.

(iii) U is a set of rigid individual concepts that is naturally determined by the set of all individual constants of the extended language $\mathcal{L}^K$. For any constant c of $\mathcal{L}^K$, let $\mathbf{u}^c$ be the corresponding individual concept in U. We require: for any world i, $\mathbf{u}^c(i) = c$. In other words, the reference of the individual concept $\mathbf{u}^c$ is, in any world,

[1] A substitution of the free variable x with the term t in $\alpha(x)$ is *legitimate* when the following condition is satisfied: if t is a variable, then t is not bound by a quantifier in the formula $\alpha(x/t)$ that is obtained by substituting every free occurrence of x in $\alpha(x)$ with t (in other words, the substitution does not transform the free variable x into a bound variable).

the constant c. We will indicate by $c^{\mathbf{u}}$ the constant corresponding to $\mathbf{u}$.

(iv) $V(a_m) = \mathbf{u}^{a_m}$;
$V(P^n_m)(\mathbf{u}^{c_1}_1, \ldots, \mathbf{u}^{c_n}_n) = \{i : P^n_m c_1, \ldots, c_n \in i\}$.
This V is well defined since one can prove for any sentence α of $\mathcal{L}^{\mathcal{K}}$:

$$i \not\vdash \alpha \;\curvearrowright\; \exists j \not\perp i : j \vdash \neg\alpha.$$

As a consequence, $V^\sigma(P^n_m t_1, \ldots, t_n)$ is a proposition.

(v) $Pr_{\mathcal{C}}$ is the set of all "meanings" of formulas (i.e., $X \in Pr^C$ iff $\exists\alpha\,\exists\sigma(X = V^\sigma(\alpha))$; C is the set of all sets $\{V^{\sigma[x/\mathbf{u}]}(\beta) : \mathbf{u} \in U\}$ for any formula β.

One can easily check that $\mathcal{K}$ is a realization with a constant domain.

LEMMA 12.1.13. Lemma of the canonical model
For any α, any $i \in I$ and any σ:

$$V^\sigma_i \models \alpha \quad \textit{iff} \quad \alpha^\sigma \in i,$$

where α^σ is the sentence obtained by substituting in α any free variable x with the constant $c^{\sigma(x)}$ corresponding to the individual concept $\sigma(x)$.

Sketch of the proof. By induction on the length of α. The cases $\alpha = P^n_m t_1, \cdots, t_n$, $\alpha = \neg\beta$, $\alpha = \beta \curlywedge \gamma$ are proved by an obvious transformation of the sentential argument. Let us consider the case $\alpha = \forall x\beta$ and suppose x occurring in β (otherwise the proof is trivial). In order to prove the left to right implication, suppose $V^\sigma_i \models \forall x\beta$. Then, for any $\mathbf{u}$ in U, $V^{\sigma[x/\mathbf{u}]} \models \beta(x)$. Hence, by inductive hypothesis, $\forall \mathbf{u} \in U$, $[\beta(x)]^{\sigma[x/\mathbf{u}]} \in i$. In other words, for any constant $c^{\mathbf{u}}$ of i: $[\beta(x)]^\sigma(x/c^{\mathbf{u}}) \in i$. And, since i is inductive and deductively closed: $\forall x\beta(x)^\sigma \in i$. In order to prove the right to left implication, suppose $[\forall x\beta(x)]^\sigma \in i$. Then, [by (PR1)], for any constant c of i: $[\beta(x/c)]^\sigma \in i$. Hence, by inductive hypothesis: for any $\mathbf{u}^c \in U$, $V^{\sigma[x/\mathbf{u}^c]}_i \models \beta(x)$, i.e., $V^\sigma_i \models \forall x\beta(x)$. Based on this, as in the sentential case, one can prove $T \vdash \alpha$ iff $T \models_{\mathcal{K}} \alpha$. □

First-order **QL** can be easily extended (in a standard way) to a first-order logic with identity. However, a critical problem is represented by the possibility of developing, within this logic, a satisfactory *theory of descriptions*. The main difficulty can be sketched as follows. A natural condition to be required in any characterization of an ι-operator is obviously the following:

$$\exists x\,\{\beta(x) \curlywedge \forall y\,[(\beta(y) \curlywedge x = y) \curlyvee (\neg\beta(y) \curlywedge \neg x = y)] \curlywedge \alpha(x)\}$$
$$\text{is true} \;\curvearrowright\; \alpha(\iota x\beta(x)) \text{ is true.}$$

However, in **QL**, the truth of the antecedent of this implication does not generally guarantee the existence of a particular individual such that $\iota x\beta$ can be regarded as a name for such an individual. As a counterexample, let us consider the following case (in the algebraic semantics): let $\mathcal{A}$ be $\langle \mathcal{B}, D, v\rangle$ where $\mathcal{B}$ is the complete orthomodular lattice based on the set of all closed subspaces of the plane $\mathbb{R}^2$, and D contains exactly two elements $\mathbf{d}_1, \mathbf{d}_2$. Let

P be a monadic predicate and X, Y two orthogonal 1-dimensional subspaces of B such that $v(P)(\mathbf{d}_1) = X$, $v(P)(\mathbf{d}_2) = Y$. If the equality predicate $=$ is interpreted as the standard identity relation (i.e., $v^\sigma(t_1 = t_2) = \mathbf{1}$, if $v^\sigma(t_1) = v^\sigma(t_2)$; $v^\sigma(t_1 = t_2) = \mathbf{0}$, otherwise), one can easily calculate:

$$v\left(\exists x\left[Px \curlywedge \forall y((Py \curlywedge x = y) \vee (\neg Py \curlywedge \neg x = y))\right]\right) = \mathbf{1}.$$

However, for both elements $\mathbf{d_1}, \mathbf{d_2}$ of the domain, we have:

$$v^{\sigma[x/\mathbf{d_1}]}(Px) \neq \mathbf{1}, \quad v^{\sigma[x/\mathbf{d_2}]}(Px) \neq \mathbf{1}.$$

In other words, there is no precise individual in the domain that satisfies the property expressed by predicate P!

12.2. Quantum set theories

An important application of **QL** to set theory has been developed by Takeuti (1981). We will sketch here only the fundamental idea of this application. Let $\mathcal{L}$ be a standard set-theoretical language. One can construct *ortho-valued models* for $\mathcal{L}$, which are formally very similar to the usual *Boolean-valued models* for standard set-theory, with the following difference: the set of truth-values is supposed to have the algebraic structure of a complete orthomodular lattice, instead of a complete Boolean algebra. Let $\mathcal{B}$ be a complete orthomodular lattice, and let ν, λ,... represent ordinal numbers. An *ortho-valued (set-theoretical) universe* V is constructed as follows:

$$U^{\mathcal{B}} = \bigcup_{\nu \in On} U(\nu),$$

where,

- $U(0) = \emptyset$;
- $U(\nu{+}1) = \{g : g \text{ is a function and } Dom(g) \subseteq U(\nu) \text{ and } Ran(g) \subseteq B\}$;
- $U(\lambda) = \bigvee_{\nu<\lambda} U(\nu)$, for any limit-ordinal λ.[2]

Given an orthovalued universe $U^{\mathcal{B}}$ one can define for any formula of $\mathcal{L}$ the truth-value $[\![\alpha]\!]^\sigma$ in $\mathcal{B}$ induced by any interpretation σ of the variables into the universe $U^{\mathcal{B}}$.

$$\begin{aligned}
[\![x \in y]\!]^\sigma &= \textstyle\bigvee_{g \in Dom(\sigma(y))} \left\{\sigma(y)(g) \wedge [\![x = z]\!]^{\sigma[z/g]}\right\} \\
[\![x = y]\!]^\sigma &= \textstyle\bigwedge_{g \in Dom(\sigma(x))} \left\{\sigma(x)(g) \rightsquigarrow [\![z \in y]\!]^{\sigma[z/g]}\right\} \wedge \\
&\quad \textstyle\bigwedge_{g \in Dom(\sigma(y))} \left\{\sigma(y)(g) \rightsquigarrow [\![z \in x]\!]^{\sigma[z/g]}\right\}.
\end{aligned}$$

where $\rightsquigarrow$ is the quantum logical conditional operation ($a \rightsquigarrow b := a' \vee (a \wedge b)$, for any $a, b \in B$).

A formula α is called *true* in the universe $V^{\mathcal{B}}$ ($\models_{V^{\mathcal{B}}} \alpha$) iff $[\![\alpha]\!]^\sigma = \mathbf{1}$, for any σ.

Interestingly enough, the segment $U(\omega)$ of $U^{\mathcal{B}}$ turns out to contain some important mathematical objects, that we might call *quantum-logical natural numbers*.

---

[2] $Dom(g)$ and $Ran(g)$ are the *domain* and the *range* of function g, respectively.

The standard axioms of set-theory hold in $\mathcal{B}$ only in a restricted form. An extremely interesting property of $U^{\mathcal{B}}$ is connected with the notion of identity. Unlike the case of Boolean-valued models, the identity relation in $U^{\mathcal{B}}$ turns out to be non-Leibnizian. For, one can choose an orthomodular lattice $\mathcal{B}$ such that:

$$\not\models_{U^{\mathcal{B}}} x = y \to \forall z(x \in z \leftrightarrow y \in z).$$

According to our semantic definitions, the relation $=$ represents a kind of "extensional equality." As a consequence, one may conclude that two quantum-sets that are extensionally equal do not necessarily share all the same properties. Such a failure of the Leibniz-substitutivity principle in quantum set theory might perhaps find interesting applications in the field of intensional logics.

A completely different approach is followed in the framework of the theories of *quasisets* (or *quasets*). The basic aim of these theories is to provide a mathematical description for collections of micro-objects, which seem to violate some characteristic properties of classical identity.

In some of his general writings, Schrödinger discussed the inconsistency between the classical concept of physical object (conceived as an individual entity) and the behavior of particles in quantum mechanics. Quantum particles – he noticed – lack individuality and the concept of identity cannot be applied to them, unlike the case of classical objects.

One of the aims of the *theories of quasisets* (proposed by da Costa, French and Krause (1992)) is to describe formally the following idea defended by Schrödinger: identity is generally not defined for micro-objects. As a consequence, one cannot even assert that an "electron is identical with itself." In the realm of micro-objects only an *indistinguishability relation* (an equivalence relation that may violate the substitutivity principle) makes sense.

Because of this, different formal systems have been proposed. Generally, these systems represent convenient generalizations of a Zermelo-Fraenkel-like set theory with *urelements*. Unlike the classical case, an urelement may be either a *macro*- or a *micro-object*. Collections are represented by *quasisets* and classical sets turn out to be limit cases of quasisets.

A somewhat different approach has been followed in the *theory of quasets* (proposed by Dalla Chiara and Toraldo di Francia (1993)).

The starting point is based on the following observation: collections of objects and compound systems in QT seem to share some features that are characteristic of *intensional entities*, which give rise to violations of the *extensionality-principle*.

Furthermore, the relation between *intensions* and *extensions* turns out to behave quite differently from the classical semantic situations. Generally, one cannot say that a *quantum intensional notion* uniquely determines a corresponding *extension*. For instance, take the notion of *electron*, whose intension is well defined by the following physical property: mass $= 9.1 \times 10^{-28}$g, electron charge $= 4.8 \times 10^{-10}$e.s.u., spin $= 1/2$. Does this property determine a corresponding *set*, whose elements should be precisely all the physical objects that satisfy this property at a certain time interval? The answer is negative. In fact, physicists have the possibility of recognizing, by theoretical or experimental means, whether a given physical system is an electron system or not. If yes, they can also enumerate all the quantum states available within it. But they can do so in a number of different ways. For example, take the spin. One can choose the x-axis and state how many electrons have spin up and how many have spin down. However, we could instead refer to the z-axis or any other direction, obtaining *different collections* of quantum states, all having the same cardinality. This seems to suggest that micro-object systems present an irreducibly *intensional behavior*: generally they do not determine precise extensions and are not determined thereby. Accordingly, a basic feature of the theory is a strong violation of the extensionality principle.

Quasets are convenient generalizations of classical sets, for which both the extensionality axiom and Leibniz' principle of indiscernibles are violated. Generally a quaset has only a cardinal but not an ordinal number, since it cannot be well ordered.

CHAPTER 13

Partial classical logic, the Lindenbaum property and the hidden variable problem

Orthologic and orthomodular quantum logic are examples of *total logics*: their language is syntactically closed under the logical connectives and any molecular sentence always has a meaning both in the algebraic and in the Kripkean semantics. We will now investigate a form of quantum logic (first proposed by Kochen and Specker (1965b)) based on a different idea. From the semantic point of view, the crucial relation will be represented by a *compatibility* relation, that may hold between the *meanings* of two sentences. As expected, the intended physical interpretation of the compatibility relation will be the following: two sentences α and β will have *compatible meanings* iff α and β can be *simultaneously tested.*

On this basis, one can characterize a logic whose main semantic features are the following:

- meanings of molecular sentences are not necessarily defined;
- the logic is locally classical in the sense that certain groupings of sentences whose meanings are defined exhibit classical behavior. This is the reason why such a logic has been termed *partial classical logic* (abbreviated **PaCL**).

As a consequence, some characteristic classical laws (like distributivity) that are violated in standard quantum logic, represent logical truths of **PaCL**. In spite of this, one can prove that **PaCL** is strictly weaker than classical logic.

13.1. Partial classical logic

We will study **PaCL** only at the sentential level. First, we will characterize this logic in the framework of an algebraic semantics, based on the notion of *partial Boolean algebra*.

DEFINITION 13.1.1. *Partial Boolean algebra*
A *partial Boolean algebra* (PBA) is a structure

$$\mathcal{B} = \langle B \, , \heartsuit \, , \veebar \, , \barwedge \, , ' \, , \mathbf{0} \, , \mathbf{1} \rangle \, ,$$

where $\heartsuit$ is a binary relation on B (called *compatibility*) , $\veebar$ and $\barwedge$ are two partial binary operations whose domain is $\heartsuit$, $'$ is a unary operation, $\mathbf{0}$ and $\mathbf{1}$ are special distinct elements of B. The following conditions are required:

(i) $\heartsuit$ is reflexive and symmetric;

(ii) $\forall a \in B[a \heartsuit \mathbf{0}$ and $a \heartsuit \mathbf{1}]$;
(iii) $\forall a, b, c \in B$:
if $a \heartsuit b$, $a \heartsuit c$ and $b \heartsuit c$, then $(a \veebar b) \heartsuit c$, $(a \barwedge b) \heartsuit c$ and $a' \heartsuit b$;
(iv) $\forall a, b, c \in B$: if $a \heartsuit b$, $a \heartsuit c$ and $b \heartsuit c$, then the Boolean polynomials in a, b, c form a Boolean algebra with maximum $\mathbf{1}$, minimum $\mathbf{0}$. In other words, the set $\{a, b, c\}$ generates a Boolean algebra with respect to the operations $\veebar, \barwedge, '$.

One can easily check that the set $\Pi(\mathcal{H})$ of all projections of a Hilbert space $\mathcal{H}$ gives rise to the following partial Boolean algebra,

$$\mathcal{B}^{\mathcal{H}} = \langle \Pi(\mathcal{H}), \heartsuit, \veebar, \barwedge, ', \mathbf{0}, \mathbf{1} \rangle,$$

where:

- $P \heartsuit Q$ iff $PQ = QP$;
- If $P \heartsuit Q$, then $P \veebar Q = P + Q - PQ$;
- If $P \heartsuit Q$, then $P \barwedge Q = PQ$;
- $P' = \mathbb{1} - P$;
- $\mathbf{0}$ and $\mathbf{1}$ are the zero ($\mathbb{O}$) and the identity ($\mathbb{1}$) operators, respectively.

Hence, partial Boolean algebras also represent interesting abstractions from the sharp events that may occur to a quantum system, as happens in the case of complete orthomodular lattices. In this framework, *compatibility* is defined as *commutativity* between projections.[1]

DEFINITION 13.1.2. *Homomorphism, weak embedding, embedding*
Let $\mathcal{B}_1 = \langle B_1, \heartsuit_1, \veebar_1, \barwedge_1, '^1, \mathbf{0}_1, \mathbf{1}_1 \rangle$ and $\mathcal{B}_2 = \langle B_2, \heartsuit_2, \veebar_2, \barwedge_2, '^2, \mathbf{0}_2, \mathbf{1}_2 \rangle$ be two partial Boolean algebras.

(i) A *homomorphism* of $\mathcal{B}_1$ into $\mathcal{B}_2$ is a map $h : B_1 \mapsto B_2$ such that $\forall a, b \in B_1$, the following conditions are satisfied:
(iia) if $a \heartsuit_1 b$, then $h(a) \heartsuit_2 h(b)$;
(iib) if $a \heartsuit_1 b$, then $h(a \veebar_1 b) = h(a) \veebar_2 h(b)$;
(iic) $h(a'^1) = h(a)'^2$.
(ii) A *weak embedding* of $\mathcal{B}_1$ into $\mathcal{B}_2$ is a homomorphism of $\mathcal{B}_1$ into $\mathcal{B}_2$ such that $\forall a, b \in B_1$: if $a \neq b$ and $a \heartsuit b$, then $h(a) \neq h(b)$.
(iii) An *embedding* of $\mathcal{B}_1$ into $\mathcal{B}_2$ is a homomorphism of $\mathcal{B}_1$ into $\mathcal{B}_2$ such that $\forall a, b \in B_1$: if $a \neq b$, then $h(a) \neq h(b)$.

We will now give an algebraic semantic characterization of **PaCL** (following the work of Kochen and Specker (1965a)).

The language of **PaCL** consists of a set of atomic sentences, of a distinguished atomic sentence **f** (the *falsity*) and of one primitive connective: the

[1]Recall that in orthodox QT any two (physically) compatible observables (which can be simultaneously measured) are represented by two projection-valued measures M_1 and M_2 that satisfy the following condition: $\forall \Delta, \Gamma \in \mathcal{B}(\mathbb{R})[M_1(\Delta)M_2(\Gamma) = M_2(\Gamma)M_1(\Delta)]$ (see Definition 1.3.3).

implication $\rightarrow$. The connectives *negation* ($\neg$), *conjunction* ($\curlywedge$) and *disjunction* ($\curlyvee$) are metalinguistically defined as follows:

$$\neg\alpha := \alpha \rightarrow \mathbf{f}.$$

$$\alpha \curlyvee \beta := \neg\alpha \rightarrow \beta.$$

$$\alpha \curlywedge \beta := \neg(\neg\alpha \curlyvee \neg\beta).$$

The notion of sentence (formula) is defined in the expected way. Notice that, according to this approach, the compatibility relation has no syntactical counterpart. Compatibility only concerns the meanings of our sentences.[2]

DEFINITION 13.1.3. *Algebraic realization for* **PaCL**
An *algebraic realization for* **PaCL** is a pair $\mathcal{A} = \langle \mathcal{B}, v \rangle$, where:

(i) $\mathcal{B} = \langle B, \heartsuit, \veebar, \barwedge, ', \mathbf{0}, \mathbf{1} \rangle$ is a partial Boolean algebra;
(ii) v is a partial valuation-function from the set of all sentences into B. The domain of v ($dom(v)$) is the smallest set that satisfies the following conditions:
- all atomic sentences belong to $dom(v)$;
- $v(\mathbf{f}) = \mathbf{0}$;
- If $\beta, \gamma \in dom(v)$ and $v(\beta) \heartsuit v(\gamma)$, then $\beta \rightarrow \gamma \in dom(v)$. Furthermore, $v(\beta \rightarrow \gamma) = v(\beta)' \veebar v(\gamma)$.

In other words, atomic sentences have always a meaning. A molecular sentence $\alpha \rightarrow \beta$ has a meaning only if both subformulas (α and β) have a meaning and further these two meanings are compatible.

Since in this chapter we will always refer to algebraic realizations, by simplicity we will more briefly speak of *realizations* (instead of *algebraic realizations*).

Given a partial Boolean algebra $\mathcal{B}$, the set of all valuation-functions of the language into B will be denoted by $Val(\mathcal{B})$. The set of all realizations based on $\mathcal{B}$ will be denoted by $R(\mathcal{B})$, while for any nonempty class Ω of partial Boolean algebras, $R(\Omega)$ will represent the class of all realizations based on elements of Ω.

DEFINITION 13.1.4. *Truth*
A sentence α is *true* in a realization $\mathcal{A} = \langle \mathcal{B}, v \rangle$ (abbreviated $\models_{\mathcal{A}} \alpha$) iff $\alpha \in dom(v)$ and $v(\alpha) = \mathbf{1}$.

In other words a sentence α is true iff

- α has a meaning;
- this meaning is $\mathbf{1}$.

[2]According to other approaches, the compatibility relation is supposed to be expressed in the object-language (see, for instance, Kochen and Specker (1965b) and Czelakowski (1975)). In this situation, one can obtain a calculus of **PaCL** that turns out to be very natural but somewhat heavy (one has systematically to distinguish sentences according to whether or not the compatibility relation occurs).

DEFINITION 13.1.5. *$\mathcal{B}$-validity*
Let $\mathcal{B}$ be a partial Boolean algebra. A sentence α is *$\mathcal{B}$-valid* (abbreviated $\models_{\mathcal{B}} \alpha$) iff, for any valuation function $v \in Val(\mathcal{B})$ such that $\alpha \in dom(v)$, $v(\alpha) = \mathbf{1}$.

DEFINITION 13.1.6. *Logical truth*
A sentence α is a *logical truth of* **PaCL** ($\models_{\mathbf{PCL}} \alpha$) iff, for every partial Boolean algebra $\mathcal{B}$, $\models_{\mathcal{B}} \alpha$.[3]

DEFINITION 13.1.7. *Γ-validity*
Let Γ be a class of realizations. A sentence α is *Γ-valid* ($\models_{\Gamma} \alpha$) iff, for any realization $\langle \mathcal{B}, v \rangle$ of Γ, $\alpha \in dom(v) \curvearrowright v(\alpha) = 1$.

PQL can be axiomatized. We will present here the calculus proposed by Kochen and Specker (where the language does not express the compatibility relation). The intuitive idea is the following:

- a sentence α "makes sense" if and only if it is implied by the false ($\mathbf{f} \rightarrow \alpha$). In fact, from a semantic point of view, asserting that a sentence α makes sense in a realization $\langle \mathcal{B}, v \rangle$ amounts to showing that $\alpha \in dom(v)$. Now, if $\alpha \in dom(v)$, then also $\mathbf{f} \rightarrow \alpha \in dom(v)$ and therefore $v(\mathbf{f} \rightarrow \alpha) = \mathbf{1}$. Vice versa, if $\mathbf{f} \rightarrow \alpha$ is true in $\langle \mathcal{B}, v \rangle$, then by definition, $\alpha \in dom(v)$.
- Two sentences α and β are "compatible" iff $\mathbf{f} \rightarrow (\alpha \rightarrow \beta)$ makes sense. In fact, suppose $v(\alpha) \heartsuit v(\beta)$ in a realization $\langle \mathcal{B}, v \rangle$. Then, $\mathbf{f} \rightarrow (\alpha \rightarrow \beta)$ is true in $\langle \mathcal{B}, v \rangle$. Vice versa, if $\mathbf{f} \rightarrow (\alpha \rightarrow \beta)$ is true in $\langle \mathcal{B}, v \rangle$, then $f \rightarrow (\alpha \rightarrow \beta) \in dom(v)$ so that $(\alpha \rightarrow \beta) \in dom(v)$. Thus, $v(\neg\alpha) \heartsuit v(\beta)$ and therefore $v(\alpha) \heartsuit v(\beta)$.

As happens in the case of **OL** and **OQL**, the **PaCL**-calculus is also determined as a set of *rules*. However here, unlike the case of **OL** and **OQL**, any rule will have the following simpler form:

$$\frac{\alpha_1, \ldots, \alpha_n}{\beta}$$

(if $\alpha_1, \ldots, \alpha_n$ are proved, then also β is proved).
The sentences $\alpha_1, \ldots, \alpha_n$ represent the *premisses*, while β is the *conclusion* of the rule.

Rules of **PaCL**

(PaCL1) $$\frac{\alpha}{\mathbf{f} \rightarrow \alpha}$$

(PaCL2) $$\frac{\mathbf{f} \rightarrow (\alpha \rightarrow \beta)}{\mathbf{f} \rightarrow \beta}$$

[3]An alternative semantic characterization of partial classical logic can be given in terms of *matrix-semantics*. Such semantics turns out to be very flexible and especially adequate for the analysis of the logical consequence relation (Hughes, 1987).

(PaCL3) $$\frac{(\alpha \to \mathbf{f}) \to \mathbf{f}}{\alpha}$$

(PaCL4) $$\frac{\mathbf{f} \to (\alpha \to \beta)}{\alpha \to (\beta \to \alpha)}$$

(PaCL5) $$\frac{\mathbf{f} \to (\alpha \to \beta),\ \mathbf{f} \to (\alpha \to \gamma),\ \mathbf{f} \to (\beta \to \gamma)}{(\alpha \to (\beta \to \gamma)) \to ((\alpha \to \beta) \to (\alpha \to \gamma))}$$

(PaCL6) $$\frac{\alpha,\ \alpha \to \beta}{\beta}$$

(PaCL7) $$\frac{\mathbf{f} \to (\alpha \to \beta),\ \beta \to \gamma,\ \gamma \to \beta}{\mathbf{f} \to (\alpha \to \gamma)}$$

DEFINITION 13.1.8. *Derivability*
A *derivable rule* of **PaCL** is a rule

$$\frac{\alpha_1, \cdots, \alpha_n}{\beta},$$

that satisfies the following condition: there is a sequence of sentences $\gamma_1, \cdots, \gamma_m$ such that

- $\beta = \gamma_m$;
- any γ_i $(1 \leq i < m)$ is either one of the premisses $\alpha_1, \cdots, \alpha_n$ or the conclusion of one of the **PaCL** rules, whose premisses are sentences γ_k that precede γ_i in the sequence $\gamma_1, \cdots, \gamma_m$;
- γ_m is the conclusion of a **PaCL** rule, whose premisses are in the sequence $\gamma_1, \cdots, \gamma_{m-1}$.

EXAMPLE 13.1.9. The rule $\dfrac{f \to (\alpha \to \beta)}{f \to (\beta \to \alpha)}$ is derivable in **PaCL**.

PROOF.

$$\cfrac{\cfrac{\cfrac{f \to (\alpha \to \beta)}{\alpha \to (\beta \to \alpha)}}{f \to (\alpha \to (\beta \to \alpha))}}{f \to (\beta \to \alpha)} \quad \begin{matrix}(\text{PaCL4})\\(\text{PaCL1})\\(\text{PaCL2})\end{matrix}$$

□

DEFINITION 13.1.10. *Logical theorem*
A sentence α is a *logical theorem* of **PaCL** ($\models_{\mathbf{PCL}} \alpha$) iff $\dfrac{f \to \alpha}{\alpha}$ is a derivable rule.

One can prove that this calculus is sound and complete with respect to the algebraic semantics. We will only sketch the basic steps of the arguments (by referring to the original articles).

THEOREM 13.1.11. Soundness theorem

$$\vdash_{\mathbf{PCL}} \alpha \curvearrowright \models_{\mathbf{PCL}} \alpha.$$

PROOF. Straightforward. □

As for the completeness theorem, the usual method of constructing the Lindenbaum-Tarski algebra of *all* sentences does not work for **PaCL**. Since the compatibility relation is not transitive, one cannot define an appropriate equivalence relation on the set of all sentences. However, as Kochen and Specker have shown, it is possible to overcome such a difficulty by constructing, for any given sentence, a particular Lindenbaum-Tarski algebra.
For any sentence α, let

$$B^\alpha := \left\{ \beta : \frac{\mathbf{f} \to \alpha}{\beta} \text{ is a derivable rule} \right\}.$$

We can define a relation $\equiv_\alpha$ on B^α in the following way:

$$\beta \equiv_\alpha \gamma \text{ iff } \frac{\mathbf{f} \to \alpha}{\beta \to \gamma} \text{ and } \frac{\mathbf{f} \to \alpha}{\gamma \to \beta} \text{ are derivable rules of } \mathbf{PaCL}.$$

LEMMA 13.1.12. (Kochen and Specker, 1965a)

(i) $\mathbf{f} \in B^\alpha$; $\mathbf{f} \to \mathbf{f} \in B^\alpha$.
(ii) *If* $(\beta_1 \to \beta_2) \in B^\alpha$, *then* $\beta_1, \beta_2 \in B^\alpha$.
(iii) $\equiv_\alpha$ *is an equivalence relation.*
(iv) *If* $\beta_1 \equiv_\alpha \beta_2$, *then* $(\beta_1 \to \mathbf{f}) \equiv_\alpha (\beta_2 \to \mathbf{f})$.
(v) *If* $\beta_1 \to \beta_2 \in B^\alpha$, *then* $(((\beta_1 \to \mathbf{f}) \to \mathbf{f}) \to \beta_2) \in B^\alpha$ *and* $(((\beta_1 \to \mathbf{f}) \to \mathbf{f}) \to \beta_2) \equiv_\alpha (\beta_1 \to \beta_2)$.
(vi) *If* $\beta_1 \equiv_\alpha \gamma_1$ *and* $\beta_2 \equiv_\alpha \gamma_2$, *then,* $(\beta_1 \to \beta_2) \in B^\alpha$ *iff* $\gamma_1 \to \gamma_2 \in B^\alpha$.
(vii) *If* $\beta_1 \equiv_\alpha \gamma_1$, $\beta_2 \equiv_\alpha \gamma_2$, *and* $(\beta_1 \to \beta_2) \in B^\alpha$, *then* $((\beta_1 \to \mathbf{f}) \to \beta_2) \equiv_\alpha ((\gamma_1 \to \mathbf{f}) \to \gamma_2)$.
(viii) $\beta \equiv_\alpha (\mathbf{f} \to \mathbf{f})$ *iff* $\frac{\mathbf{f} \to \alpha}{\beta}$ *is a derivable rule of* **PaCL**.

Since $\equiv_\alpha$ is an equivalence relation, we can consider the *partition* induced by $\equiv_\alpha$ on the set B^α. Let $|B^\alpha| := B^\alpha/_{\equiv_\alpha}$ be the set of all equivalence classes $[\beta]_{\equiv_\alpha}$ (where each equivalence class $[\beta]_{\equiv_\alpha}$ contains precisely all the sentences of B^α that are $\equiv_\alpha$ to β). One can define the following algebraic structure:

$$\mathcal{D}er(\alpha) = \left\langle |B^\alpha| \, , \heartsuit \, , \veebar , \barwedge , {}' \, , \mathbf{0} \, , \mathbf{1} \right\rangle ,$$

where

- for any $a_1, a_2 \in |B^\alpha|$, $a_1 \heartsuit a_2$ iff $\exists \beta_1 \in a_1, \exists \beta_2 \in a_2 : \frac{\mathbf{f} \to \alpha}{\mathbf{f} \to (\beta_1 \to \beta_2)}$ is a derivable rule of **PaCL**;
- $\mathbf{0} := [\mathbf{f}]_{\equiv_\alpha}$; $\mathbf{1} := [\mathbf{f} \to \mathbf{f}]_{\equiv_\alpha}$;

- $a' := [\beta \to \mathbf{f}]_{\equiv_\alpha}$, where $\beta \in a$;
- Let $a_1, a_2 \in |B^\alpha|$ such that $a_1 \heartsuit a_2$. Then, $a_1 \veebar a_2 := [(\beta_1 \to \mathbf{f}) \to \beta_2]_{\equiv_\alpha}$ and $a_1 \barwedge a_2 := (a_1' \veebar a_2')'$, where $\beta_1 \in a_1$ and $\beta_2 \in a_2$.

One can prove that the structure $\mathcal{D}er(\alpha)$ is a partial Boolean algebra.

THEOREM 13.1.13. (Kochen and Specker, 1965a)
Let β a sentence in B^α. If β is $\{\mathcal{D}er(\alpha)\}$-valid, then $\frac{\mathbf{f}\to\alpha}{\beta}$ is a derivable rule of **PaCL**.

On this basis, one can immediately prove the completeness theorem:

THEOREM 13.1.14. Completeness theorem

$$\models_{\mathbf{PCL}} \alpha \curvearrowright \vdash_{\mathbf{PCL}} \alpha.$$

PROOF. Suppose $\models_{\mathbf{PCL}} \alpha$. Hence, trivially, α will be $\mathcal{D}er(\alpha)$-valid. Consequently, by Theorem 13.1.13, the rule $\frac{\mathbf{f}\to\alpha}{\alpha}$ is derivable in **PaCL**, and α is a logical theorem. □

Partial Boolean algebras are not generally posets. At the same time, in any partial Boolean algebra $\mathcal{B}$, one can define a natural binary relation $\preceq$ that is "close" to a partial order. The definition is the following:

$$a \preceq b \text{ iff } a \heartsuit b \text{ and } a \veebar b = b.$$

One can easily show that such relation $\preceq$ is reflexive and antisymmetric but, generally, not transitive.[4] This is the reason why the notion of *logical consequence* cannot be adequately dealt with in the framework of the **PaCL**-semantics.

At the same time, it is quite natural to try to strengthen partial Boolean algebras, by introducing the notion of *transitive partial Boolean algebra*.

DEFINITION 13.1.15. *Transitive partial Boolean algebra*
A partial Boolean algebra $\mathcal{B}$ is *transitive* (TPBA) iff the relation $\preceq$ is a partial order.

Interestingly enough, one can show that any partial Boolean algebra $\mathcal{B}^{\mathcal{H}}$ based on the set $\Pi(\mathcal{H})$ of all projections of a Hilbert space $\mathcal{H}$ is transitive.

Thus, we can consider a natural extension of **PaCL** that we will call *transitive partial classical logic* (**TPaCL**). Semantically, **TPaCL** is characterized by the class of all transitive partial Boolean algebras (hence a sentence α will be a logical truth of **TPaCL** iff α is valid in any realization based on a transitive partial Boolean algebra). Since transitive partial Boolean algebras are partially ordered, the usual notion of *logical consequence* can be defined (as in the case of **OL** and of **OQL**).

The logic **TPaCL** can be axiomatized by adding a new rule to the **PaCL**-calculus. The rule asserts the transitivity property for the implication connective:

---

[4]A counterexample can be found in (Hughes, 1985).

(TPaCL)

$$\frac{\alpha \to \beta,\ \beta \to \gamma}{\alpha \to \gamma}.$$

The soundness and the completeness theorem for **TPaCL** can be easily proved.

13.2. Partial classical logic and the Lindenbaum property

In Chapter 9 we have seen that **OL** and **OQL** violate an important metalogical condition: the *Lindenbaum property*, according to which any semantically noncontradictory set of sentences admits a semantically noncontradictory and complete extension.

We will now study the Lindenbaum property in the framework of **PaCL**. We will first define a *relative* notion of Lindenbaum property with respect to a given class of realizations. We will show that:

(i) **PaCL** *does not satisfy* the relative Lindenbaum property with respect to the class of all realizations based on the partial Boolean algebra $\mathcal{B}^{\mathcal{H}}$ of all projections in a Hilbert space $\mathcal{H}$ of dimension at least three.

(ii) **PaCL** *does satisfy* the relative Lindenbaum property with respect to the class of all realizations based on the partial Boolean algebra $\mathcal{B}^{\mathbb{C}^2}$ of all projections in the 2-dimensional Hilbert space $\mathbb{C}^2$. As we will see in Section 13.4, the properties of such semantic realizations turn out to be correlated with the possibility of *a non-contextual hidden variable reconstruction* of spin 1/2 particles (which are described by 2-dimensional complex Hilbert spaces).

DEFINITION 13.2.1. *Realizability*
Consider a set T of sentences.

(i) Let $\langle \mathcal{B}, v\rangle$ be a realization. T is called $\langle \mathcal{B}, v\rangle$*-realizable* iff there exists an element $a \in B$ such that $a \neq \mathbf{0}$ and $\forall\beta \in T[\beta \in dom(v)$ and $a \preceq v(\beta)]$.

(ii) Let Γ be a nonempty class of realizations. T is called Γ-*realizable* (Γ-Real T) iff there exists an algebraic realization $\langle \mathcal{B}, v\rangle$ in Γ such that T is $\langle \mathcal{B}, v\rangle$-realizable.

DEFINITION 13.2.2. *Relative completeness*
Let T be a set of sentences and let $\langle \mathcal{B}, v\rangle$ be a realization. T is called $\langle \mathcal{B}, v\rangle$*-complete* iff $\forall\beta\, [\beta \in dom(v) \curvearrowright\ \beta \in T$ or $\neg\beta \in T]$.

DEFINITION 13.2.3. *Relative Lindenbaum property*
Let Γ be a nonempty class of realizations. We say that **PaCL** satisfies the *relative Lindenbaum property with respect to* Γ iff for any set T of sentences such that Γ-Real T, there exists a set T^* of sentences and a realization $\langle \mathcal{B}, v\rangle$ in Γ that satisfy the following conditions:

(i) $T \subseteq T^*$;

(ii) T^* is $\langle \mathcal{B}, v \rangle$-realizable;
(iii) T^* is $\langle \mathcal{B}, v \rangle$-complete.

The following theorem asserts a strong relation between the validity of the relative Lindenbaum property and the validity of classical logical truths.

THEOREM 13.2.4. *Let Ω be a nonempty class of partial Boolean algebras. If* **PaCL** *satisfies the Lindenbaum property with respect to $R(\Omega)$, then every logical truth of classical logic is $R(\Omega)$-valid.*

PROOF. Suppose **PaCL** satisfies the Lindenbaum property with respect to $R(\Omega)$. Let α be a logical truth of classical logic. Suppose, by contradiction, that α is not $R(\Omega)$-valid. Then, there exists an algebraic realization $\langle \mathcal{B}, v \rangle \in R(\Omega)$ such that $\alpha \in dom(v)$ and $v(\alpha) \neq \mathbf{1}$. Thus, $\neg\alpha \in dom(v)$ and $v(\neg\alpha) \neq \mathbf{0}$. Accordingly, $\{\neg\alpha\}$ is $\langle \mathcal{B}, v \rangle$-realizable. By hypothesis, **PaCL** satisfies the relative Lindenbaum property with respect to $R(\Omega)$. Thus, there is a set T^* of sentences including $\{\neg\alpha\}$ and an algebraic realization $\langle \mathcal{B}^*, v^* \rangle \in R(\Omega)$ such that T^* is $\langle \mathcal{B}^*, v^* \rangle$-realizable and $\langle \mathcal{B}^*, v^* \rangle$-complete. Define

$$B^*_{Fm} := \{v^*(\beta) : \beta \in dom(v^*)\}.$$

Let $\heartsuit, \veebar, \barwedge, '$ be the restriction to B^*_{Fm} of the corresponding relation and operations of $\mathcal{B}^*$. Since the atomic sentence $\mathbf{f}$ belongs to $dom(v^*)$ and $v^*(\mathbf{f}) = \mathbf{0}$, the element $\mathbf{0}$ belongs to B^*_{Fm}. Similarly, $\mathbf{1} = v^*(\neg\mathbf{f}) \in B^*_{Fm}$. Thus, the structure

$$\mathcal{B}^*_{Fm} = \langle B^*_{Fm}, \heartsuit, \veebar, \barwedge, ', \mathbf{0}, \mathbf{1} \rangle$$

is a partial Boolean subalgebra of $\mathcal{B}^*_{Fm}$. Let

$$F_{T^*} := \{v^*(\beta) : \beta \in T^*\}.$$

The set F_{T^*} is well defined since T^* is $\langle \mathcal{B}^*, v^* \rangle$-realizable so that every $\beta \in T^*$ belongs to $dom(v^*)$. We want to show that F_{T^*} is a complete proper filter of $\mathcal{B}^*_{Fm}$. In other words, we have to show that:

(i) F_{T^*} is a filter of $\mathcal{B}^*_{Fm}$, i.e., the following conditions are satisfied:
  (ia) $\forall a, b \in B^*_{Fm}$: if $a, b \in F_{T^*}$ and $a \heartsuit b$, then $a \barwedge b \in F_{T^*}$;
  (ib) $\forall a, b \in B^*_{Fm}$: if $a \in F_{T^*}$ and $a \heartsuit b$, then $a \veebar b \in F_{T^*}$.
(ii) F_{T^*} is proper, i.e., $\mathbf{0} \notin F^*_{T^*}$;
(iii) F_{T^*} is complete, i.e., $\forall a \in B^*_{Fm}$: $a \in B^*_{Fm}$ or $a' \in B^*_{Fm}$.

We prove only (ia). The other conditions are easily verified. Suppose $v^*(\beta), v^*(\gamma) \in B^*_{Fm}$ and $v^*(\beta) \heartsuit v^*(\gamma)$. We have to show that $v^*(\beta) \barwedge v^*(\gamma) \in F_{T^*}$. By hypothesis, $\beta, \gamma \in T^*$. Since T^* is $\langle \mathcal{B}^*, v^* \rangle$-realizable, there is an element $a \in B^*$ such that $a \neq \mathbf{0}$ and $\forall \delta \in T^* [a \preceq v^*(\delta)]$. Thus, $a \preceq v^*(\beta), v^*(\gamma)$. By definition of partial Boolean algebra,

$$a \preceq v^*(\beta) \barwedge v^*(\gamma).$$

It remains to show that $\beta \curlywedge \gamma \in T^*$. Suppose, by contradiction, that $\beta \curlywedge \gamma \notin T^*$. Now, $\beta \curlywedge \gamma \in dom(v^*)$. Thus, $\neg(\beta \curlywedge \gamma) \in dom(v^*)$. Since T^* is

$\langle \mathcal{B}^*, v^* \rangle$-complete, we obtain that $\neg(\beta \curlywedge \gamma) \in T^*$. Hence,

$$a \preceq v^*(\neg(\beta \curlywedge \gamma)) = v^*(\beta \curlywedge \gamma)'.$$

Since $v^*(\beta \curlywedge \gamma) \heartsuit v^*(\neg(\beta \curlywedge \gamma))$ and $a \preceq v^*(\beta \curlywedge \gamma)$ we can conclude that $a = \mathbf{0}$, contradiction. Thus, the set F_{T^*} is a proper and complete filter of $\mathcal{B}^*_{Fm}$.

Now let h be the map of $\mathcal{B}^*_{Fm}$ into the two-element Boolean algebra $\mathcal{B}_2$ defined as follows $\forall \beta \in dom(v^*)$:

$$h(v^*(\beta)) := \begin{cases} \mathbf{1} & \text{if } v^*(\beta) \in F_{T^*}, \\ \mathbf{0} & \text{otherwise,} \end{cases}$$

An easy computation shows that h is a homomorphism of the partial Boolean algebra $\mathcal{B}^*_{Fm}$ into the Boolean algebra $\mathcal{B}_2$. Let k be the composition of the two maps h and v^*. The map k can be extended to a valuation-function $k^\sharp$ of classical logic such that $\forall \beta \in dom(v^*)$:

$$k^\sharp(\beta) = k(\beta) := h(v^*(\beta)).$$

Now, the classical logical truth α belongs to $dom(v^*)$. Thus, $\neg\alpha \in dom(v^*)$. Furthermore, $\neg\alpha \in T^*$. Consequently, $v^*(\neg\alpha) \in F_{T^*}$. Accordingly,

$$k^\sharp(\neg\alpha) = k(\neg\alpha) = h(v^*(\neg\alpha)) = \mathbf{1}.$$

Hence, $k^\sharp(\alpha) = \mathbf{0}$, contradiction. □

THEOREM 13.2.5. Kochen-Specker-Czelakowski Theorem I (Kochen and Specker, 1967; Czelakowski, 1975)
Let $\mathcal{B}$ be a partial Boolean algebra. The following conditions are equivalent:

(i) *There is a homomorphism of $\mathcal{B}$ into a Boolean algebra;*
(ii) *Every logical truth of classical logic is* not refuted *in $\mathcal{B}$. In other words, for any logical truth α of classical logic and for any valuation function v in $Val(\mathcal{B})$ such that $\alpha \in dom(v)$, $v(\alpha) \neq \mathbf{0}$.*

THEOREM 13.2.6. Kochen-Specker-Czelakowski Theorem II (Kochen and Specker, 1967; Czelakowski, 1975)
Let $\mathcal{B}$ be a partial Boolean algebra. The following conditions are equivalent:

(i) *$\mathcal{B}$ is weakly embeddable into a Boolean algebra;*
(ii) *Every logical truth of classical logic is $\mathcal{B}$-valid.*

13.3. States on partial Boolean algebras

In Part I we have studied the abstract notion of *event-state system*, where the set of events was supposed to have the structure of an event algebra (a σ-orthocomplete orthomodular poset), while the set of states was a (nonempty σ-convex order determining) set of probability measures defined on the set of all events.

Suppose now that we have a situation in which the events are structured as a partial Boolean algebra $\mathcal{B}$. The notion of *state* on $\mathcal{B}$ will be naturally defined as follows.

DEFINITION 13.3.1. *State on a partial Boolean algebra*
Let $\mathcal{B}$ be a partial Boolean algebra. A *state* on $\mathcal{B}$ is a map $s : B \mapsto [0,1]$ that satisfies the following condition:

(i) $s(\mathbf{0}) = 0,\ s(\mathbf{1}) = 1$;
(ii) $\forall a, b \in B$ such that $a \perp b$ (i.e., $a \heartsuit b$ and $a \veebar b' = b'$):
$$s(a \veebar b) = s(a) + s(b).$$

DEFINITION 13.3.2. *Dispersion-free state*
A state s on a partial Boolean algebra $\mathcal{B}$ is called *dispersion-free* iff, $\forall a \in B$, $s(a) \in \{0,1\}$.

In other words, dispersion-free states represent dichotomous functions that assign to any event either value 1 or value 0.

LEMMA 13.3.3. *Let $\mathcal{B}$ be a partial Boolean algebra and let s be a map of B into $\{0,1\}$. The following conditions are equivalent:*

(i) *s is a dispersion-free state on $\mathcal{B}$;*
(ii) *s is a homomorphism of $\mathcal{B}$ into the two-element Boolean algebra $\mathcal{B}_2$.*

PROOF. Straightforward. □

THEOREM 13.3.4. (Kochen and Specker, 1967)
Let $\mathcal{B}$ a partial Boolean algebra. The following conditions are equivalent:

(i) *$\mathcal{B}$ is weakly embeddable into a Boolean algebra;*
(ii) *There is a weakly separating set H of homomorphisms of $\mathcal{B}$ into the two-element Boolean algebra $\mathcal{B}_2$. In other words, the set H satisfies the following condition: if $a \heartsuit b$ and $a \neq b$, then there is a map $h \in H$ such that $h(a) \neq h(b)$.*

THEOREM 13.3.5. *Let $\mathcal{H}$ be a separable Hilbert space (over the reals, the complexes or the quaternions) of dimension at least three and let $\mathcal{B}^{\mathcal{H}}$ be the partial Boolean algebra of all projections of $\mathcal{H}$. If Ω is a class of partial Boolean algebras containing $\mathcal{B}^{\mathcal{H}}$, then* **PaCL** *does not satisfy the relative Lindenbaum property with respect to $R(\Omega)$.*

PROOF. Suppose, by contradiction, that **PaCL** satisfies the Lindenbaum property with respect to $R(\Omega)$. Then, by Theorem 13.2.4, every logical truth of classical logic is $R(\Omega)$-valid. Thus, in particular, every logical truth of classical logic is $\mathcal{B}^{\mathcal{H}}$-valid. Therefore, by Kochen-Specker-Czelakowski Theorem II, there exists a Boolean algebra $\mathcal{B}$ and a weak embedding of $\mathcal{B}^{\mathcal{H}}$ into $\mathcal{B}$. By Theorem 13.3.4, there exists a weakly separating set H of homomorphisms of $\mathcal{B}^{\mathcal{H}}$ into the two-element Boolean algebra B_2. Hence, by Lemma 13.3.3, H is a weakly separating set of dispersion-free states on $\mathcal{B}^{\mathcal{H}}$. Let $h \in H$. We distinguish two cases, according to whether or not the dimension of $\mathcal{H}$ is finite.
(a) Suppose $\mathcal{H}$ is finite dimensional. Then, the dispersion-free state h is trivially σ-additive. Accordingly, h is a probability measure on the orthomodular lattice $\Pi(\mathcal{H})$ of all projections of $\mathcal{H}$. Thus, by Gleason's Theorem

(see Theorem 3.3.3), there exists a density operator ρ of $\mathcal{H}$ such that, for any projection $P \in \Pi(\mathcal{H})$, $h(P) = \mathrm{Tr}(\rho P) \in \{0,1\}$. Now, the following conditions hold:

$$\mathrm{Tr}(\rho P) = 1 \text{ iff } \rho P = \rho; \ \mathrm{Tr}(\rho P) = 0 \text{ iff } \rho P = \mathbb{O}.$$

Thus, if $\rho P \neq \rho$ and $\rho P \neq \mathbb{O}$, we will have: $h(P) \notin \{0,1\}$, contradiction.
(b) Suppose $\mathcal{H}$ is infinite dimensional. Then, the dispersion-free state h need not be σ-additive. Consequently, the map h is not, in general, a probability measure on $\Pi(\mathcal{H})$. Therefore, we cannot apply Gleason's Theorem directly. Let $\mathbf{D}$ be one of the reals, the complexes or the quaternions. One can show that the partial Boolean algebra $\mathcal{B}^{\mathbf{D}^3}$ of all projections of $\mathbf{D}^3$ can be embedded into $\mathcal{B}^{\mathcal{H}}$. Accordingly, the restriction of h to $\mathcal{B}^{\mathbf{D}^3}$ is a dispersion-free state on $\mathcal{B}^{\mathbf{D}^3}$, allowing us to use the previous argument to conclude that h cannot be a dispersion-free state on $\mathcal{B}^{\mathcal{H}}$. □

Note that the proof of Theorem 13.3.5 presented uses Gleason's Theorem in an essential way. However, the theorem can also be proved in a more direct way. As shown by Schütte, there is a logical truth of classical logic that is refuted in $\mathcal{B}^{\mathbb{R}^3}$ (Bub, 1999). By Theorem 13.2.4, this is sufficient to conclude that **PaCL** does not satisfy the relative Lindenbaum property with respect to any class of realizations that includes the class $R(\mathcal{B}^{\mathbb{R}^3})$.

THEOREM 13.3.6. *Let Ω be a nonempty class of partial Boolean algebras containing the two-element Boolean algebra $\mathcal{B}_2$. If every logical truth of classical logic is $R(\Omega)$-valid, then* **PaCL** *satisfies the relative Lindenbaum property with respect to $R(\Omega)$.*

PROOF. Let T be a $R(\Omega)$-realizable set of sentences. Then, there is an algebraic realization $\langle \mathcal{B}^*, v^* \rangle \in R(\Omega)$ such that T is $\langle \mathcal{B}^*, v^* \rangle$-realizable. Thus, there is an element $a \in B^*$ such that $a \neq \mathbf{0}$ and, $\forall \beta \in T$, $\beta \in dom(v^*)$ and $a \preceq v^*(\beta)$. By hypothesis, every logical truth of classical logic is $R(\Omega)$-valid. Therefore, by the Kochen-Specker-Czelakowski Theorem II, there exists a weak embedding h of $\mathcal{B}^*$ into a Boolean algebra $\mathcal{B}$. Let

$$h(T) := \{h(v^*(\beta)) \, : \, \beta \in T\}\,.$$

The set $h(T)$ is well-defined since, $\forall \beta \in T$, $\beta \in dom(v^*)$. We want to show that $h(T)$ has the finite intersection property. In other words, if $\{a_1, \ldots, a_n\}$ is a finite subset of $h(T)$, then $a_1 \wedge \ldots \wedge a_n \neq \mathbf{0}$. Let $\{h(v^*(\beta_1)), \ldots, h(v^*(\beta_n))\}$ be a finite subset of $h(T)$. Suppose, by contradiction, that $h(v^*(\beta_1)) \wedge \ldots \wedge h(v^*(\beta_n)) = \mathbf{0}$. Now, $\forall i (1 \leq i \leq n)$: $a \heartsuit v^*(\beta_i)$ and $a \wedge v^*(\beta_i) = a$. Hence, $h(a) \leq h(v^*(\beta_i))$. Therefore, $h(a) \leq h(v^*(\beta_1)) \wedge \ldots \wedge h(v^*(\beta_n)) = \mathbf{0}$. Thus, $h(a) = \mathbf{0}$. Since $a \heartsuit \mathbf{0}$ and h is a weak embedding of $\mathcal{B}^*$ into the Boolean algebra $\mathcal{B}$, we obtain $a = \mathbf{0}$, contradiction. Thus, $h(T)$ has the finite intersection property. Therefore, $h(T)$ can be extended to a proper filter of $\mathcal{B}$. By Stone's Theorem, there exists a proper and complete filter F of $\mathcal{B}$ such that $h(T) \subseteq F$. Accordingly, the map $k : \mathcal{B} \to \{0,1\}$ such that $\forall a \in B$:

$$k(a) := \begin{cases} 1 & \text{if } a \in F, \\ 0 & \text{otherwise,} \end{cases}$$

is a homomorphism of $\mathcal{B}$ into $\mathcal{B}_2$. Let $v^\sharp$ be the map that associates to any atomic sentence $\mathbf{p}$ the element $k(h(v^*(\mathbf{p}))) \in \{0,1\}$. The map $v^\sharp$ can be extended to a valuation-function $v^\flat$ of classical logic such that $\forall \beta \in dom(v^*)$:

$$v^\flat(\beta) = k(h(v^*(\beta))).$$

Let

$$T^* := \left\{\gamma : v^\flat(\gamma) = 1\right\}.$$

It is easy to see that $T \subseteq T^*$. Furthermore, T^* is both $\langle \mathcal{B}_2 \,, v^\flat \rangle$-realizable and $\langle \mathcal{B}_2 \,, v^\flat \rangle$-complete. Consequently, **PaCL** satisfies the Lindenbaum property with respect to $R(\Omega)$. □

As a consequence of Theorems 13.2.4 and 13.3.6, for any class Ω of partial Boolean algebras containing the two-element Boolean algebra $\mathcal{B}_2$, the following two conditions are equivalent:

- the relative Lindenbaum property holds with respect to $R(\Omega)$;
- all classical logical truths are valid in $R(\Omega)$.

THEOREM 13.3.7. *The partial Boolean algebra $\mathcal{B}^{\mathbb{C}^2}$ of all projections of the 2-dimensional Hilbert space $\mathbb{C}^2$ admits a separating set of dispersion-free states (homomorphisms into $\mathcal{B}_2$).*

PROOF. Let $X = \left\{\{P,Q\} : P \in \Pi(\mathbb{C}^2) \text{ and } Q = \mathbb{1} - P\right\}$. Let σ be a selection function $\sigma : X \to \bigcup X$, i.e., $\sigma(\{P,Q\}) \in \{P,Q\}$, with $\sigma(\{\mathbb{0},\mathbb{1}\}) = \mathbb{1}$. Let $s_\sigma : \Pi(\mathbb{C}^2) \to [0,1]$ be defined by $s_\sigma(P) = 1$ if $\sigma(\{P, \mathbb{1} - P\}) = P$ and 0 otherwise. It is clear that each s_σ is a dispersion-free state on $\Pi(\mathbb{C}^2)$ and that these are all of them. Moreover, it is also clear that they form a separating set of dispersion-free states on $\mathcal{B}^{\mathbb{C}^2}$. □

THEOREM 13.3.8. *Consider the partial Boolean algebra $\mathcal{B}^{\mathbb{C}^2}$ and the two-element Boolean algebra $\mathcal{B}_2$.* **PaCL** *satisfies the Lindenbaum property with respect to $R(\mathcal{B}^{\mathbb{C}^2}) \cup R(\mathcal{B}_2)$.*

PROOF. By Theorem 13.3.7, $\mathcal{B}^{\mathbb{C}^2}$ admits a separating set of homomorphisms into $\mathcal{B}_2$. Thus, by Theorem 13.2.6, $\mathcal{B}^{\mathbb{C}^2}$ is weakly embeddable into a Boolean algebra. By the Kochen-Specker-Czelakowski Theorem II, every logical truth of classical logic is $R(\mathcal{B}^{\mathbb{C}^2})$-valid. Therefore, by Theorem 13.3.6, **PaCL** satisfies the Lindenbaum property with respect $R(\mathcal{B}^{\mathbb{C}^2}) \cup R(\mathcal{B}_2)$. □

13.4. The Lindenbaum property and the hidden variable problem

The debate concerning the question whether QT can be considered a *physically complete* account of microphenomena has a long and deep history. A turning point in this discussion has been the celebrated Einstein-Bohr debate, with the ensuing charge of incompleteness raised by the Einstein-Podolsky-Rosen argument (EPR).

As we already know, in the framework of orthodox QT, physical systems can be prepared in *pure states* that have, in general, positive dispersion for most physical quantities. In the EPR argument, the attention is focused on the question whether the account of the microphysical phenomena provided by QT is to be regarded as an *exhaustive* description of the physical reality to which those phenomena are supposed to refer, a question to which Einstein himself answered in the negative.

There is a mathematical side of the completeness issue: the question becomes whether *states with positive dispersion* can be represented as a different, *dispersion-free*, kind of states in a way that is *consistent* with the mathematical constraints of the quantum theoretical formalism. In his book on the mathematical foundations of quantum mechanics, John von Neumann proved a celebrated "No go theorem" asserting the logical incompatibility between the quantum formalism and the existence of dispersion free states (satisfying some general conditions). Already in the preface, von Neumann anticipates the program and the conclusion concerning the possibility of 'neutralizing' the statistical character of QT:

> There will be a detailed discussion of the problem as to whether it is possible to trace the statistical character of quantum mechanics to an ambiguity (i.e., incompleteness) in our description of nature. Indeed, such an interpretation would be a natural concomitant of the general principle that each probability statement arises from the incompleteness of our knowledge. This explanation "by hidden parameters" [...] has been proposed more than once. However, it will appear that this can scarcely succeed in a satisfactory way, or more precisely, such an explanation is incompatible with certain qualitative fundamental postulates of quantum mechanics.

According to the advocates of hidden variables, QT is a *physically incomplete theory*. The intuitive idea that represents the common background to almost all hidden variable theories can be described in the following way:

(I) the reason why a physical theory is *statistical* depends on the fact that the description provided by the states is *incomplete*.

(II) It is possible to add a set Ξ of parameters (*hidden variables*) in such a way that

- for every state s and for every $\omega \in \Xi$, there exists a dispersion-free (dichotomous) state s_ω which semantically decides every *property* (*event*) of the physical system at issue;
- the statistical predictions of the original theory should be recovered by averaging over these dichotomous states;
- the algebraic structures determined by the properties (events) of the system should be preserved in the hidden variable extension.

The hidden variable theories based on the assumptions (I) and (II) are usually called *non-contextual*, because they require the existence of a single space Ξ of hidden variables determining dispersion-free states. A weaker position is represented by the *contextual hidden variable theories*, according to which the choice of the hidden variable space depends on the physical quantity to be dealt with. As pointed out by Beltrametti and Cassinelli (1981):

> Despite the absence of mathematical obstacles against contextual hidden variable theories, it must be stressed that their calling for completed states that are probability measures not on the whole proposition [event] lattice $\mathcal{E}$ but only on a subset of $\mathcal{E}$ is rather far from intuitive physical ideas of what a state of a physical system should be. Thus, contextual hidden variable theorists, in their search for the restoration of some classical deterministic aspects, have to pay, on other sides, in quite radical departures from properties of classical states.

Von Neumann's proof of his "No go theorem" was based on a general assumption that has been, later, considered too strong. The condition asserts the following:

Let s_ω be a dispersione-free state and let A, B be two (possibly noncompatible) observables. Then,

$$Exp(A + B, s_\omega) = Exp(A, s_\omega) + Exp(B, s_\omega).$$

In other words, the expectation functional determined by the completed state s_ω is linear.

Interestingly enough, **PaCL** permits us to develop a purely logical argument for a "No go theorem," such that von Neumann's strong assumption can be relaxed. In fact, Theorem 13.3.5 and Theorem 13.3.8 have a deep impact on the possibility of a non-contextual hidden variable reconstruction of QT. Suppose a quantum system $\mathfrak{S}$ admits of a non-contextual hidden variable theory and let $\mathcal{H}$ be the Hilbert space associated to $\mathfrak{S}$. Then, the partial Boolean algebra $\mathcal{B}^{\mathcal{H}}$ of the projections of $\mathcal{H}$ shall admit a set of dispersion-free states. As a consequence, one obtains that **PaCL** should satisfy the relative Lindenbaum property with respect to the class of all algebraic realizations based on the two-element Boolean algebra and on the partial Boolean algebra of all projections of $\mathcal{H}$. However, we have proved

(Theorem 13.3.5) that this is impossible, whenever the dimension of $\mathcal{H}$ is at least 3.

On the other hand, 2-dimensional complex Hilbert spaces turn out to admit non-contextual hidden variable models. In fact, we have proved that **PaCL** does satisfy the Lindenbaum property with respect to the class of all algebraic realizations based on the two-element Boolean algebra and on the partial Boolean algebra $\mathcal{B}^{\mathbb{C}^2}$.

Thus, we can conclude that there is a *deep logical connection* between the two following questions:

- does a quantum system $\mathfrak{S}$ admit a non-contextual hidden variable theory?
- Does **PaCL** satisfy the relative Lindenbaum property with respect to the algebraic realizations concerning the events that may occur to the system $\mathfrak{S}$?

CHAPTER 14

Unsharp quantum logics

The quantum logics we have studied so far are all examples of *sharp logics*. Both the *logical* and the *semantic* version of the noncontradiction principle hold:

- any contradiction $\alpha \curlywedge \neg\alpha$ is always false;[1]
- a sentence α and its negation $\neg\alpha$ cannot both be true.

We will now study *unsharp* forms of quantum logic; these may be regarded as natural logical abstractions from the *effect-state systems* that we have investigated in Part I.[2] The first logic we are going to present is *paraconsistent quantum logic.*

14.1. Paraconsistent quantum logic

Paraconsistent quantum logic (**PQL**) represents the most obvious unsharp weakening of orthologic. In the algebraic semantics, this logic is characterized by the class of all realizations based on a bounded involution lattice, where the noncontradiction principle ($a \wedge a' = \mathbf{0}$) is possibly violated.

In the Kripkean semantics, instead, **PQL** is characterized by the class of all realizations $\mathcal{K} = \langle I\,, \mathcal{R}\,, Pr\,, V\rangle$, where the accessibility relation $\mathcal{R}$ is symmetric (but not necessarily reflexive), while Pr behaves as in the **OL** case (i.e., Pr is a set of propositions that contains I, $\emptyset$ and is closed under the operations $\cap$ and $'$). Any pair $\langle I\,, \mathcal{R}\rangle$, where R is a symmetric relation on I, will be called a *symmetric frame*. All the other semantic definitions are given as in the case of **OL**, *mutatis mutandis*. On this basis, one can show that our algebraic and Kripkean semantics characterize the same logic.

Unlike **OL** and **OQL**, a world i of a **PQL** realization may verify a contradiction. Since $\mathcal{R}$ is generally not reflexive, it may happen that $i \in V(\alpha)$ and $i \perp V(\alpha)$. Hence, $i \models \alpha \curlywedge \neg\alpha$.

In spite of this a contradiction cannot be verified by all worlds of a realization K.

LEMMA 14.1.1. *For any realization $\mathcal{K} = \langle I\,, \mathcal{R}\,, Pr\,, V\rangle$ and any sentence α, $\mathcal{K} \not\models \alpha \curlywedge \neg\alpha$.*

[1]Of course, in the case of **PaCL**, contradictions are false only if defined.

[2]These logics will be only studied at the sentential level. First-order extensions of unsharp quantum logics present an interesting subject for future research.

PROOF. Suppose, by contradiction, $\mathcal{K} \models \alpha \curlywedge \neg\alpha$. Hence, for any $i \in I$, $i \models \alpha$ and $\forall j \not\perp i [j \not\models \alpha]$. Since $\emptyset$ is a proposition, for any $i \in I$, we have:

$$i \notin \emptyset \curvearrowright \exists j \not\perp i \forall k \not\perp j \, [k \notin \emptyset].$$

Whence $\forall i \in I \, \exists j \in I \, [i \not\perp j]$. Consequently, $\exists j \in I \, [j \models \alpha$ and $j \not\models \alpha]$. □

An axiomatization of **PQL** can be obtained by dropping the *absurdity rule* and the *Duns Scotus rule* in the **OL** calculus. As with **OL**, the logic **PQL** satisfies the finite model property and is consequently decidable.

Hilbert-space realizations for **PQL** can be constructed, in a natural way, both in the algebraic and in the Kripkean semantics. In the algebraic semantics, take the realizations based on the Mac Neille completion of a bounded involution poset having the form

$$\langle \mathcal{E}(\mathcal{H}) \, , \leq \, , ' \, , \mathbf{0} \, , \mathbf{1} \rangle \, ,$$

where $\mathcal{E}(\mathcal{H})$ is the set of all effects of a Hilbert space $\mathcal{H}$. In the Kripkean semantics, consider the realizations based on the following frames

$$\langle \mathcal{E}(\mathcal{H}) - \{\mathbf{0}\} \, , \not\perp \rangle \, ,$$

where $\not\perp$ represents the nonorthogonality relation between effects ($E \not\perp F$ iff $E \not\leq F'$). Unlike the corresponding case involving projections, in this situation the accessibility relation is symmetric but generally nonreflexive. For instance, the semi-transparent effect $\frac{1}{2}\mathbb{I}$ (representing the prototypical ambiguous property) is a fixed point of the generalized complement $'$. Hence,

$$\frac{1}{2}\mathbb{I} \perp \frac{1}{2}\mathbb{I} \text{ and } (\frac{1}{2}\mathbb{I})' \perp (\frac{1}{2}\mathbb{I})'.$$

From the physical point of view, possible worlds are here identified with possible pieces of information about the physical system under investigation. Any information may correspond to:

- a pure state (a maximal information);
- a proper mixture (a non-maximal information);
- a projection (a sharp property);
- a proper effect (an unsharp property).

Thus, unlike the sharp situations considered in Chapter 8, here possible worlds do not always correspond to states of the quantum system under investigation. As expected, violations of the noncontradiction principle will be determined by unsharp (ambiguous) pieces of knowledge.

PQL represents a somewhat rough logical abstraction from the class of all effect-structures. As we already know, a characteristic condition that holds in all effect-structures is the *regularity property* (which may fail in a generic **PQL**-realization).

DEFINITION 14.1.2. An algebraic **PQL** realization $\langle \mathcal{B} \, , v \rangle$ is called *regular* iff the bounded involution lattice $\mathcal{B}$ is regular ($a \wedge a' \leq b \vee b'$).

It is easy to see that, in the setting of an involution lattice, this property is equivalent to the definition given in 1.1.6. The regularity property can be naturally formulated also in the framework of the Kripkean semantics.

DEFINITION 14.1.3. A **PQL** Kripkean realization $\langle I\,,\,\mathcal{R}\,,\,Pr\,,\,V\rangle$ is *regular* iff its frame $\langle I\,,\,\mathcal{R}\rangle$ is *regular*. In other words, $\forall i,j \in I$: $i \perp i$ and $j \perp j \curvearrowright i \perp j$.

One can prove that a symmetric frame $\langle I\,,\,\mathcal{R}\rangle$ is regular iff the bounded involution lattice of all propositions of $\langle I\,,\,\mathcal{R}\rangle$ is regular. As a consequence, an algebraic realization is regular iff its Kripkean transformation is regular, where the Kripkean (resp., algebraic) transformation of an algebraic (resp., Kripkean) realization is defined as we did for **OL**.

Thus, one can introduce a proper extension of **PQL**: *regular paraconsistent quantum logic* (**RPQL**). Semantically **RPQL** is characterized by the class of all regular realizations (both in the algebraic and in the Kripkean semantics). The calculus for **RPQL** is obtained by adding to the **PQL**-calculus the following rule:

$$\alpha \curlywedge \neg\alpha \vdash \beta \curlyvee \neg\beta \qquad (\textit{Kleene rule})$$

A completeness theorem for both **PQL** and **RPQL** can be proved, similarly to the case of **OL**. Both logics **PQL** and **RPQL** admit a natural modal translation (as happens in the case of **OL**).

An interesting question concerns the relation between **PQL** and the orthomodular property.

Let $\mathcal{B} = \langle B\,,\,\leq\,,\,'\,,\,\mathbf{0}\,,\mathbf{1}\rangle$ be an ortholattice. By Lemma 8.1.14 the following three conditions (expressing possible definitions of the orthomodular property) turn out to be equivalent:

(i) $\forall a,b \in B$: $a \leq b \curvearrowright b = a \vee (b \wedge a')$;
(ii) $\forall a,b \in B$: $a \leq b$ and $b \wedge a' = \mathbf{0} \curvearrowright a = b$;
(iii) $\forall a,b \in B$: $a \wedge (a' \vee (a \wedge b)) \leq b$.

However, this equivalence breaks down in the case of bounded involution lattices. One can only prove:

LEMMA 14.1.4. *Let $\mathcal{B}$ be a bounded involution lattice. If $\mathcal{B}$ satisfies condition (i), then $\mathcal{B}$ satisfies conditions (ii) and (iii).*

PROOF. (i) implies (ii): trivial. Suppose (i); we want to show that (iii) holds. Now, $a' \leq a' \vee b' = (a \wedge b)'$. Therefore, by (i), $(a \wedge b)' = a' \vee (a \wedge (a \wedge b)')$. By the de Morgan law, $a \wedge b = (a \wedge (a' \vee (a \wedge b)) \leq b$. □

LEMMA 14.1.5. *Any bounded involution lattice $\mathcal{B}$ that satisfies condition (iii) is an ortholattice.*

PROOF. Suppose (iii). It is sufficient to prove that, $\forall a,b \in B$, $a \wedge a' \leq b$. Now, $a \wedge a' \leq a$ and $a' \leq a' \vee (a \wedge b)$; hence, by (iii), $a \wedge a' \leq a \wedge (a' \vee (a \wedge b)) \leq b$. Thus, $\forall a \in B$, $a \wedge a' = \mathbf{0}$. □

As a consequence, we can conclude that there exists no proper orthomodular paraconsistent quantum logic when orthomodularity is understood in the sense (i) or (iii). A residual possibility for a proper paraconsistent quantum logic to be orthomodular is orthomodularity in the sense (ii). In fact, the 3-element chain is a bounded involution lattice which is orthomodular (ii), but neither orthomodular (i) nor orthomodular (iii).

Hilbert space realizations for orthomodular paraconsistent quantum logic can be constructed in the algebraic semantics by taking as support the following proper subset of the set of all effects:

$$\mathcal{E}_c(\mathcal{H}) := \{a\mathbb{I} \, : \, a \in [0,1]\} \cup \Pi(\mathcal{H}).$$

In other words, a possible meaning of a sentence is either a sharp event (projection) or an unsharp effect that can be represented as a multiple of the universal event ($\mathbb{I}$). Hence, all proper unsharp effects are supposed to have a very special form. We will call the elements of $\mathcal{E}_c(\mathcal{H})$ *eccentric effects*.

Unlike the case of $\mathcal{E}(\mathcal{H})$, which is not a lattice, the set $\mathcal{E}_c(\mathcal{H})$ of all eccentric effects is closed under $\wedge$ and $\vee$. As a consequence, $\mathcal{E}_c(\mathcal{H})$ determines a regular lattice which satisfies the condition we have called orthomodular (ii), where the partial order is the partial order of $\mathcal{E}(\mathcal{H})$ restricted to $\mathcal{E}_c(\mathcal{H})$, while the involution is defined as in the class of all effects ($E' := \mathbb{I} - E$).

From the logical point of view, an interesting feature of **PQL** is represented by the fact that this logic is a common sublogic in a wide class of important logics. In particular, **PQL** is a sublogic of Girard's linear logic, of Łukasiewicz' infinitely many-valued logic and of some relevant logics.

14.2. ε-Preclusivity spaces

Interesting physical models for **RPQL** can be constructed by using the notion of ε-preclusivity space (Cattaneo and Giuntini, 1995), where the preclusivity relation is accompanied by a characteristic approximation degree ε. Consider a Hilbert effect-state system $(\mathcal{E}(\mathcal{H}), \mathcal{S}(\mathcal{H}))$, and let $\varepsilon \in [0,1/2)$. In other words, the approximation-degree is supposed to be smaller than 1/2. Define now the following relation $\perp_\varepsilon$ on the set of states: $\forall m_\rho, m_\sigma \in \mathcal{S}(\mathcal{H})$:

$$m_\rho \perp_\varepsilon m_\sigma \;\text{ iff }\; \exists E \in \mathcal{E}(\mathcal{H})[m_\rho(E) \geq 1-\varepsilon \;\text{ and }\; m_\sigma(E) \leq \varepsilon].$$

Clearly, for $\varepsilon = 0$, we have: $\perp_\varepsilon = \perp$.

LEMMA 14.2.1. $\forall \varepsilon \in [0,1/2)$, *the pair* $(\mathcal{S}(\mathcal{H}), \perp_\varepsilon)$ *is an orthoframe.*

PROOF. Straightforward. □

Thus, for any $\varepsilon \in [0,1/2)$, the ε-preclusivity relation will give rise to Kripkean realizations for **OQL**.

Consider now a situation in which the approximation-degree is greater than or equal to 1/2, in other words, $\varepsilon \in [1/2,1]$. Define the relation $\perp_\varepsilon$ as follows $\forall m_\rho, m_\sigma \in \mathcal{S}(\mathcal{H})$: $m_\rho \perp_\varepsilon m_\sigma$ iff

$$\exists E \in \mathcal{E}(\mathcal{H}) - \{\lambda\mathbb{I} \, : \, \lambda \in [\varepsilon, 1-\varepsilon]\} \, [m_\rho(E) \geq 1-\varepsilon \;\text{ and }\; m_\sigma(E) \leq \varepsilon].$$

Notice that if the set $\{\lambda \mathbb{I} : \lambda \in [\varepsilon, 1-\varepsilon]\}$ is not excluded, the relation $\perp_\varepsilon$ would have the unpleasant feature of being reflexive.

LEMMA 14.2.2. *$\forall \varepsilon \in [1/2, 1]$, the pair $(\mathcal{S}(\mathcal{H}), \perp_\varepsilon)$ is a symmetric and regular frame.*

PROOF. That $\perp_\varepsilon$ is symmetric is trivial. We prove that $\perp_\varepsilon$ is regular. Let m_ρ, m_σ be two states in $\mathcal{S}(\mathcal{H})$ such that

$$m_\rho \perp_\varepsilon m_\rho \text{ and } m_\sigma \perp_\varepsilon m_\sigma.$$

We want to show that $m_\rho \perp_\varepsilon m_\sigma$. If $\varepsilon = 1$, the proof is trivial. Thus, we may suppose $\varepsilon \neq 1$. By hypothesis there are two effects $E_1, E_2 \notin \{\lambda \mathbb{I} : \lambda \in [1-\varepsilon, \varepsilon]\}$ such that

$$1 - \varepsilon \leq k_1 := m_\rho(E_1) \leq \varepsilon \text{ and } 1 - \varepsilon \leq k_2 := m_\sigma(E_2) \leq \varepsilon.$$

Since $\varepsilon \neq 1$, we have that $k_1, k_2 \notin \{0, 1\}$. Suppose, without loss of generality, that $k_1 \leq k_2$. It is easy to see that the following inequalities hold:

$$k_1 + 1 - \frac{k_1}{k_2} \leq 1 \tag{a}$$

$$1 - \frac{k_1}{k_2} \geq 0 \tag{b}$$

Let us the define the following linear operator:

$$F = k_1 \mathbb{I} + \left(1 - \frac{k_1}{k_2}\right) E_2.$$

One can easily show that F does not belong to $\{\lambda \mathbb{I} : \lambda \in [1-\varepsilon, \varepsilon]\}$. In order to show that F is an effect, it is sufficient to prove that F is positive and dominated by $\mathbb{I}$; but the first follows from (b) and the second follows from (a).

We want to show that $m_\rho(F) \geq 1 - \varepsilon$ and $m_\sigma(F) \leq \varepsilon$, so that $m_\rho \perp_\varepsilon m_\sigma$. Now, $m_\rho(F) = k_1 + (1 - \frac{k_1}{k_2}) m_\rho(E_2)$. Since $(1 - \frac{k_1}{k_2}) m_\rho(E_2) \geq 0$, we have $m_\rho(F) \geq k_1 = m_\rho(E_1) \geq 1 - \varepsilon$. Now, $m_\sigma(F) = k_1 + (1 - \frac{k_1}{k_2}) k_2 = k_2 = m_\sigma(E_2) \leq \varepsilon$. Hence, $m_\rho \perp_\varepsilon m_\sigma$. □

On this basis, one immediately obtains genuine Kripkean realizations for **RPQL** where:

- possible worlds are represented by pure and mixed states of a quantum system;
- the accessibility relation $\not\perp_\varepsilon$ is generally not reflexive. Hence, the noncontradiction principle will be violated.

Interestingly enough, the dividing line between the sharp and the unsharp case is given by the value of ε. As we have seen:

(i) if $\varepsilon \in [0, \frac{1}{2})$, one obtains Kripkean realizations for the sharp logic **OQL**;

(ii) if $\varepsilon \in [\frac{1}{2}, 1]$, one obtains Kripkean realization for the unsharp **RPQL**.

14.3. An aside: similarities of PQL and historiography

The Kripkean semantics for **PQL** focuses a kind of semantic behavior, that can be also recognized in other domains of experience (*prima facie*, quite far from microphysics). In fact, the notion of possible world can be successfully applied not only by logicians and by physicists. For instance, also historians naturally interact with possible worlds, at least implicitly. The role of possible worlds in historical research has been investigated by Toraldo di Francia (1990). We will use here a slightly different semantic approach.

As in the case of physical states and of physical effects, even *historical sources* determine pieces of information about possible states of affairs. From an abstract point of view, in the simplest case, a source can be idealized as a set of sentences:

a source i will *assert* a sentence α when α is contained in i.

For instance the source *The Lives of the Twelve Caesars* by Svetonius asserts the sentence:

Evident miracles had announced to Julius Caesar his violent death.

More generally a source may be represented also by some non-linguistic objects (monuments, tools, tombs, ...). As a consequence, the relation "i asserts α" might have a weaker meaning, such as "α is *confirmed*, *testified* by i."

As in a quantum logical situation, two sources may be either compatible or incompatible. We will say that two sources i and j are (logically) compatible when it is not the case that i asserts a sentence α whereas j asserts the negation $\neg\alpha$. Since a source is not necessarily consistent, we will have that some sources might be incompatible with themselves. Apparently, the presence of a local contradiction is not a sufficient reason that renders a given source completely unreliable. Consequently, as in the case of unsharp quantum logic, the accessibility relation $\not\perp$ between pieces of information is symmetric, but generally not reflexive.

Let I represent a set of sources that are available to a given historian. When will our historian accept the historical truth of a given statement? At first sight, one is tempted to take into consideration the two following possibilities:

I) All sources assert α.

II) At least one source asserts α.

However, our first choice seems too strong: many events that are sometimes considered as "historical facts" are not asserted by all sources. At the same time, our second choice appears too weak: some sources might be not completely reliable. For instance, why should we trust Svetonius when he asserts that "Evident miracles had announced to Julius Caesar his violent death"?

An intermediate choice between I) and II), which seems quite reasonable from an intuitive point of view, naturally leads to a quantum-logic-like truth condition:

> a source i asserts the truth of an atomic sentence **p** if and only if for any source j compatible with i, there exists a compatible source k that asserts **p**.
>
> In other words:

$$i \models \mathbf{p} \text{ iff } \forall j \not\perp i \exists k \not\perp j (k \models \mathbf{p}).$$

From an intuitive point of view, this typical quantum-logical condition seems to assert that **p** represents a *stable assertion*. As to the case of compound sentences, again the quantum logical truth conditions appear to be quite natural. In particular, one might mention a number of examples in historical contexts, where an alternative is true, whereas both members are strongly indeterminate. Hence, it seems reasonable to require:

$$i \models \neg\alpha \text{ iff } \forall j \not\perp i \, [j \not\models \alpha].$$

$$i \models \alpha \curlywedge \beta \text{ iff } i \models \alpha \text{ and } i \models \beta.$$

Finally let us ask: when will a historian accept the historical truth of a given sentence? The natural answer, from the abstract semantics perspective, is the following: a historian will accept the truth of α, when α is asserted by a sufficiently dominant set of sources that are judged reliable by the historian. In this framework, the set of all reliable sources can be modeled as the set of all possible worlds of a **PQL**-realization. As a consequence, by Lemma 14.1.1, we obtain the following situation: even if some sources are contradictory, our historian will not assert contradictory sentences. On this basis we can conclude that truths discovered by historians and by quantum physicists turn out to have some strong formal similarities!

CHAPTER 15

The Brouwer Zadeh logics

We now study some stronger examples of unsharp quantum logic, that have been called *Brouwer Zadeh logics* (also *fuzzy intuitionistic logics*). These logics represent natural abstractions from the class of all BZ-lattices (defined in Chapter 4). As expected, a characteristic property of Brouwer Zadeh logics is a splitting of the connective "not" into two forms of negation: a *fuzzy-like* negation, that gives rise to a *paraconsistent* behavior and an *intuitionistic-like* negation. The fuzzy "not" represents a weak negation, that inverts the two extreme truth-values (truth and falsity), satisfies the double negation principle but generally violates the noncontradiction principle. The second "not" is a stronger negation, a kind of necessitation of the fuzzy "not".

We consider two different forms of Brouwer Zadeh logic: **BZL** (*weak Brouwer Zadeh logic*) and $\mathbf{BZL^3}$ (*strong Brouwer Zadeh logic*). The language of both logics is an extension of the language of **QL**. The primitive connectives are: the *conjunction* ($\curlywedge$), the *fuzzy* negation ($\neg$), the *intuitionistic* negation ($\sim$).

Disjunction is metatheoretically defined in terms of conjunction and of the fuzzy negation:

$$\alpha \curlyvee \beta := \neg(\neg\alpha \curlywedge \neg\beta)\,.$$

A *necessity* operator is defined in terms of the intuitionistic and of the fuzzy negation:

$$L\alpha := \sim \neg\alpha\,.$$

A *possibility* operator is defined in terms of the necessity operator and of the fuzzy negation:

$$M\alpha := \neg L \neg\alpha\,.$$

15.1. The weak Brouwer Zadeh logic

Let us first consider the weaker logic **BZL**. As happens with **OL**, **OQL** and **PQL**, also **BZL** can be characterized by an algebraic and by a Kripkean semantics.

DEFINITION 15.1.1. *Algebraic realization for* **BZL**
An *algebraic realization* of **BZL** is a pair $\langle \mathcal{B}\,,\, v\rangle$, consisting of a BZ-lattice $\langle B\,, \leq\,, '\,, {}^{\sim}\,, \mathbf{0}\,, \mathbf{1}\rangle$ and a valuation-function v that associates to any sentence α an element in B, satisfying the following conditions:

(i) $v(\neg\beta) = v(\beta)'$;

(ii) $v(\sim \beta) = v(\beta)^{\sim}$;
(iii) $v(\beta \curlywedge \gamma) = v(\beta) \wedge v(\gamma)$.

The definitions of truth, consequence in an algebraic realization for **BZL**, logical truth and logical consequence are given as in the case of **OL**.

A Kripkean semantics for **BZL** has been first proposed in (Giuntini, 1991b). A characteristic feature of this semantics is the use of frames with two accessibility relations.

DEFINITION 15.1.2. *Kripkean realization for* **BZL**
A *Kripkean realization* for **BZL** is a system $\mathcal{K} = \langle I\,, \not\perp\,, \not\underset{\sim}{\perp}\,, Pr\,, V\rangle$ where:

(i) $\langle I\,, \not\perp\,, \not\underset{\sim}{\perp}\rangle$ is a frame with a nonempty set I of possible worlds and two accessibility relations: $\not\perp$ (the *fuzzy accessibility* relation) and $\not\underset{\sim}{\perp}$ (the *intuitionistic accessibility* relation).
Two worlds $i\,, j$ are called *fuzzy-accessible* iff $i \not\perp j$. They are called *intuitionistically-accessible* iff $i \not\underset{\sim}{\perp} j$. We write $i \perp j$ and $i \underset{\sim}{\perp} j$ for not $(i \not\perp j)$ and not $(i \not\underset{\sim}{\perp} j)$, respectively.
The following conditions are required for the two accessibility relations:

  (ia) $\langle I\,, \not\perp\rangle$ is a regular symmetric frame;
  (ib) any world is fuzzy-accessible to at least one world:
  $$\forall i\, \exists j : i \not\perp j.$$
  (ic) $\langle I\,, \not\underset{\sim}{\perp}\rangle$ is an orthoframe;
  (id) Fuzzy accessibility implies intuitionistic accessibility:
  $$i \not\perp j \quad \curvearrowright \quad i \not\underset{\sim}{\perp} j.$$
  (ie) Any world i has a kind of "twin-world" j such that for any world k:
    (a) $i \not\underset{\sim}{\perp} k$ iff $j \not\underset{\sim}{\perp} k$
    (b) $i \not\underset{\sim}{\perp} k \quad \curvearrowright \quad j \not\perp k$.

  For any set X of worlds, the *fuzzy orthogonal* set $X^{\perp}$ is defined as in **OL**:
  $$X^{\perp} = \{i \in I \,:\, \forall j \in X[i \perp j]\}\,.$$
  Similarly, the *intuitionistic orthogonal* set $X^{\sim}$ is defined as follows:
  $$X^{\sim} = \{i \in I \,:\, \forall j \in X[i \underset{\sim}{\perp} j]\}\,.$$
  The notion of *proposition* is defined as in **OL**: a set of worlds X is a proposition iff $X = X^{\perp\perp}$.
  One can prove that for any set of worlds X, both $X^{\perp}$ and $X^{\sim}$ are propositions. Furthermore, as in **OL**, $X \wedge Y$ (the greatest proposition included in the propositions X and Y) is $X \cap Y$, while $X \vee Y$ (the smallest proposition including X and Y) is $(X \cup Y)^{\perp\perp}$.

(ii) Pr is a set of propositions that contains I, and is closed under $^{\perp}$, $^{\sim}$, and $\wedge$.

(iii) V associates to any sentence a proposition in Pr according to the following conditions:
$V(\neg\beta) = V(\beta)^{\perp}$;
$V(\sim\beta) = V(\beta)^{\sim}$;
$V(\beta \curlywedge \gamma) = V(\beta) \wedge V(\gamma)$.

THEOREM 15.1.3. *Let* $\langle I\,, \not\perp\,, \not\perp\!\!\!\perp \rangle$ *be a BZ-frame (i.e., a frame satisfying the conditions of Definition 15.1.2) and let* Pr^0 *be the set of all propositions of the frame. Then, the structure* $\langle Pr^0\,, \subseteq\,, {}^{\perp}\,, {}^{\sim}\,, \emptyset\,, I \rangle$ *is a complete BZ-lattice such that for any set* $\Gamma \subseteq Pr^0$*:*

$$\bigwedge \Gamma = \bigcap \Gamma \quad \textit{and} \quad \bigvee \Gamma = \left(\bigcup \Gamma\right)^{\perp\perp}.$$

As a consequence, the proposition-structure $\langle Pr\,, \subseteq\,, {}^{\perp}\,, {}^{\sim}\,, \emptyset\,, I \rangle$ of a **BZL** realization, is a BZ-lattice.

The definitions of truth, consequence in a Kripkean realization, logical truth and logical consequence, are given as in the case of **OL**.

One can prove, with standard techniques, that the algebraic and the Kripkean semantics for **BZL** characterize the same logic.

We now introduce a calculus that represents an adequate axiomatization for the logic **BZL**. The most intuitive way to formulate this calculus is to present it as a *modal extension* of the axiomatic version of regular paraconsistent quantum logic **RPQL**. (Recall that the modal operators of **BZL** are defined as follows: $L\alpha :=\sim \neg\alpha$; $M\alpha := \neg L \neg\alpha$).

Rules of **BZL.**

The BZL-calculus includes, besides the rules of **RPQL** the following modal rules:

(BZ1) $L\alpha \vdash \alpha$

(BZ2) $L\alpha \vdash LL\alpha$

(BZ3) $ML\alpha \vdash L\alpha$

(BZ4) $\dfrac{\alpha \vdash \beta}{L\alpha \vdash L\beta}$

(BZ5) $L\alpha \curlywedge L\beta \vdash L(\alpha \curlywedge \beta)$

(BZ6) $\emptyset \vdash \neg(L\alpha \curlywedge \neg L\alpha)$

The rules (BZ1)-(BZ5) give rise to an $\mathbf{S}_5$–like modal behavior. The rule (BZ6) (the noncontradiction principle for necessitated sentences) is, of course, trivial in any classical modal system.

One can prove a soundness and completeness theorem with respect to the Kripkean semantics (by an appropriate modification of the corresponding proofs for **OL**).

THEOREM 15.1.4. Soundness theorem

$$T \vdash_{\mathbf{BZL}} \alpha \curvearrowright T \models_{\mathbf{BZL}} \alpha.$$

THEOREM 15.1.5. Completeness theorem

$$T \models_{\mathbf{BZL}} \alpha \curvearrowright T \vdash_{\mathbf{BZL}} \alpha.$$

Characteristic logical properties of **BZL** are the following:

(a) as in **PQL**, the distributive principles, Duns Scotus, the non-contradiction and the excluded middle principles break down for the fuzzy negation;

(b) as in intuitionistic logic:

$$\models_{\mathbf{BZL}} \sim (\alpha \curlywedge \sim \alpha); \quad \not\models_{\mathbf{BZL}} \alpha \curlyvee \sim \alpha; \quad \alpha \models_{\mathbf{BZL}} \sim\sim \alpha; \quad \sim\sim \alpha \not\models_{\mathbf{BZL}} \alpha;$$

$$\sim\sim\sim \alpha \models_{\mathbf{BZL}} \sim \alpha; \; \alpha \models_{\mathbf{BZL}} \beta \curvearrowright \sim \beta \models_{\mathbf{BZL}} \sim \alpha;$$

(c) furthermore:

$$\sim \alpha \models_{\mathbf{BZL}} \neg \alpha; \quad \neg \alpha \not\models_{\mathbf{BZL}} \sim \alpha; \quad \neg \sim \alpha \models_{\mathbf{BZL}} \sim\sim \alpha;$$

One can prove that **BZL** has the finite model property; as a consequence it is decidable (Giuntini, 1992).

15.2. The pair semantics and the strong Brouwer Zadeh logic

The stronger logic $\mathbf{BZL}^3$ represents a natural abstraction from the structure of all *pair-propositions* that arise in the framework of concrete effect-state systems (see Chapter 4). Consider an effect-state system $(\mathcal{E}(\mathcal{H}), \mathcal{S}(\mathcal{H}))$. As we have learned, for any effect E, the pair-proposition associated to E is the pair $(\mathbf{Yes}(E), \mathbf{No}(E))$, where both $\mathbf{Yes}(E)$ and $\mathbf{No}(E)$ are closed subsets of $\mathcal{S}(\mathcal{H})$. Unlike the sharp case, the positive proposition $\mathbf{Yes}(E)$ does not determine the negative proposition $\mathbf{No}(E)$ (and vice versa). Of course, the notion of pair-proposition can be naturally generalized to the case of abstract effect-state systems.

This situation has suggested a new form of quantum logical semantics (first studied in (Cattaneo and Nisticò, 1986). The intuitive idea, underlying this semantics (which has some features in common with Klaua's *partielle Mengen* and with Dunn's *polarities*) can be sketched as follows: one supposes that interpreting a language means associating to any sentence two *domains of certainty*: the domain of the situations where our sentence certainly holds, and the domain of the situations where our sentence certainly does not hold. As happens in the case of Kripkean semantics, the situations we are referring to can be thought of as *possible worlds*. However, unlike the standard Kripkean behavior, the positive domain of a given sentence does not generally determine the negative domain of the same sentence. As a consequence, propositions are here identified with particular *pairs* of sets of worlds, rather than with particular sets of worlds.

Let us assume again that we are dealing with the **BZL** language. We now define the notion of *pair-realization* for a **BZL** language.

DEFINITION 15.2.1. *Pair-realization*
A *pair-realization* is a system $\mathcal{O} = \langle I, \mathcal{R}, \Omega, V \rangle$, where:

(i) $\langle I, \mathcal{R} \rangle$ is an orthoframe.

(ii) Let Pr^0 be the set of all propositions of the orthoframe $\langle I, \mathcal{R} \rangle$. As we already know, this set gives rise to an ortholattice with respect to the operations $\wedge, \vee$ and ${}^{\perp}$ (where $\wedge$ is the set-theoretic intersection).
A *pair-proposition* of $\langle I, \mathcal{R} \rangle$ is any pair $\langle A_1, A_0 \rangle$, where A_1, A_0 are propositions in Pr^0 such that $A_1 \perp A_0$. The following operations and relations can be defined on the set of all orthopairpropositions:

(iia) The fuzzy complement:

$$\langle A_1, A_0 \rangle^{\circlearrowleft} := \langle A_0, A_1 \rangle .$$

(iib) The intuitionistic complement:

$$\langle A_1, A_0 \rangle^{\sim} := \left\langle A_0, A_0^{\perp} \right\rangle .$$

(iic) The pair-propositional conjunction:

$$\langle A_1, A_0 \rangle \sqcap \langle B_1, B_0 \rangle := \langle A_1 \wedge B_1, A_0 \vee B_0 \rangle .$$

(iid) The pair-propositional disjunction:

$$\langle A_1, A_0 \rangle \sqcup \langle B_1, B_0 \rangle := \langle A_1 \vee B_1, A_0 \wedge B_0 \rangle .$$

(iie) The (possibly) infinitary conjunction:

$$\bigsqcap_{j \in J} \{\left\langle A_1^j, A_0^j \right\rangle\} := \left\langle \bigcap_{j \in J} \{A_1^j\}, \bigvee_{j \in J} \{A_0^j\} \right\rangle$$

(where J is any set of indices).

(iif) The (possibly) infinitary disjunction:

$$\bigsqcup_{j \in J} \{\left\langle A_1^j, A_0^j \right\rangle\} := \left\langle \bigvee_{j \in J} \{A_1^j\}, \bigcap_{j \in J} \{A_0^j\} \right\rangle$$

(where J is any set of indices).

(iig) The necessity operator:

$$\Box(\langle A_1, A_0 \rangle) := \left\langle A_1, A_1^{\perp} \right\rangle .$$

(iih) The possibility operator:

$$\Diamond(\langle A_1, A_0 \rangle) := (\Box(\langle A_1, A_0 \rangle^{\circlearrowleft}))^{\circlearrowleft}.$$

(iik) The order-relation:

$$\langle A_1, A_0 \rangle \leq \langle B_1, B_0 \rangle \text{ iff } A_1 \subseteq B_1 \text{ and } B_0 \subseteq A_0.$$

(iii) Ω is a set of pair-propositions, that is closed under ${}^{\circlearrowleft}$, ${}^{\sim}$, $\sqcap$ and $\sqcup$, while $\mathbf{0} := \langle \emptyset, I \rangle$.

(iv) V is a valuation-function that maps sentences into pair-propositions according to the following conditions:
$V(\neg\beta) = V(\beta)^{\textcircled{\prime}}$;
$V(\sim\beta) = V(\beta)^{\textcircled{\sim}}$;
$V(\beta \curlywedge \gamma) = V(\beta) \sqcap V(\gamma)$.

The other basic semantic definitions are given as in the algebraic semantics. One can prove the following theorem:

THEOREM 15.2.2. *Let $\langle I\,,\,\mathcal{R}\rangle$ be an orthoframe and let Ω^0 be the set of all pair-propositions of $\langle I\,,\,\mathcal{R}\rangle$. Then, the structure*

$$\left\langle \Omega^0\,,\,\leq\,,\,^{\textcircled{\prime}}\,,\,^{\textcircled{\sim}}\,,\,\langle\emptyset, I\rangle\,,\,\langle I,\emptyset\rangle\right\rangle$$

is a complete BZ-lattice (with respect to the infinitary conjunction and disjunction defined above). Furthermore, the following conditions are satisfied: for any $\langle A_0, A_1\rangle\,,\langle B_0\,,B_1\rangle \in \Omega^0$:

(i) $\Box\,\langle A_1\,,A_0\rangle = \langle A_1\,,A_0\rangle^{\textcircled{\prime}\textcircled{\sim}}$;
(ii) $\langle A_1\,,A_0\rangle^{\textcircled{\sim}} = \Box(\langle A_1\,,A_0\rangle^{\textcircled{\prime}})$;
(iii) $\Diamond\,\langle A_1\,,A_0\rangle = \langle A_1\,,A_0\rangle^{\textcircled{\sim}\textcircled{\prime}}$;
(iv) $(\langle A_1, A_0\rangle \sqcap \langle B_1, B_0\rangle)^{\textcircled{\sim}} = \langle A_1, A_0\rangle^{\textcircled{\sim}} \sqcup \langle B_1, B_0\rangle^{\textcircled{\sim}}$
(Strong de Morgan law)
(v) $(\langle A_1, A_0\rangle \sqcap \langle B_1, B_0\rangle^{\textcircled{\sim}\textcircled{\sim}}) \subseteq (\langle A_1, A_0\rangle^{\textcircled{\prime}\textcircled{\sim}} \sqcup \langle B_1, B_0\rangle)$.

Accordingly, in any pair-realization the set of all pair-propositions Ω^0 gives rise to a BZ-lattice. As a consequence, one can immediately prove a soundness theorem with respect to the pair-semantics.

Does, perhaps, the pair-semantics characterizes the logic **BZL**? The answer to this question is negative. As a counterexample, let us consider an instance of the fuzzy excluded middle and an instance of the intuitionistic excluded middle applied to the same sentence α:

$$\alpha \curlyvee \neg\alpha \quad \text{and} \quad \alpha \curlyvee \sim\alpha.$$

One can easily check that they are logically equivalent in the pair-semantics. For, given any pair-realization $\mathcal{O}$, we have:

$$\alpha \curlyvee \neg\alpha \models_{\mathcal{O}} \alpha \curlyvee \sim\alpha \quad \text{and} \quad \alpha \curlyvee \sim\alpha \models_{\mathcal{O}} \alpha \curlyvee \neg\alpha\,.$$

However, generally

$$\alpha \curlyvee \neg\alpha \not\models_{\mathbf{BZL}} \alpha \curlyvee \sim\alpha\,.$$

For instance, consider the following algebraic **BZL**–realization $\mathcal{A} = \langle\mathcal{B}\,,\,v\rangle$, where the support of $\mathcal{B}$ is the real interval $[0\,,1]$ and the algebraic structure on $\mathcal{B}$ is defined as follows:

$\leq$ is the natural order defined on $[0,1]$;
$a' = 1 - a$;

$$a^{\sim} = \begin{cases} 1 & \text{if } a = 0, \\ 0 & \text{otherwise.} \end{cases}$$

$$\mathbf{1} = 1; \quad \mathbf{0} = 0.$$

Suppose that for a given atomic sentence $\mathbf{p}$: $0 < V(\mathbf{p}) < 1/2$. We have: $V(\mathbf{p} \curlyvee \sim \mathbf{p}) = \max(V(\mathbf{p}), 0) = V(\mathbf{p}) < 1/2$. At he same time, $V(\mathbf{p} \curlyvee \neg \mathbf{p}) = \max(V(\mathbf{p}), 1 - V(\mathbf{p})) = 1 - V(\mathbf{p}) \geq 1/2$. Hence, $V(\mathbf{p} \curlyvee \sim \mathbf{p}) < V(\mathbf{p} \curlyvee \neg \mathbf{p})$. As a consequence, the pair-semantics characterizes a logic stronger than **BZL**. We call this logic $\mathbf{BZL^3}$. The name is due to the characteristic three-valued features of the pair-semantics. Since the meaning of any sentence α is identified with the pair consisting of its positive and of its negative domain, only three cases turn out to be semantically relevant (for any possible world): 1) α is true; 2) α is false; 3) α is indeterminate.

The logic $\mathbf{BZL^3}$ is axiomatizable. A suitable calculus can be obtained by adding to the **BZL**-calculus the following rules.

Rules of $\mathbf{BZL^3}$.

(BZ^{3}1) $\quad L(\alpha \curlyvee \beta) \vdash L\alpha \curlyvee L\beta$

(BZ^{3}2) $\quad \dfrac{L\alpha \vdash \beta,\ \alpha \vdash M\beta}{\alpha \vdash \beta}$

The following rules are derivable:

(DR1) $\quad \dfrac{L\alpha \vdash \beta, M\alpha \vdash M\beta}{\alpha \vdash \beta}$

(DR2) $\quad M\alpha \curlywedge M\beta \vdash M(\alpha \curlywedge \beta)$

(DR3) $\quad \sim(\alpha \curlywedge \beta) \vdash \sim\alpha \curlyvee \sim\beta$

The validity of a strong de Morgan's principle for the connective $\sim$ (DR3) shows that this connective represents, in this logic, a kind of strong "superintuitionistic" negation. This is in contrast with the case of **BZL**, where the strong de Morgan law fails, as it does in intuitionistic logic.

One can prove a soundness and a completeness theorem of this calculus with respect to the pair-semantics.

THEOREM 15.2.3. Soundness theorem

$$T \vdash_{\mathbf{BZL^3}} \alpha \;\curvearrowright\; T \models_{\mathbf{BZL^3}} \alpha.$$

PROOF. By routine techniques. □

THEOREM 15.2.4. Completeness theorem

$$T \models_{\mathbf{BZL^3}} \alpha \;\curvearrowright\; T \vdash_{\mathbf{BZL^3}} \alpha.$$

Sketch of the proof. Instead of $T \models_{\mathbf{BZL^3}} \alpha$ and $T \vdash_{\mathbf{BZL^3}} \alpha$, we will write, more briefly, $T \models \alpha$ and $T \vdash \alpha$, respectively. It is sufficient to construct a canonical model $\mathcal{O} = \langle I, \mathcal{R}, \Omega, V \rangle$ such that:

$$T \models_{\mathcal{O}} \alpha \;\curvearrowright\; T \vdash \alpha.$$

(The converse follows from the soundness theorem).

Definition of a canonical model for $\mathbf{BZL^3}$

(i) I is the set of all possible sets i of sentences satisfying the following conditions:
  (ia) i is *noncontradictory* with respect to the fuzzy negation: for any α, if $\alpha \in i$, then $\neg\alpha \notin i$;
  (ib) i is *L-closed*: for any α, if $\alpha \in i$, then $L\alpha \in i$;
  (ic) i is *deductively closed*: for any α, if $i \vdash \alpha$, then $\alpha \in i$.

(ii) The accessibility relation $\mathcal{R}$ is defined as follows:
  $i\mathcal{R}j$ iff, for any sentence α, $\alpha \in i \curvearrowright \neg\alpha \notin j$.
  (In other words, i and j are not contradictory with respect to the fuzzy negation).
  Instead of not $i\mathcal{R}j$, we write $i \perp j$.

(iii) Ω is the set of all pair-propositions of $\langle I\,, \mathcal{R}\rangle$.

(iv) For any atomic sentence $\mathbf{p}$:

$$V(\mathbf{p}) = \langle V_1(\mathbf{p})\,, V_0(\mathbf{p})\rangle\ ,$$

where:

$$V_1(\mathbf{p}) = \{i\,:\,i \vdash \mathbf{p}\} \quad\text{and}\quad V_0(\mathbf{p}) = \{i\,:\,i \vdash \neg\mathbf{p}\}\,.$$

$\mathcal{O}$ is well defined since one can prove the following lemmas:

LEMMA 15.2.5. *R is reflexive and symmetric.*

LEMMA 15.2.6. *For any $\alpha\,, \{i\,:\,i \vdash \alpha\}$ is a proposition of the orthoframe $\langle I\,, \mathcal{R}\rangle$.*

LEMMA 15.2.7. *For any α,* $\{i\,:\,i \vdash \alpha\} \perp \{i\,:\,i \vdash \neg\alpha\}$.

Furthermore, one can prove:

LEMMA 15.2.8. *For any α,* $V(\alpha) = \langle V_1(\alpha)\,, V_0(\alpha)\rangle$, *where:*

$$V_1(\alpha) = \{i\,:\,i \vdash \alpha\}\,;$$
$$V_0(\alpha) = \{i\,:\,i \vdash \neg\alpha\}\,.$$

PROOF. By induction on the length of α. □

LEMMA 15.2.9. *For any sentence α:*

$$\mathbf{0} := \langle \emptyset, I\rangle = \langle \{i\,:\,i \vdash L\alpha \curlywedge \neg L\alpha\}\,, \{i\,:\,i \vdash \neg(L\alpha \curlywedge \neg L\alpha)\}\rangle\,.$$

LEMMA 15.2.10. *Let $T = \{\alpha_1\,, \ldots\,, \alpha_n\,, \ldots\}$ be a set of sentences and let α be any sentence. Then,*

$$\bigcap\{V_1(\alpha_n)\,:\,\alpha_n \in T\} \subseteq V_1(\alpha) \curvearrowright L\alpha_1\,, \ldots\,, L\alpha_n\,, \ldots \vdash \alpha.$$

As a consequence, one can prove:

LEMMA 15.2.11. Lemma of the canonical model

$$T \models_{\mathcal{O}} \alpha \curvearrowright T \vdash \alpha.$$

Suppose $T \models_{\mathcal{O}} \alpha$. Hence, (by definition of consequence in a given realization): for any pair-proposition $\langle A_1, A_0\rangle \in \Omega$, if for all $\alpha_n \in T$, $\langle A_1, A_0\rangle \leq V(\alpha_n)$, then $\langle A_1, A_0\rangle \leq V(\alpha)$.

The propositional lattice, consisting of all pair-propositions of $\mathcal{O}$ is complete (see Theorem 15.2.2). Hence, $\bigsqcap_n \{V(\alpha_n) : \alpha_n \in T\} \leq V(\alpha)$. In other words, by definition of $\leq$:

(i) $\bigcap\{V_1(\alpha_n) : \alpha_n \in T\} \subseteq V_1(\alpha)$;
(ii) $V_0(\alpha) \subseteq \bigvee\{V_0(\alpha_n) : \alpha_n \in T\}$.

Thus, by (i) and by Lemma 15.2.10: $L\alpha_1, \ldots, L\alpha_n, \ldots \vdash \alpha$. Consequently, there exists a finite subset $\{\alpha_{n_1}, \ldots, \alpha_{n_k}\}$ of T such that $L\alpha_{n_1} \curlywedge \ldots \curlywedge L\alpha_{n_k} \vdash \alpha$. Hence, by the rules for $\curlywedge$ and L: $L(\alpha_{n_1} \curlywedge \ldots \curlywedge \alpha_{n_k}) \vdash \alpha$. At the same time, we obtain from (ii) and by Lemma 15.2.8: $V_1(\neg\alpha) \leq \bigvee\{V_1(\neg\alpha_n) : \alpha_n \in T\}$.
Whence, by de Morgan,

$$V_1(\neg\alpha) \subseteq \left[\bigcap\left\{(V_1(\neg\alpha_n))^{\perp} : \alpha_n \in T\right\}\right]^{\perp}.$$

Now, one can easily check that in any realization: $V_1(\neg\alpha)^{\perp} = V_1(M\alpha)$. As a consequence: $V_1(\neg\alpha) \subseteq [\bigcap\{(V_1(M\alpha_n) : \alpha_n \in T\}]^{\perp}$. Hence, by contraposition:

$$\bigcap\{V_1(M\alpha_n) : \alpha_n \in T\} \subseteq (V_1(\neg\alpha))^{\perp}$$

and

$$\bigcap\{V_1(M\alpha_n) : \alpha_n \in T\} \subseteq V_1(M\alpha).$$

Consequently, by Lemma 15.2.10 and by the $\mathbf{S_5}$-rules:

$$LM\alpha_1, \ldots, LM\alpha_n, \ldots \vdash M\alpha, \qquad M\alpha_1, \ldots, M\alpha_n, \ldots \vdash M\alpha.$$

By syntactical compactness, there exists a finite subset $\{\alpha_{m_1}, \ldots, \alpha_{m_h}\}$ of T such that $M\alpha_{m_1}, \ldots, M\alpha_{m_h} \vdash M\alpha$. Whence, by the rules for $\curlywedge$ and M: $M(\alpha_{m_1} \curlywedge \ldots \curlywedge \alpha_{m_h}) \vdash M\alpha$. Let us put $\gamma_1 = \alpha_{n_1} \curlywedge \ldots \curlywedge \alpha_{n_k}$ and $\gamma_2 = \alpha_{m_1} \curlywedge \ldots \curlywedge \alpha_{m_h}$. We have obtained: $L\gamma_1 \vdash \alpha$ and $M\gamma_2 \vdash M\alpha$. Whence, $L\gamma_1 \curlywedge L\gamma_2 \vdash \alpha$, $L(\gamma_1 \curlywedge \gamma_2) \vdash \alpha$, $M\gamma_1 \curlywedge M\gamma_2 \vdash M\alpha$ and $M(\gamma_1 \curlywedge \gamma_2) \vdash M\alpha$. From $L(\gamma_1 \curlywedge \gamma_2) \vdash \alpha$ and $M(\gamma_1 \curlywedge \gamma_2) \vdash M\alpha$ we obtain, by the derivable rule (DR1), $\gamma_1 \curlywedge \gamma_2 \vdash \alpha$. Consequently, $T \vdash \alpha$. □

As happens with other forms of quantum logic, $\mathbf{BZL^3}$ admits of an algebraic semantic characterization (Giuntini, 1993) based on the notion of BZ^3-lattice.

DEFINITION 15.2.12. A *BZ^3-lattice* (BZL^3) is a BZ-lattice

$$\mathcal{B} = \langle B, \leq, ', ^{\sim}, \mathbf{0}, \mathbf{1}\rangle,$$

that satisfies the following conditions:

(i) $(a \wedge b)^{\sim} = a^{\sim} \vee b^{\sim}$;
(ii) $a \wedge b^{\sim\sim} \sqsubseteq a'^{\sim} \vee b$.

By Theorem 15.2.2, the set of all pair-propositions of an orthoframe determines a complete BZ^3-lattice.

The following representation theorem holds:

THEOREM 15.2.13. *Every BZ^3-lattice is embeddable into the (complete) BZ^3-lattice of all pair-propositions of an orthoframe.*

One can prove that the pair-semantics and the algebraic semantics strongly characterize the same logic.[1]

Some problems concerning the Brouwer Zadeh logics still remain open:

1) Is there any Kripkean characterization of the logic that is algebraically characterized by the class of all de Morgan BZ-lattices? In this framework, the problem can be reformulated in this way: is the (strong) de Morgan law elementary?
2) Is it possible to axiomatize a logic based on an infinite many-valued generalization of the ortho-pair semantics?
3) Find possible conditional connectives in $\mathbf{BZL^3}$.
4) Find an appropriate orthomodular extension of $\mathbf{BZL^3}$.

15.3. $\mathbf{BZL^3}$-effect realizations

As we have seen, the logic $\mathbf{BZL^3}$ has been suggested by the structure of all pair-propositions arising from (concrete and abstract) effect-state systems. On this basis, the construction of effect-realizations for $\mathbf{BZL^3}$ turns out to be quite natural. Suppose that $(\mathcal{E}(\mathcal{H}), \mathcal{S}(\mathcal{H}))$ is a concrete effect-state system associated to a quantum system $\mathfrak{S}$. As we already know, a pair-proposition of $(\mathcal{E}(\mathcal{H}), \mathcal{S}(\mathcal{H}))$ is a pair $(\mathbf{Yes}(E), \mathbf{No}(E))$, where E is an effect belonging to $\mathcal{E}(\mathcal{H})$.

The pair-realization associated to the quantum system $\mathfrak{S}$ can be now defined as follows:

$$\mathcal{O}^{\mathfrak{S}} = \langle I\,,\, \mathcal{R}\,, \Omega\,, V\rangle\,,$$

where:

- I is the set $\mathcal{S}(\mathcal{H})$ of all states of $\mathfrak{S}$;
- $\mathcal{R}$ is the nonorthogonality relation $\not\perp$ defined on $\mathcal{S}(\mathcal{H})$. Hence, we have:

$$\begin{aligned} i \not\perp j \ \text{ iff not } \exists E \in \mathcal{E}(\mathcal{H})[i(E) = 1 \text{ and } j(E) = 0] \\ \text{iff not } \exists P \in \Pi(\mathcal{H})[i(P) = 1 \text{ and } j(P) = 0]; \end{aligned}$$

- Ω is the smallest set that contains all the possible pair-propositions $(\mathbf{Yes}(E), \mathbf{No}(E))$ of the effect system $(\mathcal{E}(\mathcal{H}), \mathcal{S}(\mathcal{H}))$, and is closed under the operations $^{\circlearrowleft}$, $^{\circleddash}$, $\sqcap$ and $\sqcup$;
- $V(\mathbf{p})$ is an effect-pair proposition $(\mathbf{Yes}(E), \mathbf{No}(E))$, where E is an effect in $\mathcal{E}(\mathcal{H})$.

One can easily show that $\mathcal{O}^{\mathfrak{S}}$ is a pair-realization because:

[1] $\mathbf{BZL^3}$ can be also characterized by means of a non-standard version of Kripkean semantics (Giuntini, 1993).

(i) $\mathcal{R}$ is a similarity relation defined on I;

(ii) For any effect E, both $\mathbf{Yes}(E)$ and $\mathbf{No}(E)$ are propositions of the orthoframe $\langle I\,,\,\mathcal{R}\rangle$, since they are closed subsets of I (see Section 4.4);

(iii) The set Ω of all pair-propositions of $(\mathcal{E}(\mathcal{H}), \mathcal{S}(\mathcal{H}))$ contains the impossible proposition $\mathbf{0} = (\mathbf{Yes}(\mathbb{O}), \mathbf{No}(\mathbb{O}))$, and the certain proposition $\mathbf{1} = (\mathbf{Yes}(\mathbb{1}), \mathbf{No}(\mathbb{1}))$. Furthermore, Ω is, by definition, closed under the operations $^{\circlearrowleft}$, $^{\circleddash}$, $\sqcap$ and $\sqcup$;

(iv) $V(\mathbf{p})$ is by definition a pair-proposition.

Notice that, unlike the sharp case, in this setting the atomic sentences do not generally correspond to sharp questions (A, Δ). Whenever $V(\mathbf{p})$ is the pair-proposition $(\mathbf{Yes}(E), \mathbf{No}(E))$ of a proper effect E, we may say that E has an "ambiguous meaning".

Let us now compare the BZ-poset $\langle \mathcal{E}(\mathcal{H})\,,\,\le\,,\,'\,,\,^{\sim}\,,\,\mathbf{0}\,,\mathbf{1}\rangle$ and the proposition-BZ lattice $\langle \Omega\,,\,\le\,,\,^{\circlearrowleft}\,,\,^{\circleddash}\,,\,\mathbf{0}\,,\mathbf{1}\rangle$. One can naturally define a map h that associates to any effect $E \in \mathcal{E}(\mathcal{H})$ a pair-proposition in Ω:

$$h : E \rightarrowtail (\mathbf{Yes}(E), \mathbf{No}(E)).$$

The following conditions hold:

- h is order-preserving. Suppose $E \le F$. Then,

$$\mathbf{Yes}(E) \subseteq \mathbf{Yes}(F), \quad \mathbf{No}(F) \subseteq \mathbf{No}(E).$$

  Hence, $(\mathbf{Yes}(E), \mathbf{No}(E)) \le (\mathbf{Yes}(F), \mathbf{No}(F))$.

- h preserves both the Zadeh and the Brouwer complements:

$$h(E') = (\mathbf{Yes}(E'), \mathbf{No}(E')) = (\mathbf{No}(E), \mathbf{Yes}(E)) = [h(E)]^{\circlearrowleft};$$

$$\begin{aligned} h(E^{\sim}) &= h(P_{Ker(E)}) = (\mathbf{Yes}(P_{Ker(E)}), \mathbf{No}(P_{Ker(E)})) \\ &= (\mathbf{No}(E), \mathbf{No}(E)^{\perp}) = (\mathbf{Yes}(E), \mathbf{No}(E))^{\circleddash} = [h(E)]^{\circleddash}. \end{aligned}$$

- h is not injective. Unlike the case of projections, it may happen that two different effects E,F have one and the same positive and negative proposition:

$$\mathbf{Yes}(E) = \mathbf{Yes}(F), \quad \mathbf{No}(E) = \mathbf{No}(F).$$

  As an example take $\frac{1}{2}\mathbb{1}$ and $\frac{3}{4}\mathbb{1}$.

  Hence, moving from effects to pair-propositions clearly determines a loss of information. In fact, pair-propositions are only concerned with the two extreme probability-values 0 and 1, a situation that corresponds to a three-valued semantics.

Unlike the effect-structure, our proposition-structure is a lattice. It may happen that the conjunction between two effects E and F is not defined; but the conjunction between the two corresponding propositions is always defined. We have:

$$(\mathbf{Yes}(E), \mathbf{No}(E)) \sqcap (\mathbf{Yes}(F), \mathbf{No}(F)) = (\mathbf{Yes}(E) \wedge \mathbf{Yes}(F), \mathbf{No}(E) \vee \mathbf{No}(F)).$$

As a consequence, there is an important logical difference between the sharp and the unsharp situation. As we have seen, in the sharp case, events and propositions give rise to isomorphic structures and to one and the same form of quantum logic. In the unsharp case, however, we have obtained different algebraic structures and different logics. The "metaphysical disaster" (mentioned by Foulis and Randall (1983)) seems to be, to a certain extent, avoided. We have no longer an *extensional collapse*, because effects cannot be identified with their *extensions*.

CHAPTER 16

Partial quantum logics and Łukasiewicz' quantum logic

16.1. Partial quantum logics

In Chapter 5, we have considered examples of partial algebraic structures, where the basic operations are not always defined. How does one give a semantic characterization for the different forms of quantum logic corresponding, respectively, to the class of all effect algebras, of all orthoalgebras and of all orthomodular posets? We will call these logics: *unsharp partial quantum logic* (**UPaQL**), *weak partial quantum logic* (**WPaQL**) and *strong partial quantum logic* (**SPaQL**), respectively. By **PaQL** we will denote any instance of our three logics. Since **PaQL** as well as *partial classical logic* **PaCL** (studied in Chapter 13) are examples of *partial logics*, one could expect some natural semantic connections between the two different cases. However, we will see that the basic idea of the semantic characterization of **PaQL** is somewhat different with respect to the **PaCL**-semantics.

The language of **PaQL** consists of a set of atomic sentences and of two primitive connectives: the *negation* $\neg$ and the *exclusive disjunction* $\veebar$ (aut).

The set of sentences is defined in the usual way. A *conjunction* is metalinguistically defined, via de Morgan law:

$$\alpha \barwedge \beta := \neg(\neg\alpha \veebar \neg\beta).$$

The intuitive idea underlying our semantics for **PaQL** is the following: disjunctions and conjunctions are always considered "legitimate" from a mere linguistic point of view. However, semantically, a disjunction $\alpha \veebar \beta$ will have the intended meaning only in the "appropriate cases:" where the values of α and β are orthogonal in the corresponding algebra. Otherwise, $\alpha \veebar \beta$ will have any meaning whatsoever (generally not connected with the meanings of α and β). As is well known, a similar semantic "trick" is used in some classical treatments of the description operator ι ("the unique individual satisfying a given property"; for instance, "the present king of Italy"). Apparently one is dealing with a different idea with respect to the **PaCL**-semantics, where the meaning of a sentence is not necessarily defined (hence, the valuation-function v is a partial function).

We will first consider the case of **UPaQL**, that represents the "logic of effect algebras" (Dalla Chiara and Giuntini, 1995a).

DEFINITION 16.1.1. *Realization for* **UPaQL**
A *realization* for **UPaQL** is a pair $\mathcal{A} = \langle \mathcal{B}\,,\, v \rangle$, where $\mathcal{B} = \langle B\,,\, \boxplus\,,\, \mathbf{0}\,,\mathbf{1} \rangle$ is an effect algebra (see Definition 5.1.1); v (the valuation-function) associates to any sentence α an element of B, satisfying the following conditions:

(i) $v(\neg\beta) = v(\beta)'$, where $'$ is the generalized complement (defined in $\mathcal{B}$).

(ii)

$$v(\beta \curlyvee \gamma) = \begin{cases} v(\beta) \boxplus v(\gamma) & \text{if } v(\beta) \boxplus v(\gamma) \text{ is defined in } \mathcal{B}, \\ \text{any element} & \text{otherwise.} \end{cases}$$

The other semantic definitions (truth, consequence in a given realization, logical truth, and logical consequence) are given as in the **QL**-case.

Weak partial quantum logic (**WPaQL**) and strong partial quantum logic (**SPaQL**) will be naturally characterized *mutatis mutandis*. It will be sufficient to replace, in the definition of realization, the notion of effect algebra with the notion of orthoalgebra and of orthomodular poset (see Definition 5.1.6 and Definition 1.1.11). Of course, **UPaQL** is weaker than **WPaQL**, which is, in turn, weaker than **SPaQL**.

Partial quantum logics are axiomatizable. We will first present a calculus for **UPaQL**, which is obtained as a natural transformation of the calculus for orthologic.

Unlike **QL**, the rules of our calculus will always have the form:

$$\frac{\alpha_1 \vdash \beta_1, \ldots, \alpha_n \vdash \beta_n}{\alpha \vdash \beta}$$

In other words, we will consider only inferences from single sentences.

Rules of **UPaQL**

(UPa1) $\alpha \vdash \alpha$ (identity)

(UPa2) $$\frac{\alpha \vdash \beta, \quad \beta \vdash \gamma}{\alpha \vdash \gamma}$$ (transitivity)

(UPa3) $\alpha \vdash \neg\neg\alpha$ (weak double negation)

(UPa4) $\neg\neg\alpha \vdash \alpha$ (strong double negation)

(UPa5) $$\frac{\alpha \vdash \beta}{\neg\beta \vdash \neg\alpha}$$ (contraposition)

(UPa6) $\beta \vdash \alpha \Cup \neg\alpha$ (excluded middle)

(UPa7) $$\frac{\alpha \vdash \neg\beta, \quad \alpha \Cup \neg\alpha \vdash \alpha \Cup \beta}{\neg\alpha \vdash \beta}$$ (uniqueness of negation)

(UPa8) $$\frac{\alpha \vdash \neg\beta,\ \alpha \vdash \alpha_1, \quad \alpha_1 \vdash \alpha, \quad \beta \vdash \beta_1, \quad \beta_1 \vdash \beta}{\alpha \Cup \beta \vdash \alpha_1 \Cup \beta_1}$$ (weak substitutivity)

(UPa9) $$\frac{\alpha \vdash \neg\beta}{\alpha \Cup \beta \vdash \beta \Cup \alpha}$$ (weak commutativity)

(UPa10) $$\frac{\beta \vdash \neg\gamma, \quad \alpha \vdash \neg(\beta \Cup \gamma)}{\alpha \vdash \neg\beta}$$ (weak associativity)

(UPa11) $$\frac{\beta \vdash \neg\gamma, \quad \alpha \vdash \neg(\beta \Cup \gamma)}{\alpha \Cup \beta \vdash \neg\gamma}$$ (weak associativity)

(UPa12) $$\frac{\beta \vdash \neg\gamma, \quad \alpha \vdash \neg(\beta \Cup \gamma)}{\alpha \Cup (\beta \Cup \gamma) \vdash (\alpha \Cup \beta) \Cup \gamma}$$ (weak associativity)

(UPa13) $$\frac{\beta \vdash \neg\gamma, \quad \alpha \vdash \neg(\beta \Cup \gamma)}{(\alpha \Cup \beta) \Cup \gamma) \vdash \alpha \Cup (\beta \Cup \gamma)}$$ (weak associativity)

The concepts of *derivation* and of *derivability* are defined in the expected way. In order to axiomatize weak partial quantum logic (**WPaQL**) it is sufficient to add a rule, which asserts a *Duns Scotus principle*:

(WPaQL) $$\frac{\alpha \vdash \neg\alpha}{\alpha \vdash \beta}$$ (Duns Scotus)

Clearly, the Duns Scotus rule corresponds to the strong consistency condition in our definition of orthoalgebra (see Definition 5.1.6). In other words, unlike **UPaQL**, the logic **WPaQL** forbids paraconsistent situations.

Finally, an axiomatization of strong partial quantum logic (**SPaQL**) can be obtained, by adding the following rule to (UPa1)-(UPa13), and (WPa):

(SPaQL) $$\frac{\alpha \vdash \neg\beta, \quad \alpha \vdash \gamma, \quad \beta \vdash \gamma}{\alpha \Cup \beta \vdash \gamma}$$

In other words, (**SPaQL**) requires that the disjunction $\Cup$ behaves like a supremum, whenever it has the "right meaning".

Let **PaQL** represent any instance of these three calculi. We will use the following abbreviations. Instead of $\alpha \vdash_{\mathbf{PaQL}} \beta$ we will write $\alpha \vdash \beta$. When α and β are logically equivalent ($\alpha \vdash \beta$ and $\beta \vdash \alpha$) we will write $\alpha \equiv \beta$.

Let **p** represent a particular atomic sentence of the language: **t** will be an abbreviation for $\mathbf{p} \mathbin{\forall\!\!\!\!\vee} \neg\mathbf{p}$; while **f** will be an abbreviation for $\neg(\mathbf{p} \mathbin{\forall\!\!\!\!\vee} \neg\mathbf{p})$.

Some important derivable rules of all calculi are the following:

(D1) $$\mathbf{f} \vdash \beta, \quad \beta \vdash \mathbf{t}$$ (Weak Duns Scotus)

(D2) $$\frac{\alpha \vdash \neg\beta}{\alpha \vdash \alpha \mathbin{\forall\!\!\!\!\vee} \beta}$$ (weak sup rule)

(D3) $$\frac{\alpha \vdash \beta}{\beta \equiv \alpha \mathbin{\forall\!\!\!\!\vee} \neg(\alpha \mathbin{\forall\!\!\!\!\vee} \neg\beta)}$$ (orthomodular-like rule)

(D4) $$\frac{\alpha \vdash \neg\gamma, \quad \beta \vdash \neg\gamma, \quad \alpha \mathbin{\forall\!\!\!\!\vee} \gamma \equiv \beta \mathbin{\forall\!\!\!\!\vee} \gamma}{\alpha \equiv \beta}$$ (cancellation)

As a consequence, the following syntactical lemma holds:

LEMMA 16.1.2. *For any* α, β: $\alpha \vdash \beta$ *iff there exists a formula* γ *such that*

(i) $\alpha \vdash \neg\gamma$;
(ii) $\beta \equiv \alpha \mathbin{\forall\!\!\!\!\vee} \gamma$.

In other words, the logical implication behaves like the partial order relation in the effect algebras.

The following derivable rule holds for **WPaQL** and for **SPaQL**:

(D5) $$\frac{\alpha \vdash \neg\beta, \quad \alpha \vdash \gamma, \quad \beta \vdash \gamma, \quad \gamma \vdash \alpha \mathbin{\forall\!\!\!\!\vee} \beta}{\alpha \mathbin{\forall\!\!\!\!\vee} \beta \vdash \gamma}$$

Our calculi turn out to be adequate with respect to the corresponding semantic characterizations. Soundness proofs are straightforward. Let us sketch the proof of the completeness theorem for our weakest calculus (**UPaQL**).

THEOREM 16.1.3. Completeness theorem

$$\alpha \models \beta \;\curvearrowright\; \alpha \vdash \beta.$$

PROOF. Following the usual procedure, it is sufficient to construct a canonical model $\mathcal{A} = \langle \mathcal{B}, v \rangle$ such that for any sentences α, β:

$$\alpha \vdash \beta \;\curvearrowright\; \alpha \models_{\mathcal{A}} \beta.$$

Definition of a canonical model for **UPaQL**.

(i) The algebra $\mathcal{B} = \langle B, \boxplus, \mathbf{0}, \mathbf{1} \rangle$ is determined as follows:
- (ia) B is the class of all equivalence classes of logically equivalent sentences: $B := \{[\alpha]_\equiv : \alpha \text{ is a sentence}\}$. (In the following, we write $[\alpha]$ instead of $[\alpha]_\equiv$).
- (ib) $[\alpha] \boxplus [\beta]$ is defined iff $\alpha \vdash \neg\beta$. If defined, $[\alpha] \boxplus [\beta] := [\alpha \mathbin{\forall\!\!\!\!\vee} \beta]$.

(ic) $\mathbf{1} := [\mathbf{t}]$; $\mathbf{0} := [\mathbf{f}]$.

(ii) The valuation function v is defined as follows: $v(\alpha) = [\alpha]$.

One can easily check that $\mathcal{A}$ is a realization for our logic. The operation $\boxplus$ is well defined (by the transitivity, contraposition and weak substitutivity rules). Furthermore, $\mathcal{B}$ is an effect algebra: $\boxplus$ is weakly commutative and weakly associative, because of the corresponding rules of our calculus. The strong excluded middle axiom holds, by definition of $\boxplus$ and in virtue of the following rules: excluded middle, uniqueness of negation, and double negation. Finally, the weak consistency axiom holds by weak Duns Scotus (D1) and by definition of $\boxplus$.

LEMMA 16.1.4. Lemma of the canonical model

$$[\alpha] \leq [\beta] \ \textit{iff} \ \alpha \mid\!\!\!- \beta.$$

Sketch of the proof. By definition of $\leq$ (in any effect algebra) one has to prove:

$$\alpha \mid\!\!\!- \beta \ \text{ iff for a given } \gamma \text{ such that } \ [\alpha] \perp [\gamma] : \ [\alpha] \boxplus [\gamma] = [\beta].$$

This equivalence holds by Lemma 16.1.2 and by definition of $\boxplus$.

Finally, let us check that v is a valuation-function. In other words:

(i) $v(\neg\beta) = v(\beta)'$

(ii) $v(\beta \uplus \gamma) = v(\beta) \boxplus v(\gamma)$, if $v(\beta) \boxplus v(\gamma)$ is defined.

(i) By definition of v, we have to show that $[\neg\beta]$ is the unique $[\gamma]$ such that $[\beta] \boxplus [\gamma] = \mathbf{1} := [\mathbf{t}]$. In other words,

(ia) $[\mathbf{t}] \leq [\beta] \boxplus [\neg\beta]$.

(ib) $[\mathbf{t}] \leq [\beta] \boxplus [\gamma] \ \curvearrowright \ \neg\beta \equiv \gamma$.

This holds by definition of the canonical model, by definition of $\boxplus$ and by the following rules: double negation, excluded middle, and uniqueness of negation.

(ii) Suppose $v(\beta) \boxplus v(\gamma)$ is defined. Then, $\beta \mid\!\!\!- \neg\gamma$. Hence, by definition of $\boxplus$ and of v: $v(\beta) \boxplus v(\gamma) = [\beta] \boxplus [\gamma] = [\beta \uplus \gamma] = v(\beta \uplus \gamma)$.

As a consequence, we obtain:

$$\alpha \mid\!\!\!- \beta \ \text{ iff } \ [\alpha] \leq [\beta] \ \text{ iff} \, v(\alpha) \leq v(\beta) \ \text{ iff } \ \alpha \models_{\mathcal{A}} \beta$$

□

The completeness argument can be easily transformed, *mutatis mutandis* for the logics **WPaQL** and **SPaQL**.

16.2. Łukasiewicz quantum logic

As we have seen in Chapter 5, the class $\mathcal{E}(\mathcal{H})$ of all effects on a Hilbert space $\mathcal{H}$ determines a quasi-linear QMV algebra. The theory of QMV algebras suggests, in a natural way, the semantic characterization of a new form of quantum logic (called *Łukasiewicz quantum logic* (**ŁQL**)), which generalizes both **OQL** and $\mathbf{L}_{\aleph}$ (Łukasiewicz' infinite many valued logic). The language of **ŁQL** contains the same primitive connectives as **UPaQL** ($\uplus$, $\neg$).

The conjunction ($\curlywedge$) is defined via the de Morgan law (as with**UPaQL**). Furthermore, a new pair of conjunction ($\Cap$) and disjunction ($\Cup$) connectives are defined as follows:

$$\alpha \Cap \beta := (\alpha \curlyvee \neg\beta) \curlywedge \beta$$

$$\alpha \Cup \beta := \neg(\neg\alpha \Cap \neg\beta)$$

DEFINITION 16.2.1. *Realization for* **ŁQL**
A *realization* of **ŁQL** is a pair $\mathcal{A} = \langle \mathcal{M}, v \rangle$, where

(i) $\mathcal{M} = \langle M, \oplus, ', \mathbf{0}, \mathbf{1} \rangle$ is a QMV algebra.
(ii) v (the valuation-function) associates to any sentence α an element of M, satisfying the following conditions:
$v(\neg\beta) = v(\beta)'$.
$v(\beta \curlyvee \gamma) = v(\beta) \oplus v(\gamma)$.

The other semantic definitions (truth, consequence in a given realization, logical truth, and logical consequence) are given as in the **QL**-case.

ŁQL can be easily axiomatized by means of a calculus that simply mimics the axioms of QMV algebras.

As we have learned in Section 5.4, the QMV algebras of concrete effects satisfies the quasi-linearity property (Definition 5.4.1). Thus, the following question naturally arises: is **ŁQL** characterized by the class of all quasi-linear QMV algebras (qLQMV)? In the case of $\mathbf{L}_{\aleph}$, Chang (1959) has proved that $\mathbf{L}_{\aleph}$ is characterized by the MV algebra determined by the real interval $[0, 1]$. This MV algebra is clearly quasi-linear, being totally ordered.

The relation between **ŁQL** and QMV algebras turns out to be much more complicated. In fact on can show that **ŁQL** cannot be characterized even by the class of all *weakly linear* QMV algebras (WLQMV). Since WLQMV is strictly contained in qLQMV, it follows that **ŁQL** is not characterized by qLQMV. To obtain these results, something stronger is proved. In particular, we can show (Giuntini, 1995a) that:

- the variety generated by the class all QMV algebras ($\mathcal{QMV}$) strictly includes the variety generated by the class of all weakly linear QMV algebras ($\mathbb{HSP}$(WLQMV)).[1]
- $\mathbb{HSP}$(WLQMV) strictly includes the variety generated by the class of all quasi-linear QMV-algebras ($\mathbb{HSP}$(qLQMV)).

So far, little is known about the axiomatizability of the logic based on $\mathbb{HSP}$(qLQMV). In the case of $\mathbb{HSP}$(WLQMV), instead, one can prove that

[1]A *variety* is a class of algebraic structures closed under the formation of products, subalgebras and homomorphic images. Thus, for any class of algebras $\mathcal{C}$, $\mathbb{P}(\mathcal{C})$ is the class of all products of algebras in the class $\mathcal{C}$, $\mathbb{SP}(\mathcal{C})$ is the class of all subalgebras of that, and $\mathbb{HSP}(\mathcal{C})$ is that class of all homomorphic images of $\mathbb{SP}(\mathcal{C})$. Birkhoff (1967) proved that varieties are precisely the classes of algebras that can be defined by universally quantified equations.

this variety is generated by the QMV-axioms together with the following axiom:

$$a = (a \oplus c \odot b') \Cap (a \oplus c' \odot b).$$

The problem of the axiomatizability of the logic based on $\mathbb{HSP}$(qLQMV) is complicated by the fact that not every (quasi-linear) QMV algebra $\mathcal{M} = \langle M, \oplus, ', \mathbf{0}, \mathbf{1} \rangle$ admits of a well behaved polynomial conditional, i.e., a polynomial binary operation $\circ$ such that

$$a \circ b = 1 \text{ iff } a \leq b.$$

Thus, it might happen that the notion of logical truth of the logic based on $\mathbb{HSP}$(qLQMV) is axiomatizable, while the notion of logical consequence ($\alpha \models \beta$) is not.

We will now show that the "diamond" $\mathcal{M}_4$ (see Figure 5.3.1) does not admit any well behaved polynomial conditional.

Let us consider the three-valued MV-algebra $\mathcal{M}_3$. It is easy to see that the map $h : \mathcal{M}_4 \mapsto \mathcal{M}_3$ such that, $\forall x \in M_4$,

$$h(x) := \begin{cases} 0 & \text{if } x = 0, \\ \frac{1}{2} & \text{if } x = a \text{ or } x = b, \\ 1 & \text{otherwise,} \end{cases}$$

is a homomorphism of $\mathcal{M}_4$ into $\mathcal{M}_3$.

Suppose, by contradiction, that $\mathcal{M}_4$ admits of a well behaved polynomial conditional $\rightarrow_{\mathcal{M}_4}$. Since $a \not\leq b$, we have $h(a \rightarrow_{\mathcal{M}_4} b) \neq 1$. Thus,

$$1 \neq h(a \rightarrow_{\mathcal{M}_4} b) = h(a) \rightarrow_{\mathcal{M}_3} h(b) = \frac{1}{2} \rightarrow_{\mathcal{M}_3} \frac{1}{2} = 1,$$

contradiction.

16.3. The intuitive meaning of the Łukasiewicz' quantum logical connectives

In the previous chapters we have often discussed the intuitive meaning of the quantum logical connectives in physical contexts. Let us now ask: how shall we interpret the new connectives arising in the framework of Łukasiewicz' quantum logic?

An interesting intuitive interpretation of the connectives of Łukasiewicz' many valued logics has been proposed by Mundici (1992). The basic idea is the use of *Ulam games*, where players are supposed to *lie* a certain number of times (the number may be either determined or indeterminate). Instead of *lying players*, one may also think of physical situations represented by *information sources* that are disturbed by a certain *noise*. However, such interpretation cannot be simply transferred to our weaker **ŁQL** (semantically characterized by the category of all QMV algebras, which are not generally distributive lattices).

So far we have seen how the structure of a QMV algebra can be naturally induced on the set $\mathcal{E}(\mathcal{H})$ of all effects in a Hilbert space $\mathcal{H}$ (Section 5.3). Let us now consider again the subset $\Pi(\mathcal{H})$ of $\mathcal{E}(\mathcal{H})$ consisting of all sharp events (projections). As we know, unlike effects, projections have a lattice-structure (with respect to the natural order). However, the meet $\wedge$ and the join $\vee$ do not coincide with the operations $\Cap$ and $\Cup$, which generally preserve their noncommunicative even in the "smaller" structure based on $\Pi(\mathcal{H})$.

At the same time, the lattice structure of the projections permits us to induce a QMV algebra on $\Pi(\mathcal{H})$, according to an alternative method: it is sufficient to identify the sum $\oplus$ with the join operation ($\vee$) of the lattice.

Hence, we obtain the following structure:

$$\mathcal{M}^{\Pi(\mathcal{H})} = \langle \Pi(\mathcal{H}) \, , \oplus \, , ' \, , \mathbf{0} \, , \mathbf{1} \rangle \, ,$$

where $\oplus = \vee$, while $'$ are the projection-orthocomplement, the null and the identity projection, respectively.

One can easily show that $\mathcal{M}^{\Pi(\mathcal{H})}$ is a QMV algebra (which, obviously, will not be a subalgebra of the QMV algebra based on $\mathcal{E}(\mathcal{H})$).

Thus, we obtain, for any $P, Q \in \Pi(\mathcal{H})$:

(1) $P \Cap Q = (P \vee Q') \wedge Q$;
(2) $P \Cup Q = (P \wedge Q') \vee Q$.

How shall we interpret $\Cap$ and $\Cup$ in the framework of $\mathcal{M}^{\Pi(\mathcal{H})}$?

For the sake of simplicity, let us refer to the isomorphic structure, based on the set $\mathcal{C}(\mathcal{H})$ of all closed subspaces:

$$\mathcal{M}^{\mathcal{C}(\mathcal{H})} = \langle \mathcal{C}(\mathcal{H}) \, , \oplus \, , ' \, , \mathbf{0} \, , \mathbf{1} \rangle \, .$$

By definition of $\mathcal{M}^{\mathcal{C}(\mathcal{H})}$, the operations $\oplus$ and $\odot$ will represent here the usual quantum logical disjunction and conjunction. What about the operations $\Cap$ and $\Cup$?

From our definitions we obtain:

(1) $X \Cap Y = (X \vee Y') \wedge Y$;
(2) $X \Cup Y = (X \wedge Y') \vee Y$.

In other words, $\Cap$ corresponds to the so called *Sasaki projection* of X into Y.[2] The following theorems hold:

THEOREM 16.3.1. *Let $\mathcal{H}$ be a Hilbert space. For any vector $\psi \in \mathcal{H}$ and for any two closed subspaces X, Y of $\mathcal{H}$:*

$$\psi \in Y \text{ and } \exists \varphi \in X (\psi = P_Y \varphi) \curvearrowright \psi \in X \Cap Y.$$

PROOF. Suppose that $\psi \in Y$ and $\exists \varphi \in X$ such that $\psi = P_Y \varphi$. We have to prove that $\psi \in X \Cap Y$. By the projection theorem, $\varphi = \varphi_1 + \varphi_2$, where $\varphi_1 \in Y$ and $\varphi_2 \in Y'$. Thus, $\psi = P_Y \varphi = \varphi_1$. Hence, $\psi = \varphi - \varphi_2 \in X + Y' \subseteq X \vee Y'$ □

THEOREM 16.3.2. *Let $\mathcal{H}$ be a finite dimensional Hilbert space. For any vector $\psi \in \mathcal{H}$ and for any two closed subspaces X, Y of $\mathcal{H}$:*

$$\psi \in X \Cap Y \curvearrowright \psi \in Y \text{ and } \exists \varphi \in X [\psi = P_Y \varphi].$$

[2]See Definition 11.3.2.

PROOF. Suppose that $\psi \in X \Cap Y$. Then, $\psi \in Y$ and $\psi \in X \vee Y'$. Since $\mathcal{H}$ has finite dimension, $X \vee Y' = X + Y'$. Thus, $\psi = \psi_1 + \psi_2$, where $\psi_1 \in X$ and $\psi_2 \in Y'$, so that $P_Y \psi_1 = P_Y \psi - P_Y \psi_2 = P_Y \psi = \psi$. Thus, $\varphi = \psi_1$ satisfies the conditions of the theorem. □

Theorem 16.3.2 cannot be generalized to infinite dimensional Hilbert spaces as the following result shows.

THEOREM 16.3.3. *Let $\mathcal{H}$ be a separable infinite dimensional Hilbert space. Then, there are two closed subspaces X, Y of $\mathcal{H}$ and a vector ψ such that the following conditions are satisfied:*

(1) $\psi \in (X \Cap Y')$.
(2) $\forall \varphi \in X$: $\psi \neq P_{Y'} \varphi$.

PROOF. Let $B = \{\phi_n\}$ and $C = \{\psi_n\}$ be two infinite orthonormal sets of vectors of $\mathcal{H}$ such that $\forall \phi \in B, \forall \psi \in C$: $\langle \phi | \psi \rangle = 0$.
Let X be the closed subspace generated by

$$\left\{ \delta_n = cos\left[\frac{1}{n}\right] \phi_n + sin\left[\frac{1}{n}\right] \psi_n \mid n = 1, 2, \cdots \right\}$$

and let Y be the closed subspace generated by B.
i) Let $\psi := \sum_{n=1}^{\infty} sin\left[\frac{1}{n}\right] \psi_n$. By (Halmos, 1951), $\psi \in X \vee Y$ and $\psi \notin X + Y$. We want to show that $\psi \in Y'$. Let ϕ be any vector of Y. We have to prove that $\langle \psi | \phi \rangle = 0$. First, we show that $\forall n$: $\langle \psi_n | \phi \rangle = 0$.

$$\begin{aligned} \langle \psi_n | \phi \rangle = \left\langle \psi_n \middle| \sum_{m=1}^{\infty} (\phi, \phi_m) \phi_m \right\rangle &= \sum_{m=1}^{\infty} \langle \psi_n | \langle \phi | \phi_m \rangle \phi_m \rangle \\ &= \sum_{m=1}^{\infty} \langle \phi | \phi_m \rangle \langle \psi_n | \phi_m \rangle \\ &= 0. \end{aligned}$$

Thus, $\langle \psi | \phi \rangle = \langle \sum_{n=1}^{\infty} sin\left[\frac{1}{n}\right] \psi_n | \phi \rangle = \sum_{n=1}^{\infty} \langle sin\left[\frac{1}{n}\right] \psi_n | \phi \rangle = 0$. Consequently, $\psi \in (X \vee Y) \wedge Y'$.

ii) Let $\varphi \in X$. Let us suppose, by contradiction, that $\psi = P_{Y'} \varphi$. By the projection theorem, $\varphi = \varphi_1 + \varphi_2$, where $\varphi_1 \in Y$ and $\varphi_2 \in Y'$. Hence, $\psi = P_{Y'} \varphi = \varphi_2$. By hypothesis, $\varphi \in X$. Thus, $\psi = \varphi - \varphi_1 \in X + Y$, contradiction. □

Theorems 16.3.1, 16.3.2, 16.3.3 suggest a natural physical interpretation for the noncommutative conjunction $\Cap$. Suppose one has a physical system $\mathfrak{S}$, with the associated Hilbert space $\mathcal{H}$, and suppose $Dim(\mathcal{H}) < \infty$. Let $P_{[\psi]}$ represent a pure state of $\mathfrak{S}$ in $\mathcal{H}$. The following relation holds:
$P_{[\psi]}$ *verifies* the conjunctive property $X \Cap Y$ (i.e., $\psi \in X \Cap Y$) iff there exists a pure state $P_{[\varphi]}$ of $\mathfrak{S}$ such that:

(1) $P_{[\psi]}$ is the result of a transformation of $P_{[\varphi]}$ after the performance of a Y-measurement (by application of the collapse of the wave-function);
(2) $P_{[\varphi]}$ *verifies* X ($\varphi \in X$);
(3) $P_{[\psi]}$ *verifies* Y ($\psi \in Y$).

In other words, one could say that: *the system in state* $P_{[\psi]}$ *is* $X \Cap Y$ *iff the system* was X *and* is now Y, *after the performance of a* Y*-measurement.* Owing to Theorem 16.3.3, this equivalence breaks down in the infinite dimensional case. However, Theorem 16.3.1 guarantees that $\Cap$ may still represent a *temporal* noncommutative conjunction, in a weaker sense.

Let us now consider our second way of inducing a quantum MV structure on the set $\Pi(\mathcal{H})$ of the sharp properties of $\mathfrak{S}$. As we already know, the QMV disjunction $\oplus$ is, in this case, defined as follows:

$$P \oplus Q = \begin{cases} P + Q & \text{if } P + Q \in \Pi(\mathcal{H}), \\ \mathbb{1} & \text{otherwise.} \end{cases}$$

As a consequence one obtains:

$$P \odot Q = (P' \oplus Q')' = \begin{cases} P \wedge Q & \text{if } P' \leq Q, \\ \mathbb{O} & \text{otherwise.} \end{cases}$$

This justifies the following intuitive interpretation: a state $P_{[\psi]}$ *verifies* the property $P \odot Q$ iff $P_{[\psi]}$ *verifies* both P and Q and further the pair P, Q gives rise to a *strongly stable alternative* (whenever not-P then Q).

A similar interpretation can be assumed also in the case of the standard QMV algebra, based on $\mathcal{E}(\mathcal{H})$. For, the following holds:

$$E \odot F = \begin{cases} (E' \oplus F')' & \text{if } E' \leq F, \\ \mathbb{O} & \text{otherwise.} \end{cases}$$

Whence, by additivity of the probability measure m_ψ (determined by the state $P_{[\psi]}$):

$$m_\psi(E \odot F) = 1 \text{ iff } E' \leq F \text{ and } m_\psi \left[E' \oplus F'\right]' = 1 \text{ iff}$$
$$E' \leq F \text{ and } m_\psi(E) = m_\psi(F) = 1.$$

In other words, $P_{[\psi]}$ verifies $E \odot F$ iff $P_{[\psi]}$ verifies both E and F and further the pair E, F gives rise to a strongly stable alternative (whenever not-E then F).

Finally, let us discuss the question concerning the intuitive meaning of $\Cap$. One can easily show that both projections and effects satisfy the following relation:

$$E \Cap F = \begin{cases} E & \text{if } E \leq F, \\ F & \text{otherwise.} \end{cases}$$

As a consequence, we will have: a state $P_{[\psi]}$ verifies $E \Cap F$ iff (i) $P_{[\psi]}$ verifies the second member F *and also* (ii) $P_{[\psi]}$ verifies the first member E, provided that E implies F. In other words, generally, a conjunction $E \Cap F$ "looses the memory" of the first member. This memory is preserved, just in the case in which the first member implies the second one. From an intuitive point of view, this seems to be connected with the reason why, for QMV algebras, *associativity* (which generally fails for $\Cap$) is preserved only in the following weaker form

$$(E \Cap F) \Cap G = (E \Cap F) \Cap (F \Cap G).$$

The repetition of the middle F seems to guarantee a kind of "conservation of memory".

CHAPTER 17

Quantum computational logic

Finally we study a new form of *unsharp quantum logic* that has been naturally suggested by the theory of *quantum computation*. One of the most interesting logical proposals that arise from quantum computation is the idea to use the quantum theoretical formalism in order to represent parallel reasoning.[1] As is well known, the unit of measurement in classical information theory is the *bit*: one *bit* measures the information quantity that can be either transmitted or received whenever one chooses one element from a set consisting of two elements, say, from the set $\{0, 1\}$. From the intuitive point of view, both the objects 0 and 1 can be imagined as a well determined *state* of a classical physical system, for instance, the state of a tape cell in a given machine.

Let us now refer to a quantum physical system. As we already know, the *pure states* of our system are mathematically represented by unit vectors in a convenient Hilbert space $\mathcal{H}$. Let us suppose we have the simplest situation, where our Hilbert space $\mathcal{H}$ has dimension 2; hence $\mathcal{H} = \mathbb{C}^2$. In such a case $\mathcal{H}$ has a *basis* consisting of two unit elements; and any vector of the space will be representable as a *superposition* of the two basis-elements. In quantum computation, it is customary to use Dirac's notation. Accordingly, the vectors of $\mathcal{H}$ are indicated by $|\psi\rangle$, $|\varphi\rangle$,... ; while the basis-elements are denoted by $|0\rangle$, $|1\rangle$. As a consequence, for any unit $|\psi\rangle$ we will have:

$$|\psi\rangle = a_0|0\rangle + a_1|1\rangle,$$

where the coefficients a_0, a_1 are complex numbers such that:

$$|a_0|^2 + |a_1|^2 = 1.$$

Let us now try to apply such a formalism to a quantum information theory. The basic idea is to generalize the concept of *bit*, by introducing the notion of *qubit* or *quantum bit*. A *qubit* is any unit vector in the Hilbert space $\mathbb{C}^2$. Thus, any *qubit* will have the form:

$$|\psi\rangle = a_0|0\rangle + a_1|1\rangle.$$

The interpretation is determined by the Born rule. Suppose that (as in the classical case) the pure states $|0\rangle$ and $|1\rangle$ represent two maximal (and precise) pieces of information. Then, the superposition-state

[1] See for instance (Nielsen and Chuang, 2000; Ekert, Hayden and Inamori, 2000; Gruska, 1999).

$$|\psi\rangle = a_0|0\rangle + a_1|1\rangle$$

represents an information that involves a certain degree of *uncertainty*. In particular, the number $|a_0|^2$ will correspond to the probability-value of the information described by the basic state $|0\rangle$; while $|a_1|^2$ will correspond to the probability-value of the information described by the basic state $|1\rangle$.

In this context, it makes sense to imagine $|\psi\rangle$ as an *epistemic state* that "stocks" two precise pieces of information *in parallel*: the information $|0\rangle$ and the information $|1\rangle$. As a consequence, unlike the classical *bit*, the *qubit* permits us to represent two atomic pieces of information, *in parallel*.

Let us now consider a situation characterized by many *bits* or *qubits*. As is well known, in classical information theory, a system consisting of n *bits* is naturally represented by a sequence of n elements belonging to the set $\{0,1\}$ (i.e., as an element of the set $\{0,1\}^n$). In the framework of quantum computation, it is convenient to adopt the tensor product formalism, which is used in quantum theory in order to represent compound physical systems. Suppose we have a two-particle quantum system:

$$\mathfrak{S} = \mathfrak{S}_1 + \mathfrak{S}_2.$$

For instance, $\mathfrak{S}_1$ and $\mathfrak{S}_2$ might correspond respectively to the two electrons in a given helium atom. In such a case, the Hilbert space $\mathcal{H}^{\mathfrak{S}}$ associated to $\mathfrak{S}$ will be the tensor product $\mathcal{H}^{\mathfrak{S}_1} \otimes \mathcal{H}^{\mathfrak{S}_2}$ of the two Hilbert spaces $\mathcal{H}^{\mathfrak{S}_1}$ and $\mathcal{H}^{\mathfrak{S}_2}$, which are associated to $\mathfrak{S}_1$ and $\mathfrak{S}_2$, respectively (see Section 1.3). Thus, any pure state of $\mathfrak{S}$ will be a unit vector in the space $\mathcal{H}^{\mathfrak{S}}$.

A particularly interesting case is represented by those vectors $|\psi\rangle$ of $\mathcal{H}^{\mathfrak{S}}$ that can be expressed as the tensor product of two vectors $|\psi_1\rangle$ and $|\psi_2\rangle$, belonging to $\mathcal{H}^{\mathfrak{S}_1}$ and to $\mathcal{H}^{\mathfrak{S}_2}$, respectively. In other words:

$$|\psi\rangle = |\psi_1\rangle \otimes |\psi_2\rangle.$$

In such cases, one usually speaks of *factorized states*. It is worthwhile recalling that not all vectors of $\mathcal{H}^{\mathfrak{S}}$ can be expressed in such a simple form.

How does one represent, in this framework, a system consisting of n *qubits*? It seems quite natural to describe our system as the pure state of a compound physical system consisting of n quantum objects. Thus, the *n-qubit* system can be identified with a unit vector of the product space

$$\otimes^n \mathbb{C}^2 = \underbrace{\mathbb{C}^2 \otimes \ldots \otimes \mathbb{C}^2}_{n-times}.$$

Particularly interesting examples will be represented by the vectors of $\otimes^n \mathbb{C}^2$ of the form:

$$|x_1\rangle \otimes \ldots \otimes |x_n\rangle,$$

where each $|x_i\rangle$ is an element of the given basis of $\mathbb{C}^2$ (i.e., $|x_i\rangle = |0\rangle$ or $|x_i\rangle = |1\rangle$). As we have learned in Section 1.2, the set of all vectors having this form represents a basis for the product space $\otimes^n \mathbb{C}^2$.

Evidently, this computational basis of $\otimes^n\mathbb{C}^2$ can be labelled by binary strings such as

$$|\underbrace{011\ldots10}_{n-times}\rangle.$$

Since any such string represents an integer number $j \in [0, 2^n - 1]$ in binary notation, any unit vector of $\otimes^n\mathbb{C}^2$ can be briefly expressed in the following form:

$$\sum_{j=0}^{2^n-1} a_j \, \|j\rangle\rangle,$$

where $0 \leq j \leq 2^n - 1$ and $\|j\rangle\rangle$ is the basis-element corresponding to j.

How can we deal with the concept of *quantum computation*, in this framework? The basic idea is to describe a computation by means of that kind of process that corresponds to the dynamic evolution of a quantum system. Suppose we have a physical system $\mathfrak{S}$, whose pure state at time t_0 is the vector $|\psi(t_0)\rangle$ (where $|\psi(t_0)\rangle$ belongs to the Hilbert space $\mathcal{H}^{\mathfrak{S}}$ associated to $\mathfrak{S}$). As we have seen, owing to Schrödinger equation, for any time t (where either $t \leq t_0$ or $t_0 \leq t$), there exists an operator $U_{[t_0,t]}$ that determines the state of the system at time t as a function of the state of the system at time t_0. In other words:

$$|\psi(t)\rangle = U_{[t_0,t]}|\psi(t_0)\rangle.$$

Any operator $U_{[t_0,t]}$ of this kind is *unitary*. Hence, our operator will preserve the length of the vectors and the orthogonality relation. Furthermore, it is *reversible*: one can go from $|\psi(t_0)\rangle$ to $|\psi(t)\rangle$ and vice versa, without any dissipation of information.

Thus, it makes sense to represent a quantum computation by means of convenient unitary operators assuming arguments and values in particular sets of *qubit* systems. As we have seen, *qubits* are generally superposition-states; as a consequence, we will obtain some typical *parallel configurations*.

Let us describe a particular example. Consider the basis $\{|0\rangle, |1\rangle\}$ of $\mathbb{C}^2$. The quantum theoretical formalism permits us to transform each element of the basis (which represents a precise information) into a *qubit*, which represents a "maximally uncertain information." It is sufficient to apply the following transformations:

$$|0\rangle \rightarrowtail \frac{1}{\sqrt{2}}|0\rangle + \frac{1}{\sqrt{2}}|1\rangle;$$

$$|1\rangle \rightarrowtail \frac{1}{\sqrt{2}}|0\rangle - \frac{1}{\sqrt{2}}|1\rangle.$$

In both cases, the probability that the transformed *qubit* satisfies those properties that are certain for $|0\rangle$ and are not certain for $|1\rangle$ (respectively, certain for $|1\rangle$ and not certain for $|0\rangle$) is $\frac{1}{2}$. Let us call such a transformation H (for Hadamard).

Unlike classical theory, in the framework of quantum computation, one can transform a certain information into an uncertain information in a *reversible* way! What seems to be *prima facie* a disadvantage, turns out to be instead one of the most powerful resources of quantum computation.

The transformation H represents an example of a unitary operator. We shall isolate some important unitary operators that turn out to have a special role in quantum computation. Such operators have been called *logical gates*.

All this suggests, in a natural way, a new form of quantum logical semantics, whose main features are quite different from the sharp and unsharp logics we have studied in the previous chapters.

17.1. Quantum logical gates

Consider the 2-dimensional Hilbert space $\mathbb{C}^2$. Any vector $|\psi\rangle$ will be represented as a pair of complex numbers. Let $B = \{|0\rangle, |1\rangle\}$ be the *orthonormal basis* for $\mathbb{C}^2$ such that

$$|0\rangle = (1, 0); \quad |1\rangle = (0, 1).$$

Thus, the elements of B are two particular unit vectors that are mutually orthogonal.

DEFINITION 17.1.1. *Qubit*
A *qubit* is an unit vector $|\psi\rangle$ of the space $\mathbb{C}^2$.

Hence, any *qubit* will have the following form:

$$|\psi\rangle = a_0|0\rangle + a_1|1\rangle,$$

where $a_0, a_1 \in \mathbb{C}$ and $|a_0|^2 + |a_1|^2 = 1$.

Let us first fix some notation. We use $x, y, \ldots$ as variables ranging over the set $\{0, 1\}$, while $|x\rangle, |y\rangle, \ldots$ range over the vectors $\{|0\rangle, |1\rangle\}$.

The set of all vectors having the form $|x_1\rangle \otimes \ldots \otimes |x_n\rangle$ is an orthonormal basis for $\otimes^n\mathbb{C}^2$ (sometimes called a *computational basis*). We will also write $|x_1, \ldots, x_n\rangle$ for $|x_1\rangle \otimes \ldots \otimes |x_n\rangle$.

DEFINITION 17.1.2. *n-qubit system (or n-quregister)*
An *n-qubit system* (or *n-quregister*) is any unit vector $|\psi\rangle$ in the product space $\otimes^n\mathbb{C}^2$. A *quregister* is a vector that is either a qubit or an n-qubit system.

We now introduce some examples of *quantum logical gates*. Generally, a quantum logical gate can be described by a unitary operator assuming arguments and values in a product-Hilbert space $\otimes^n\mathbb{C}^2$. First of all we study the so called *Toffoli gate* (Toffoli, 1980). It is expedient to start by analyzing the simplest case, in which the Hilbert space has the form:

$$\otimes^3\mathbb{C}^2 = \mathbb{C}^2 \otimes \mathbb{C}^2 \otimes \mathbb{C}^2.$$

In order to stress that this operator is defined on the product space $\otimes^3\mathbb{C}^2$, we indicate it by $T^{(1,1,1)}$. Since T is a linear operator, it is sufficient to determine its behavior on the computational basis of $\otimes^3\mathbb{C}^2$, i.e., on the elements of the form $|x\rangle \otimes |y\rangle \otimes |z\rangle$, where $x, y, z \in \{0, 1\}$.

DEFINITION 17.1.3. *The Toffoli gate* $T^{(1,1,1)}$
The*Toffoli gate* $T^{(1,1,1)}$ is the linear operator $T^{(1,1,1)} : \otimes^3\mathbb{C}^2 \mapsto \otimes^3\mathbb{C}^2$ that is defined for any element $|x\rangle \otimes |y\rangle \otimes |z\rangle$ of the basis as follows:

$$T^{(1,1,1)}(|x\rangle \otimes |y\rangle \otimes |z\rangle) = |x\rangle \otimes |y\rangle \otimes |xy \oplus z\rangle,$$

where $\oplus$ represents the sum modulo 2.

Clearly, $T^{(1,1,1,)}$ transforms any product vector $|x\rangle \otimes |y\rangle \otimes |z\rangle$ into the product that is obtained by leaving unchanged the first two factors ($|x\rangle$ and $|y\rangle$) and by transforming the third factor $|z\rangle$ as follows:

$$|z\rangle \rightarrowtail |xy \oplus z\rangle.$$

From the intuitive point of view, it seems quite natural to "see" the gate $T^{(1,1,1)}$ as a kind of "truth-table" that transforms triples of *zeros* and of *ones* into triples of *zeros* and of *ones*. The "table" we obtain is the following:

$$\begin{array}{ccc}
|0,0,0\rangle & \rightarrowtail & |0,0,0\rangle \\
|0,0,1\rangle & \rightarrowtail & |0,0,1\rangle \\
|0,1,0\rangle & \rightarrowtail & |0,1,0\rangle \\
|0,1,1\rangle & \rightarrowtail & |0,1,1\rangle \\
|1,0,0\rangle & \rightarrowtail & |1,0,0\rangle \\
|1,0,1\rangle & \rightarrowtail & |1,0,1\rangle \\
|1,1,0\rangle & \rightarrowtail & |1,1,1\rangle \\
|1,1,1\rangle & \rightarrowtail & |1,1,0\rangle
\end{array}$$

$T^{(1,1,1)}$ behaves like the identity operator on the first six basis elements and interchanges the last two basis elements by leaving the first two components fixed and interchanging the third component. Since $T^{(1,1,1)}$ maps a basis of $\otimes^3\mathbb{C}^2$ onto itself, it is not only linear but unitary.

The matrix representation of $T^{(1,1,1)}$ relative to the computational basis is the following:

$$\begin{pmatrix}
1 & 0 & 0 & 0 & 0 & 0 & 0 & 0 \\
0 & 1 & 0 & 0 & 0 & 0 & 0 & 0 \\
0 & 0 & 1 & 0 & 0 & 0 & 0 & 0 \\
0 & 0 & 0 & 1 & 0 & 0 & 0 & 0 \\
0 & 0 & 0 & 0 & 1 & 0 & 0 & 0 \\
0 & 0 & 0 & 0 & 0 & 1 & 0 & 0 \\
0 & 0 & 0 & 0 & 0 & 0 & 0 & 1 \\
0 & 0 & 0 & 0 & 0 & 0 & 1 & 0
\end{pmatrix}$$

By using $T^{(1,1,1)}$, we can introduce a convenient notion of *conjunction*. This conjunction, which will be indicated by AND, is characterized as a function whose arguments are pairs of vectors in $\mathbb{C}^2$ and whose values are vectors of the product space $\otimes^3\mathbb{C}^2$.

DEFINITION 17.1.4. AND
For any $|\varphi\rangle \in \mathbb{C}^2$ and any $|\psi\rangle \in \mathbb{C}^2$:

$$\mathtt{AND}\,(|\varphi\rangle, |\psi\rangle) := T^{(1,1,1)}\,(|\varphi\rangle \otimes |\psi\rangle \otimes |0\rangle)\,.$$

Clearly, $|0\rangle$ represents an *ancilla* in the Definition of AND.

Let us check that AND represents a good generalization of the corresponding classical truth-function. For the arguments $|0\rangle$ and $|1\rangle$ we will obtain the following "truth-table:"

$$\begin{array}{lcl}
(|0\rangle, |0\rangle) & \rightarrowtail & T^{(1,1,1)}(|0\rangle \otimes |0\rangle \otimes |0\rangle) = |0\rangle \otimes |0\rangle \otimes |0\rangle \\
(|0\rangle, |1\rangle) & \rightarrowtail & T^{(1,1,1)}(|0\rangle \otimes |1\rangle \otimes |0\rangle) = |0\rangle \otimes |1\rangle \otimes |0\rangle \\
(|1\rangle, |0\rangle) & \rightarrowtail & T^{(1,1,1)}(|1\rangle \otimes |0\rangle \otimes |0\rangle) = |1\rangle \otimes |0\rangle \otimes |0\rangle \\
(|1\rangle, |1\rangle) & \rightarrowtail & T^{(1,1,1)}(|1\rangle \otimes |1\rangle \otimes |0\rangle) = |1\rangle \otimes |1\rangle \otimes |1\rangle
\end{array}$$

One immediately realizes how this differs from the classical case. The corresponding classical truth-table represents a typical irreversible transformation, since it is many-to-one:

$$\begin{array}{lcl}
(0,0) & \rightarrowtail & 0 \\
(0,1) & \rightarrowtail & 0 \\
(1,0) & \rightarrowtail & 0 \\
(1,1) & \rightarrowtail & 1
\end{array}$$

The arguments of the function determine the value, but not the other way around. As is well known, irreversibility generally brings about dissipation of information. Mathematically, however, any Boolean function $f : \{0,1\}^n \mapsto \{0,1\}^m$ can be transformed into a *reversible* function $\hat{f} : \{0,1\}^n \times \{0,1\}^m \mapsto \{0,1\}^n \times \{0,1\}^m$ in the following way:

$$\forall u \in \{0,1\}^n\ \forall v \in \{0,1\}^m : \hat{f}((u,v)) = (u, v \oplus f(u)),$$

where $\oplus$ is the sum modulo 2, pointwise defined in each coordinate. One of the functions that can be obtained by making reversible the irreversible classical "and" is precisely the Toffoli gate. The classical "and" is then recovered by restricting the third input bit to 0.

Accordingly, the three arguments $(0,0)$, $(0,1)$, $(1,0)$ turn out to correspond to three distinct values, represented by the triples $(0,0,0)$, $(0,1,0)$, $(1,0,0)$. The price we have paid in order to obtain a reversible situation is the increasing of the complexity of the Hilbert space. The function AND associates to pairs of arguments, belonging to the 2-dimensional space $\mathbb{C}^2$, values belonging to the 8-dimensional space $\otimes^3\mathbb{C}^2$.

All this happens in the simplest situation, when one is only dealing with elements of the basis (in other words, with precise pieces of information). Let us examine the case in which the function AND is applied to arguments that are superpositions of the basis elements in the space $\mathbb{C}^2$. Consider the following *qubit* pair:

$$|\psi\rangle = a_0|0\rangle + a_1|1\rangle, \ |\varphi\rangle = b_0|0\rangle + b_1|1\rangle.$$

By applying the definitions of AND and of $T^{(1,1,1)}$ in the manner indicated, we obtain:

$$\texttt{AND}(|\psi\rangle, |\varphi\rangle) = a_1b_1|1,1,1\rangle + a_1b_0|1,0,0\rangle + a_0b_1|0,1,0\rangle + a_0b_0|0,0,0\rangle.$$

This result suggests a quite natural logical interpretation. The four basis elements that occur in the superposition vector correspond to the four cases of the truth-table for the classical conjunction:

$$(1,1,1), (1,0,0), (0,1,0), (0,0,0).$$

However here, unlike the classical situation, each case is accompanied by a complex number, which represents a characteristic quantum *amplitude*. By applying the "Born rule" we obtain the following interpretation: $|a_1b_1|^2$ represents the probability-value that both the *qubit*-arguments are equal to $|1\rangle$, and consequently their conjunction is $|1\rangle$. The other three cases admit a similar interpretation.

The logical gate AND refers to a very special situation, having values in a Hilbert space having the form $\otimes^3\mathbb{C}^2$. However, the procedure can be easily generalized by defining the Toffoli gate in any Hilbert space having the form:

$$(\otimes^n\mathbb{C}^2) \otimes (\otimes^m\mathbb{C}^2) \otimes \mathbb{C}^2 \ (= \otimes^{n+m+1}\mathbb{C}^2).$$

DEFINITION 17.1.5. *The Toffoli gate* $T^{(n,m,1)}$
The *Toffoli gate* $T^{(n,m,1)}$ is the linear operator

$$T^{(n,m,1)} : (\otimes^n\mathbb{C}^2) \otimes (\otimes^m\mathbb{C}^2) \otimes \mathbb{C}^2 \ \mapsto \ (\otimes^n\mathbb{C}^2) \otimes (\otimes^m\mathbb{C}^2) \otimes \mathbb{C}^2,$$

that is defined for any element $|x_1,\ldots,x_n\rangle \otimes |y_1,\ldots,y_m\rangle \otimes |z\rangle$ of the computational basis of $\otimes^{n+m+1}\mathbb{C}^2$ as follows:

$$T^{(n,m,1)}(|x_1,\ldots,x_n\rangle \otimes |y_1,\ldots,y_m\rangle \otimes |z\rangle) =$$

$$|x_1,\ldots,x_n\rangle \otimes |y_1,\ldots,y_m\rangle \otimes |x_ny_m \oplus z\rangle,$$

where $\oplus$ represents the sum modulo 2.

On this basis one can immediately generalize our definition of AND.

DEFINITION 17.1.6. AND
For any $|\varphi\rangle \in \otimes^n\mathbb{C}^2$ and any $|\psi\rangle \in \otimes^m\mathbb{C}^2$:

$$\texttt{AND}(|\varphi\rangle, |\psi\rangle) := T^{(n,m,1)}(|\varphi\rangle \otimes |\psi\rangle \otimes |0\rangle).$$

How do we deal in this context with the concept of negation? A characteristic of quantum computation is the possibility of defining a plurality of negation operations, some of which represent generalizations of the classical negation. We will first consider a function that simply inverts the value of the last elements of any basis-vector.

DEFINITION 17.1.7. $\mathtt{NOT}^{(\mathtt{n})}$
The negation-gate is the linear operator $\mathtt{NOT}^{(\mathtt{n})}$ that is defined for any element $|x_1, \ldots, x_n\rangle$ of the computational basis of $\otimes^n \mathbb{C}^2$ as follows:

$$\mathtt{NOT}^{(n)}(|x_1, \ldots, x_n\rangle) = |x_1, \ldots, x_{n-1}, 1 - x_n\rangle.$$

One can immediately check that $\mathtt{NOT}^{(n)}$ represents a good generalization of the classical truth-table. Consider the basis-elements $|0\rangle$ and $|1\rangle$ of the space $\mathbb{C}^2$. In such a case we obtain:

$$\mathtt{NOT}^{(1)}(|1\rangle) = |0\rangle;$$
$$\mathtt{NOT}^{(1)}(|0\rangle) = |1\rangle.$$

The matrix corresponding to $\mathtt{NOT}^{(1)}$ is:

$$\begin{pmatrix} 0 & 1 \\ 1 & 0 \end{pmatrix}$$

Note that $\mathtt{NOT}^{(n)}$ acts on the 2^n-element computational basis

$$\{\|j\rangle\!\rangle : j = 0, 1, \ldots, 2^{n-1}\}$$

by interchanging, for each *even* j, $\|j\rangle\!\rangle$ and $\|j+1\rangle\!\rangle$. Therefore, the matrix corresponding to $\mathtt{NOT}^{(n)}$ will be a $2^n \times 2^n$ matrix of the form

$$\begin{pmatrix} 0 & 1 & . & . & . & . & 0 & 0 \\ 1 & 0 & . & . & . & . & 0 & 0 \\ . & . & 0 & 1 & . & . & . & . \\ . & . & 1 & 0 & . & . & . & . \\ . & . & . & . & . & . & . & . \\ . & . & . & . & . & . & . & . \\ 0 & 0 & . & . & . & . & 0 & 1 \\ 0 & 0 & . & . & . & . & 1 & 0 \end{pmatrix}$$

For any $|\psi\rangle \in \otimes^n \mathbb{C}^2$, we will usually write $\mathtt{NOT}(|\psi\rangle)$ instead of $\mathtt{NOT}^{(n)}(|\psi\rangle)$.

Finally, how do we introduce a reasonable disjunction? A gate $\mathtt{OR}$ can be naturally defined in terms of $\mathtt{AND}$ and $\mathtt{NOT}$ via de Morgan.

DEFINITION 17.1.8. $\mathtt{OR}$
For any $|\varphi\rangle \in \otimes^n \mathbb{C}^2$ and $|\psi\rangle \in \otimes^m \mathbb{C}^2$:

$$\mathtt{OR}(|\varphi\rangle, |\psi\rangle) = \mathtt{NOT}\left(\mathtt{AND}\left(\mathtt{NOT}(|\varphi\rangle), \mathtt{NOT}(|\psi\rangle)\right)\right).$$

The quantum logical gates we have considered so far are, in a sense, "semiclassical". A quantum logical behavior only emerges in the case in which the gates are applied to superpositions. When restricted to classical registers, the gates behave as classical truth-functions. We now investigate *genuine quantum gates* that may transform classical registers into quregisters that are superpositions.

One of the most significant genuine quantum gates is the *squareroot of the negation* NOT, which is indicated by $\sqrt{\mathtt{NOT}}$. As suggested by the name, the characteristic property of the gate $\sqrt{\mathtt{NOT}}$ is the following: for any quregister $|\psi\rangle$,

$$\sqrt{\mathtt{NOT}}(\sqrt{\mathtt{NOT}}|\psi\rangle) = \mathtt{NOT}(|\psi\rangle).$$

In other words: applying the squareroot of the negation twice "means" negating.

Interestingly enough, the gate $\sqrt{\mathtt{NOT}}$ has some significant physical models (and implementations). As an example, consider an idealized atom with a single electron and two energy levels: a *ground state* (identified with $|0\rangle$) and an *excited state* (identified with the state $|1\rangle$). By shining a pulse of light of appropriate intensity, duration and wavelength, it is possible to force the electron to change energy level. As a consequence, the state (bit) $|0\rangle$ is transformed into the state (bit) $|1\rangle$, and vice versa:

$$|0\rangle \rightarrowtail |1\rangle; \quad |1\rangle \rightarrowtail |0\rangle.$$

We have obtained a typical physical model for the gate NOT. Now, by using a light pulse of half the duration as the one needed to perform the NOT operation, we effect a half-flip between the two logical states. The state of the atom after the half pulse is neither $|0\rangle$ nor $|1\rangle$, but rather a superposition of both states. As observed by Deutsch, Ekert and Lupacchini (2000):

> Logicians are now entitled to propose a new logical operation $\sqrt{\mathtt{NOT}}$. Why? Because a faithful physical model for it exists in nature.

The physical models of the gate $\sqrt{\mathtt{NOT}}$ naturally suggest the following logical interpretation: $\sqrt{\mathtt{NOT}}$ represents a kind of "tentative negation". By applying twice our "attempt" to negate, we obtain a full negation.

Let us now give the mathematical definition of $\sqrt{\mathtt{NOT}}$.

DEFINITION 17.1.9. *The square root of the negation*
The square root of the negation on $\otimes^n\mathbb{C}^2$ is the linear operator $\sqrt{\mathtt{NOT}}^{(n)}$ such that, for every element $|x_1,\ldots,x_n\rangle$ of the computational basis,

$$\sqrt{\mathtt{NOT}}^{(n)}(|x_1,\ldots,x_n\rangle) = |x_1,\ldots,x_{n-1}\rangle \otimes (\frac{1+i}{2}|x_n\rangle + \frac{1-i}{2}|1-x_n\rangle)$$

(where, of course, $i^2 = -1$).

In other words, $\sqrt{\mathtt{NOT}}^{(n)}$ transforms the last element x_n of any basis-element $|x_1, \ldots, x_n\rangle$ into the element $\frac{1+i}{2}|x_n\rangle + \frac{1-i}{2}|1-x_n\rangle$. As a consequence, for the two bits $|0\rangle$ and $|1\rangle$ ("living" in the space $\mathbb{C}^2$), we obtain:

$$\sqrt{\mathtt{NOT}}^{(1)}(|0\rangle) = \frac{1+i}{2}|0\rangle + \frac{1-i}{2}|1\rangle;$$

$$\sqrt{\mathtt{NOT}}^{(1)}(|1\rangle) = \frac{1-i}{2}|0\rangle + \frac{1+i}{2}|1\rangle.$$

One can easily show that $\sqrt{\mathtt{NOT}}^{(n)}$ is a unitary operator, which satisfies the following condition:

$$\text{for any } |\psi\rangle \in \otimes^n\mathbb{C}^2,\ \sqrt{\mathtt{NOT}}^{(n)}(\sqrt{\mathtt{NOT}}^{(n)}(|\psi\rangle)) = \mathtt{NOT}^{(n)}(|\psi\rangle).$$

It turns out that the matrix associated to $\sqrt{\mathtt{NOT}}^{(1)}$ is

$$\begin{pmatrix} \frac{1+i}{2} & \frac{1-i}{2} \\ \frac{1-i}{2} & \frac{1+i}{2} \end{pmatrix}.$$

The matrix associated to the quantum logical gate $\sqrt{\mathtt{NOT}}^{(n)}$ relative to the computational basis is the $(2^n) \times (2^n)$ matrix of the form

$$\frac{1}{2}\begin{pmatrix} 1+i & 1-i & . & . & . & . & 0 & 0 \\ 1-i & 1+i & . & . & . & . & 0 & 0 \\ . & . & 1+i & 1-i & . & . & . & . \\ . & . & 1-i & 1+i & . & . & . & . \\ . & . & . & . & . & . & . & . \\ . & . & . & . & . & . & . & . \\ 0 & 0 & . & . & . & . & 1+i & 1-i \\ 0 & 0 & . & . & . & . & 1-1 & 1+i \end{pmatrix}.$$

We will omit the index n in $\sqrt{\mathtt{NOT}}^{(n)}$ if no confusion is possible.

THEOREM 17.1.10. *For any n, m the following properties hold:*

(i) $T^{(n,m,1)}\sqrt{\mathtt{NOT}}^{(n+m+1)} = \sqrt{\mathtt{NOT}}^{(n+m+1)}T^{(n,m,1)}$;

(ii) $\sqrt{\mathtt{NOT}}^{(n)}\mathtt{NOT}^{(n)} = \mathtt{NOT}^{(n)}\sqrt{\mathtt{NOT}}^{(n)}$.

As expected, the square root of the negation has no Boolean counterpart.

LEMMA 17.1.11. *There is no function $f : \{0,1\} \to \{0,1\}$ such that, for any $x \in \{0,1\} : f(f(x)) = 1 - x$.*

PROOF. Suppose that such a function exists and call it f. Then, f is a bijection because f^2 is a bijection (in fact, f^4 is the identity). If, $f(0) = 0$, then $1 = f(f(0)) = f(0) = 0$, contradiction. If $f(0) = 1$, then also $f(1) = f(f(0)) = 1$, contradicting the fact that f is one-to-one. Thus, $f(0) \neq 0$, and $f(0) \neq 1$; it follows that no such function exists. □

Interestingly enough, $\sqrt{\mathtt{NOT}}$ also does not have a continuous fuzzy counterpart.

LEMMA 17.1.12. *There is no continuous function $f : [0,1] \to [0,1]$ such that for any $x \in [0,1]$, $f(f(x)) = 1 - x$.*

PROOF. Suppose, by contradiction, that such a function f exists. As in the proof of the preceding lemma, $f(0) \neq 0$ and $f(0) \neq 1$. Since f is onto, there exist a and b in $[0,1]$ such that $f(a) = 0$ and $f(b) = 1$. If $a < b$ then, since f is continuous on the interval $[a,b]$ and takes on the values 0 and 1 over the interval, it must take on all intermediate values by Bolzano's Theorem,[2] so that $f([a,b]) = [0,1]$. Thus, $\exists c \in [a,b]$ such that $f(c) = f(0)$. Since f is one-to-one on $[0,1]$, we must have $c = 0$, and therefore $a = 0$. But $f(0) \neq 0$, contradiction.

A similar argument shows that if $b < a$, then $b = 0$. Thus, $f(0) = 1$, contradiction.

□

There are, however, piecewise continuous functions $f : [0,1] \to [0,1]$ satisfying $f^2(x) = 1 - x$. The graph of one is given in Figure 17.1.1. This is the graph of the function

$$f(x) = \begin{cases} 3/4 - x & \text{if } x \in [0, 1/4), \\ x - 1/4 & \text{if } x \in [1/4, 1/2), \\ 1/2 & \text{if } x = 1/2, \\ x + 1/4 & \text{if } x \in (1/2, 3/4], \\ -x + 5/4 & \text{if } x \in (3/4, 1]. \end{cases}$$

17.2. The probabilistic content of the quantum logical gates

For any quregister one can define a natural *probability-value*, which will play an important role in our quantum computational semantics.

Suppose we have a quregister

$$|\varphi\rangle = \sum_{j=0}^{2^n-1} a_j \, \|j\rangle\rangle \in \otimes^n \mathbb{C}^2.$$

Let us first define two particular sets of coefficients that occur in the superposition-vector φ:

$$C^+|\varphi\rangle = \left\{a_j \,:\, \|j\rangle\rangle = \langle x_{j1}, \ldots, x_{j_{n-1}}, 1\rangle\right\},$$

and

$$C^-|\varphi\rangle = \left\{a_j \,:\, \|j\rangle\rangle = \langle x_{j_1}, \ldots, x_{j_{n-1}}, 0\rangle\right\}.$$

Clearly, the elements of $C^+|\varphi\rangle$ ($C^-|\varphi\rangle$) represent the *amplitudes* associated to the different vector-basis of $\otimes^n \mathbb{C}^2$ ending with 1 (0, respectively).

[2]See, for instance, (Foulis and Munem, 1984).

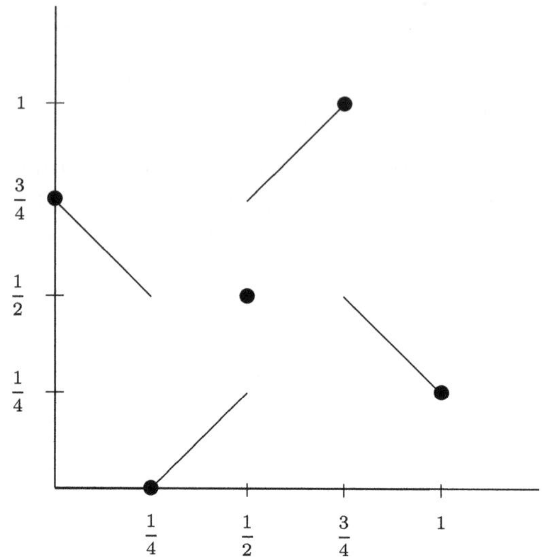

FIGURE 17.1.1. A noncontinuous fuzzy square root of the negation

On this basis, we can now define the probability-value of any vector having length less or equal than 1.

DEFINITION 17.2.1. *The probability-value of a vector*
Let $|\psi\rangle = \sum_{j=0}^{2^n-1} a_j \,\|j\rangle\!\rangle$ be any vector of $\otimes^n\mathbb{C}^2$ such that $\sum_{j=0}^{2^n-1} |a_j|^2 \leq 1$. Then, the probability-value of $|\psi\rangle$ is defined as follows:

$$\mathrm{Prob}(|\psi\rangle) := \sum_{a_j \in C^+|\psi\rangle} |a_j|^2.$$

According to our definition, in order to calculate the probability-value of such a quregister $|\psi\rangle$ one has to perform the following operations:

- consider all the amplitudes a_j that are associated to a basis-element ending with 1;
- take the squared moduli $|a_j|^2$ of all these complex numbers a_j;
- sum all the real numbers $|a_j|^2$.

One can prove the following result.

LEMMA 17.2.2.

(i) *If* $|\psi\rangle = \sum_{j=0}^{2^n-1} a_j \,\|j\rangle\!\rangle$ *is any unit vector of* $\otimes^n\mathbb{C}^2$, *then*

$$\sum_{a_j \in C^+|\psi\rangle} |a_j|^2 + \sum_{a_j \in C^-|\psi\rangle} |a_j|^2 = 1.$$

(ii) *Let* $|\psi\rangle = \sum_{j=0}^{2^n-1} a_j \,\|j\rangle\!\rangle$ *and* $|\varphi\rangle = \sum_{j=0}^{2^n-1} b_j \,\|j\rangle\!\rangle$ *be any two orthogonal vectors of* $\otimes^n\mathbb{C}^2$ *such that* $\||\psi\rangle + |\varphi\rangle\| \leq 1$ *and*

$\forall j (0 \leq j \leq 2^n - 1)$, $a_j b_j = 0$. *Then,*

$$\text{Prob}(|\psi\rangle + |\varphi\rangle) = \text{Prob}(|\psi\rangle) + \text{Prob}(|\varphi\rangle).$$

From an intuitive point of view, $\text{Prob}(|\psi\rangle)$ represents "the probability" that our quregister $|\psi\rangle$ (which is a superposition) "collapses" into a classical register whose last element is 1. The following theorem describes some interesting relations between the probability function Prob and the basic logical gates.

THEOREM 17.2.3. (Dalla Chiara, Giuntini and Leporini, 2003) *Let* $|\psi\rangle = \sum_{j=0}^{2^n-1} a_j \|j\rangle\rangle$ *and* $|\varphi\rangle = \sum_{j=0}^{2^n-1} b_k \|k\rangle\rangle$ *be two unit vectors of* $\otimes^n\mathbb{C}^2$ *and of* $\otimes^m\mathbb{C}^2$, *respectively. The following properties hold:*

(i) $\text{Prob}(\texttt{AND}(|\psi\rangle, |\varphi\rangle)) = \text{Prob}(|\psi\rangle)\text{Prob}(|\varphi\rangle)$;

(ii) $\text{Prob}(\texttt{NOT}(|\psi\rangle)) = 1 - \text{Prob}(|\psi\rangle)$;

(iii) $\text{Prob}(\texttt{OR}(|\psi\rangle, |\varphi\rangle)) = \text{Prob}(|\psi\rangle) + \text{Prob}(|\varphi\rangle) - \text{Prob}(|\psi\rangle)\text{Prob}(|\varphi\rangle)$;

(iv) $\text{Prob}(\sqrt{\texttt{NOT}}(|\psi\rangle)) = \sum_{a_j \in C^+|\psi\rangle} \mid \frac{1}{2}(1-i)a_{j-1} + \frac{1}{2}(1+i)a_j \mid^2$;

(v) $\text{Prob}(\sqrt{\texttt{NOT}}(\texttt{AND}(|\psi\rangle, |\varphi\rangle))) = \frac{1}{2}$.

Condition (i) of Theorem 17.2.3 represents a quite unusual property for probabilistic contexts: any pair of quregisters seems to behave like a pair of independent classical events (so that the probability of their conjunction is the product of their probabilities). At the same time, condition (ii) and (iii) appear to be well behaved with respect to standard probability theory. As a consequence, we obtain:

- unlike classical and quantum probability, AND, OR, NOT have a "truth-functional behavior" with respect to the function Prob: the probability of the "whole" is determined by the probabilities of the parts.
- The gate $\sqrt{\texttt{NOT}}$ is not truth-functional. It may happen at the same time that:

$$\text{Prob}(|\psi\rangle) = \text{Prob}(|\varphi\rangle) \text{ and } \text{Prob}(\sqrt{\texttt{NOT}}(\psi)) \neq \text{Prob}(\sqrt{\texttt{NOT}}(|\varphi\rangle)).$$

For example, let

$$|\psi\rangle := \frac{\sqrt{2}}{2}|0\rangle + \frac{\sqrt{2}}{2}|1\rangle \text{ and } |\varphi\rangle := \frac{\sqrt{2}}{2}|0\rangle + \frac{\sqrt{2}}{2}\left(\frac{\sqrt{2}}{2} + \frac{\sqrt{2}}{2}i\right)|1\rangle.$$

Clearly, $\text{Prob}(|\psi\rangle) = \text{Prob}(|\varphi\rangle) = \frac{1}{2}$. However, $\text{Prob}(\sqrt{\texttt{NOT}}(\psi)) = \frac{1}{2}$ and $\text{Prob}(\sqrt{\texttt{NOT}}(|\varphi\rangle)) = \frac{1}{8} + \left(\frac{1}{2} - \frac{1}{2\sqrt{2}}\right)^2 = \frac{1}{2} - \frac{\sqrt{2}}{4}$.

As one can easily see, the operators $\texttt{NOT}^{(1)}$ and $\sqrt{\texttt{NOT}}^{(1)}$ have the same set of fixed points. In other words, for any (unit) vector $|\psi\rangle \in \mathbb{C}^2$: $\texttt{NOT}^{(1)}(|\psi\rangle) = |\psi\rangle$ iff $\sqrt{\texttt{NOT}}^{(1)}(|\psi\rangle) = |\psi\rangle$. Every (unit) vector of the form $\frac{e^{i\vartheta}}{\sqrt{2}}(|0\rangle + |1\rangle)$ turns out to be a fixed point of $\texttt{NOT}^{(1)}$ and, accordingly, of $\sqrt{\texttt{NOT}}^{(1)}$.[3]

[3]Recall that any unit vector of $\mathbb{C}^2$ can be expressed in the canonical form $|\psi\rangle = cos\eta|0\rangle + e^{i\vartheta}sin\eta|1\rangle = cos\eta|0\rangle + (cos\vartheta + isin\vartheta)sin\eta|1\rangle$, where $i := \sqrt{-1}$.

17.3. Quantum computational semantics

In the previous chapters we have analyzed different semantic characterizations for different forms of sharp and unsharp quantum logic.

The starting point of the quantum computational semantics is quite different from the standard quantum logical approach. Here the meanings of the linguistic sentences are represented by quregisters. Thus, we can say that the meaning of a sentence is identified with the *information quantity* encoded by the sentence in question.

Let us now describe such a semantics technically. Consider a sentential language $\mathcal{L}$ with the following connectives: the *negation* ($\neg$), the *conjunction* ($\curlywedge$) and the *square root of the negation* ($\sqrt{\neg}$). The notion of *sentence* (or *formula*) of $\mathcal{L}$ is defined in the expected way. Let $Form^{\mathcal{L}}$ represent the set of all sentences of $\mathcal{L}$. As usual, the metavariables $\mathbf{p}, \mathbf{q}, \mathbf{r}, \ldots$ will range over atomic sentences, while α, β, $\gamma, \ldots$ will range over sentences. The connective *disjunction* ($\curlyvee$) is defined via de Morgan's law:

$$\alpha \curlyvee \beta := \neg(\neg\alpha \curlywedge \neg\beta).$$

We now introduce the basic concept of this semantics, the notion of *quantum computational realization*: an interpretation of the language $\mathcal{L}$, such that the meaning associated to any sentence is a quregister. As a consequence, here the *space of the meanings* corresponds not to a unique Hilbert space, but to varying Hilbert spaces, each one of the form $\otimes^n\mathbb{C}^2$.

DEFINITION 17.3.1. *Quantum computational realization*
A *quantum computational realization* of $\mathcal{L}$ is a function Qub associating to any sentence α a quregister in a Hilbert space $\otimes^n\mathbb{C}^2$ (where n depends on the linguistic form of α):

$$Qub : Form^{\mathcal{L}} \to \bigcup_n \otimes^n\mathbb{C}^2.$$

We will also write $|\alpha\rangle$ instead of $Qub(\alpha)$; and we will call $|\alpha\rangle$ the *information-value* of α. The following conditions are required.

(i) $|\mathbf{p}\rangle$ is a qubit.

(ii) Let $|\beta\rangle \in \otimes^n\mathbb{C}^2$. Then,
$|\neg\beta\rangle = \mathtt{NOT}(|\beta\rangle) \in \otimes^n\mathbb{C}^2$.

(iii) Let $|\beta\rangle \in \otimes^n\mathbb{C}^2$, $|\gamma\rangle \in \otimes^m\mathbb{C}^2$. Then,
$|\beta \curlywedge \gamma\rangle = \mathtt{AND}(|\beta\rangle, |\gamma\rangle) \in (\otimes^n\mathbb{C}^2) \otimes (\otimes^m\mathbb{C}^2) \otimes \mathbb{C}^2$.

(iv) Let $|\beta\rangle \in \otimes^n\mathbb{C}^2$. Then,
$|\sqrt{\neg}\beta\rangle = \sqrt{\mathtt{NOT}}(|\beta\rangle) \in \otimes^n\mathbb{C}^2$.

Our definition univocally determines, for any sentence α, the Hilbert space $\otimes^n\mathbb{C}^2$ to which $|\alpha\rangle$ belongs. Clearly, n is the number of all occurrences of atomic sentences and of the connective $\curlywedge$ in α. Since the meaning associated to a given sentence partially reflects the logical form of the sentence in question, we can say that our semantics has a typical *intensional* character.

As we have seen, a characteristic of our semantics is to identify the meanings of the linguistic sentences with unit vectors of various Hilbert spaces. As a consequence, we will obtain that the information-value of a sentence naturally determines a probability-value for that sentence.

Let *Qub* be a quantum computational realization and let α be any sentence with associated meaning $|\alpha\rangle$. Like all qubit-registers, also this $|\alpha\rangle$ will have a probability-value, which (according to Definition 17.2.1), is determined as follows:

$$\mathrm{Prob}(|\alpha\rangle) := \sum_{a_j \in C^+|\alpha\rangle} |a_j|^2.$$

On this basis, one can naturally define the probability-value of any sentence of our language.

DEFINITION 17.3.2. *The probability-value of* α

$$\mathrm{Prob}(\alpha) := \sum_{a_j \in C^+|\alpha\rangle} |a_j|^2.$$

As an example, let us first consider the simplest case, where α is an atomic sentence; in this case, its information-value will belong to the 2-dimensional space $\mathbb{C}^2$. Suppose, for instance, that $|\alpha\rangle$ has the form:

$$a_0|0\rangle + a_1|1\rangle.$$

Then, the probability-value of α will be:

$$\mathrm{Prob}(\alpha) = |a_1|^2.$$

Thus, $\mathrm{Prob}(\alpha) = |a_1|^2$ represents the probability that our *uncertain* information $|\alpha\rangle$ corresponds to the *precise* information $|1\rangle$.

From an intuitive point of view, this definition clearly attributes a privileged role to one of the two basic *qubits* (belonging to the computational basis of $\mathbb{C}^2$), the *qubit* $|1\rangle$. In such a way, $|1\rangle$ is dealt with as the truth-value *True*.

Consider now the case of a molecular sentence α. Its information-value $|\alpha\rangle$ will belong to the space $\otimes^n\mathbb{C}^2$, where n (≥ 3) depends on the linguistic complexity of α. The dimension of $\otimes^n\mathbb{C}^2$ is 2^n. Hence, $|\alpha\rangle$ will generally be a superposition of elements of the basis of $\otimes^n\mathbb{C}^2$. Thus, we will have:

$$|\alpha\rangle = \sum_{j=0}^{2^n-1} a_j \,\|j\rangle\!\rangle,$$

where $\|j\rangle\!\rangle$ ranges over the basis of $\otimes^n\mathbb{C}^2$.

From the logical point of view, any element $\|j\rangle\!\rangle$ of the computational basis of $\otimes^n\mathbb{C}^2$ represents a possible case of a "reversible truth-table" for α. For instance, suppose α has the form $\mathbf{p} \curlyvee \mathbf{q}$, where:

$$|\mathbf{p}\rangle = a_0|0\rangle + a_1|1\rangle, \; |\mathbf{q}\rangle = b_0|0\rangle + b_1|1\rangle.$$

By applying the definitions of quantum computational realization and of `OR`, we will obtain:

$$|\mathbf{p} \curlyvee \mathbf{q}\rangle = a_1 b_1|1,1,1\rangle + a_1 b_0|1,0,1\rangle + a_0 b_1|0,1,1\rangle + a_0 b_0|0,0,0\rangle.$$

We know that the number $|a_1 b_1|^2$ represents the probability that both the members of our disjunction are true and that, consequently, the disjunction is true. Similarly in the other cases. In order to calculate the probability of the truth of $\mathbf{p} \curlyvee \mathbf{q}$, it will be sufficient to sum the three probability-values corresponding to the three cases where the final result is *True* (that is, the cases of the vectors $|1,1,1\rangle$, $|1,0,1\rangle$, $|0,1,1\rangle$). Thus, we are able to assign to the disjunction $\mathbf{p} \curlyvee \mathbf{q}$ the following probability-value:

$$|a_1 b_1|^2 + |a_1 b_0|^2 + |a_0 b_1|^2.$$

We can now define the notions of *truth*, *logical truth*, *consequence* and *logical consequence* .

DEFINITION 17.3.3. *Truth and logical truth*
A sentence α is *true in a quantum computational realization Qub* ($\models_{Qub} \alpha$) iff $\text{Prob}(\alpha) = 1$.
α is a *logical truth* ($\models \alpha$) iff for any computational realization Qub, $\models_{Qub} \alpha$.

DEFINITION 17.3.4. *Consequence and logical consequence*
β is a *consequence of* α *in the quantum computational realization Qub* ($\alpha \models_{Qub} \beta$) iff $\text{Prob}(\alpha) \leq \text{Prob}(\beta)$;
β is a *logical consequence* of α ($\alpha \models \beta$) iff for any Qub: $\alpha \models_{Qub} \beta$.

Let us call the logic characterized by this semantics *quantum computational logic* (**QCL**).

Some interesting examples of logical consequences that hold in **QCL** are the following:

THEOREM 17.3.5.

(i) $\alpha \models \neg\neg\alpha, \quad \neg\neg\alpha \models \alpha;$
(double negation)
(ii) $\sqrt{\neg}\sqrt{\neg}\alpha \models \neg\alpha, \quad \neg\alpha \models \sqrt{\neg}\sqrt{\neg}\alpha;$
(iii) $\alpha \curlywedge \beta \models \beta \curlywedge \alpha, \quad \alpha \curlyvee \beta \models \beta \curlyvee \alpha;$
(commutativity)
(iv) $\alpha \curlywedge (\beta \curlywedge \gamma) \models (\alpha \curlywedge \beta) \curlywedge \gamma, \quad (\alpha \curlywedge \beta) \curlywedge \gamma \models \alpha \curlywedge (\beta \curlywedge \gamma);$
(associativity)
(v) $\alpha \curlyvee (\beta \curlyvee \gamma) \models (\alpha \curlyvee \beta) \curlyvee \gamma, \quad (\alpha \curlyvee \beta) \curlyvee \gamma \models \alpha \curlyvee (\beta \curlyvee \gamma);$
(associativity)
(vi) $\neg(\alpha \curlywedge \beta) \models \neg\alpha \curlyvee \neg\beta, \quad \neg\alpha \curlyvee \neg\beta \models \neg(\alpha \curlywedge \beta);$
(de Morgan)
(vii) $\neg(\alpha \curlyvee \beta) \models \neg\alpha \curlywedge \neg\beta, \quad \neg\alpha \curlywedge \neg\beta \models \neg(\alpha \curlyvee \beta)$
(de Morgan)
(viii) $\alpha \curlywedge \alpha \models \alpha.$
(semiidempotence 1)

(ix) $\alpha \curlywedge (\beta \curlyvee \gamma) \models (\alpha \curlywedge \beta) \curlyvee (\alpha \curlywedge \gamma)$.
(distributivity 1)

Some logical consequences and some logical truths that are violated in **QCL** are the following:

THEOREM 17.3.6.

(i) $\alpha \not\models \alpha \curlywedge \alpha$;
(semiidempotence 2)

(ii) $\not\models \alpha \curlyvee \neg\alpha$
(excluded middle)

(iii) $\not\models \neg(\alpha \curlywedge \neg\alpha)$;
(noncontradiction)

(iv) $(\alpha \curlywedge \beta) \vee (\alpha \curlywedge \gamma) \not\models \alpha \curlywedge (\beta \curlyvee \gamma)$.
(distributivity 2)

PROOF. (i)-(iii) Take $|\alpha\rangle := \frac{\sqrt{2}}{2}|0\rangle + \frac{\sqrt{2}}{2}|1\rangle$. Then, $\mathrm{Prob}(\alpha) = \frac{1}{2}$, $\mathrm{Prob}(\alpha \curlywedge \alpha) = \frac{1}{4}$, $\mathrm{Prob}(\alpha \curlyvee \neg\alpha) = \mathrm{Prob}(\neg(\alpha \curlywedge \neg\alpha)) = \frac{3}{4}$.
(iv) Take $|\alpha\rangle = |\beta\rangle := \frac{\sqrt{2}}{2}|0\rangle + \frac{\sqrt{2}}{2}|1\rangle$ and $|\gamma\rangle := \frac{\sqrt{3}}{2}|0\rangle + \frac{1}{2}|1\rangle$. Then, $\mathrm{Prob}((\alpha \wedge \beta) \vee (\alpha \wedge \gamma)) = \frac{11}{32} > \frac{10}{12} = \mathrm{Prob}(\alpha \wedge (\beta \vee \gamma))$. □

QCL turns out to be a nonstandard form of quantum logic. Conjunction and disjunction do not correspond to lattice operations, because they are not generally idempotent. Unlike the usual (sharp and unsharp) quantum logics, the weak distributivity principle ($(\alpha \curlywedge \beta) \curlyvee (\alpha \curlywedge \gamma) \models \alpha \curlywedge (\beta \curlyvee \gamma)$) breaks down. Both the excluded middle and the noncontradiction principles are violated: as a consequence, we have obtained an example of an *unsharp logic*. Interestingly enough, one is dealing with a logic that has *no logical truths*. This follows from the following.

LEMMA 17.3.7. *Let Qub be any quantum computational realization and let α be any sentence. If* $\mathrm{Prob}(Qub(\alpha)) \in \{0, 1\}$, *then there is an atomic sentence* $\mathbf{p}$ *occurring in α such that* $\mathrm{Prob}(Qub(\mathbf{p})) \in \{0, \frac{1}{2}, 1\}$.

PROOF. Suppose that $\mathrm{Prob}(Qub(\alpha)) \in \{0, 1\}$. The proof is by induction on the logical complexity of α.
(i) α is an atomic sentence. The proof is trivial.
(ii) $\alpha = \neg\beta$. By Theorem 17.2.3(ii), $\mathrm{Prob}(Qub(\alpha)) = 1 - \mathrm{Prob}(Qub(\beta)) \in \{0, 1\}$. The conclusion follows by induction hypothesis.
(iii) $\alpha = \sqrt{\neg}\beta$. By hypothesis and by Theorem 17.2.3(v), β cannot be a conjunction. Consequently, only the following cases are possible: (iiia) $\beta = \mathbf{p}$; (iiib) $\beta = \neg\gamma$; (iiic) $\beta = \sqrt{\neg}\gamma$.
(iiia) $\beta = \mathbf{p}$. By hypothesis, $\mathrm{Prob}(\sqrt{\neg}\beta) \in \{0, 1\}$. Hence, $\sqrt{\mathtt{NOT}}(Qub(\mathbf{p})) = c|x\rangle$, where $|x\rangle \in \{|0\rangle, |1\rangle\}$ and $|c| = 1$. We have:

$$\mathtt{NOT}(Qub(\mathbf{p})) = \sqrt{\mathtt{NOT}}(\sqrt{\mathtt{NOT}}(Qub(\mathbf{p}))) = \sqrt{\mathtt{NOT}}(c|x\rangle).$$

By Theorem 17.2.3(iv), $\text{Prob}(\sqrt{\texttt{NOT}}(c|x\rangle)) = \frac{1}{2}$. As a consequence,

$$\text{Prob}(Qub(\neg\mathbf{p})) = \frac{1}{2} = \text{Prob}(Qub(\mathbf{p})).$$

(iiib) $\beta = \neg\gamma$. By Theorem 17.1.10 (ii) and by Theorem 17.2.3(ii),

$$\text{Prob}(Qub(\sqrt{\neg}\neg\gamma)) = \text{Prob}(Qub(\neg\sqrt{\neg}\gamma)) = 1 - \text{Prob}(Qub(\sqrt{\neg}\gamma)).$$

The conclusion follows by induction hypothesis.
(iiic) $\beta = \sqrt{\neg}\gamma$. Then,

$$\text{Prob}(Qub(\sqrt{\neg}\sqrt{\neg}\gamma)) = \text{Prob}(Qub(\neg\gamma)) = 1 - \text{Prob}(Qub(\gamma)).$$

The conclusion follows by induction hypothesis.
(iv) $\alpha = \beta \curlywedge \gamma$. By Theorem 17.2.3(i),

$$\text{Prob}(Qub(\beta \wedge \gamma)) = \text{Prob}(Qub(\beta))\text{Prob}(Qub(\gamma)) \in \{0, 1\}.$$

The conclusion follows by induction hypothesis. □

THEOREM 17.3.8. *There exists no quantum computational logical truth.*

PROOF. Suppose, by contradiction, that α is a logical truth of **QCL**. Then, for any quantum computational realization Qub, $\text{Prob}(Qub(\alpha)) = 1$. Let $\mathbf{p}_1, \ldots, \mathbf{p}_n$ be the atomic sentences of α and let Qub be a quantum computational realization such that for any i $(1 \le i \le n)$: $\text{Prob}(Qub(\mathbf{p}_i)) \notin \{0, \frac{1}{2}, 1\}$. Then, by Lemma 17.3.7, $\text{Prob}(Qub(\alpha)) \notin \{0, 1\}$, contradiction. □

The axiomatizability of **QCL** is an open problem.

Conclusions

Some general questions that have been often discussed in connection with (or against) quantum logic are the following:

(a) Why quantum logics?
(b) Are quantum logics helpful to solve the difficulties of QT?
(c) Are quantum logics "real logics?" And how is their use compatible with the mathematical formalism of QT, based on classical logic?
(d) Does quantum logic confirm the thesis that "logic is empirical?"

Our answers to these questions are, in a sense, trivial, and close to a position that Gibbins (1987) has called a "quietist view of quantum logic." It seems to us that quantum logics are not to be regarded as a kind of "clue," capable of solving the main physical and epistemological difficulties of QT. This was perhaps an illusion of some pioneering workers in quantum logic. Let us think of the attempts to recover a *realistic interpretation* of QT based on the properties of the quantum logical connectives.[4]

Why quantum logics? Simply because "quantum logics are there!" They seem to be deeply incorporated in the abstract structures generated by QT. Quantum logics are, without any doubt, *logics*. As we have seen, they satisfy all the canonical conditions that the present community of logicians require in order to call a given abstract object *a logic*. A question that has been often discussed concerns the compatibility between quantum logic and the mathematical formalism of quantum theory, based on classical logic. Is the quantum physicist bound to a kind of "logical schizophrenia"? At first sight, the copresence of different logics in one and the same theory may give a sense of uneasiness. However, the splitting of the basic logical operations (negation, conjunction, disjunction,...) into different connectives with different meanings and uses is now a well accepted logical phenomenon, that admits consistent descriptions. Classical and quantum logic turn out to apply to different sublanguages of quantum theory, that must be carefully distinguished.

Finally, does quantum logic confirm the thesis that "logic is empirical?" At the very beginning of the contemporary discussion about the *nature of logic*, the claim that the "right logic" to be used in a given theoretical situation may also depend on experimental data appeared to be a kind of extremistic view, in contrast with a leading philosophical tradition according

[4]See for instance (Putnam, 1969).

to which a characteristic feature of logic should be its absolute independence from any content. Interestingly enough, a quite heterodox thesis, in this connection, had been defended (already in 1936) by Łukasiewicz (the "father" of fuzzy logics). The strong contrast between Łukasiewicz' position and the leading ideas of the Vienna Circle is apparent in the following quotation (Łukasiewicz, 1970a)

> I think that in Carnap the attempt to reduce certain objective problems to a linguistic one results from his erroneous interpretation of the *a priori* sciences and their role in the study of reality. That erroneous opinion was taken over by Carnap from Wittgenstein, who considers all *a priori* propositions, that is, those belonging to logic and mathematics, to be tautologies. Carnap calls such propositions analytic. I have always opposed that terminology, since the association it evokes may make it misleading. Moreover, Carnap believes, together with Wittgenstein, that *a priori* propositions do not convey anything about reality. For them the *a priori* disciplines are only instruments which facilitate the cognition of reality, but a scientific interpretation of the world could, if necessary, do without those *a priori* elements. Now, my opinion on the *a priori* disciplines and their role in the study of reality is entirely different. We know today that not only do different systems of geometry exist, but different systems of logic as well, and they have, moreover, the property that one cannot be translated into another. I am convinced that one and only one of these logical systems is valid in the real world, that is, is real, in the same way as one and only one system of geometry is real. Today, it is true, we do not yet know which system that is, but I do not doubt that empirical research will sometime demonstrate whether the space of the universe is Euclidean or non-Euclidean, and whether relationships between facts correspond to two-valued logic or to one of the many-valued logics. All *a priori* systems, as soon as they are applied to reality, become natural-science hypotheses which have to be verified by facts in a similar way as is done with physical hypotheses.

The comparison between logic and geometry was the central point of Putnam's famous article *Is Logic empirical?* (Putnam, 1969), which has highly influenced the epistemological debate about quantum logic.

These days, an empirical position in logic is generally no longer regarded as a "daring heresy" . At the same time, we are facing not only a variety of logics, but even a variety of quantum logics. The "labyrinth of quantum logics" described by van Fraassen (1974) has become more and more labyrinthine. Even the distinction between *sharp* and *unsharp* logical situations turns out to be, to a certain extent, "unsharp". We have seen, for instance, that *the logic* of *special effects* (which may be physically unsharp) corresponds to a sharp version of quantum logic (which satisfies the noncontradiction principle). At the same time, the so called *eccentric effects* (such that any unsharp property is represented as a multiple of the universal event) are adequately described by *orthomodular paraconsistent quantum logic*.

The logical behavior of standard effects (the *whole* $\mathcal{E}(\mathcal{H})$) can be represented by means of different forms of *unsharp quantum logics*. One can refer to a *partial* quantum logic (like **UPaQL**) (where conjunction and disjunction are only defined for pairs of orthogonal effects), or to a *total* logic (like *Łukasiewicz' quantum logic*, a natural logical abstraction from the QMV-structure of $\mathcal{E}(\mathcal{H})$). Another possibility is represented by different forms of *Brouwer Zadeh logics*: in the Brouwer Zadeh semantics effects are generally regarded as *possible worlds* (a kind of *unsharp* and *partial* pieces of information about possible physical situations), while the *meanings* of linguistic sentences are represented by convenient sets of effects.

A totally different situation arises in the framework of quantum computation: as we have seen, the theory of *quantum logical gates* has suggested a nonstandard version of unsharp logic (*quantum computational logic*). Unlike all other forms of quantum logic (we have investigated in this book), quantum computational logic is strongly anti-classical, since it cannot be described as a sublogic of classical logic. Strangely enough, this logic is nondistributive in the "wrong way" with respect to orthodox quantum logic (the lattice-distributivity relations are generally violated, while the strong distributivity properties, which fail in quantum logic, are valid).

In this complex situation a natural question arises: is it still reasonable to look for the "right logic" of QT?

Synoptic tables

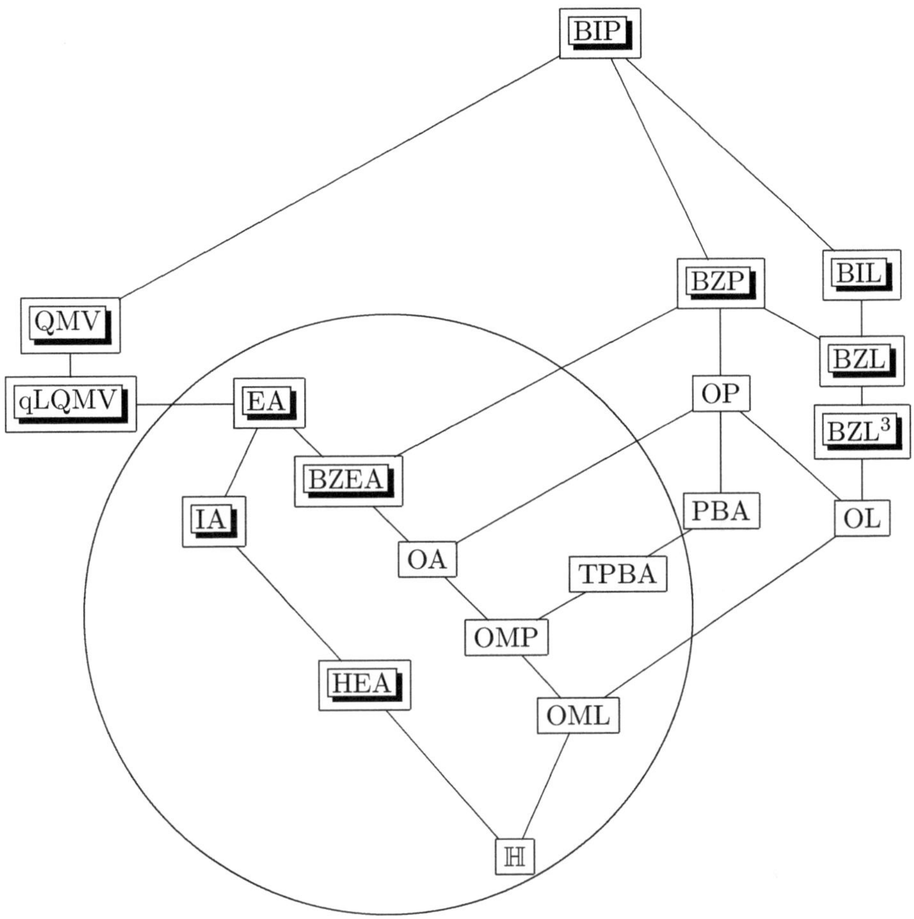

TABLE 1. The quantum structures

Legend of Table 1

X (double box) = *Unsharp* quantum structure;
X (box) = *Sharp* quantum structure.

Notes

(a) An ascending line from X to Y indicates that any structure of type X can be embedded into a structure of type Y.
(b) The one horizontal line between EA and qLQMV indicates that every effect algebra is embeddable into a quasi-linear QMV algebra, and vice versa.
(c) The structures inside the circle are "the orthomodular structures" that are mentioned in the text; these structures admit states and observables.
(d) It is also true that every EA which admits an order determining set of states can be embedded into an IA. This is the essence of the Bennett-Foulis representation theorem (Theorem 5.1.15).
(e) The reverse of the arrow from BZEA to EA also holds, for every regular EA can be embedded into a BZEA, by defining $a^{\sim}$ to be $a^{\sim} := \chi_{\mathbf{0}}(a)$, which is **1** at **0** and **0** otherwise.

BZEA := BZ-effect algebras
EA := Effect algebras
$\mathbb{H}$:= Hilbert lattices
HEA := Standard effect algebras
IA := Interval effect algebras
OA := Orthoalgebras
OML := Orthomodular lattices
OMP := Orthomodular posets
TPBA := Transitive partial Boolean algebras

BIL := Bounded involution lattices
BIP := Bounded involution posets
BZL^3 := BZ^3 lattices
BZL := BZ-lattices
BZP := BZ-posets
OL := Ortholattices
OP := Orthoposets
PBA := Partial Boolean algebras
qLQMV := Quasi-linear QMV algebras
QMV := QMV algebras

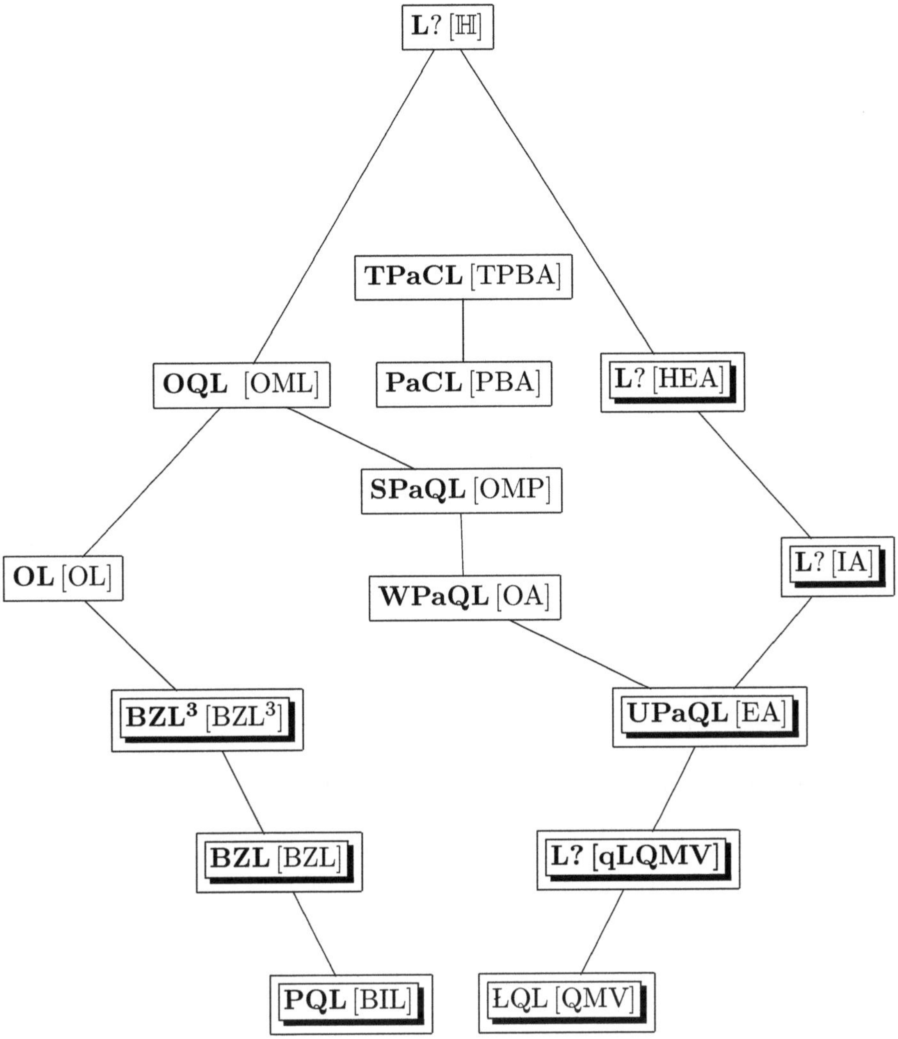

TABLE 2. The labyrinth of quantum logics

Legend of Table 2

X [Y] (double-framed box) indicates that the *unsharp* logic **X** is semantically characterized by the class Y.

X [Y] (single-framed box) indicates that the *sharp* logic **X** is semantically characterized by the class Y.

L ?[Y] (double-framed box) or **L** ?[Y] (single-framed box) means that it is not known whether the logic **L** semantically characterized by the class Y is *axiomatizable.*

An ascending line from **X**[Y] to **W**[Z] indicates that **X** is a sublogic of **W** (up to a natural linguistic translation.)

OQL := Orthomodular quantum logic
SPaQL := Strong partial quantum logic
UPaQL := Unsharp partial quantum logic
WPaQL := Weak partial quantum logic

BZL := Weak Brouwer Zadeh logic
$\mathbf{BZL^3}$:= Strong Brouwer Zadeh logic
ŁQL := Łukasiewicz quantum logic
OL := Orthologic
PaCL := Partial classical logic
PQL := Paraconsistent quantum logic
TPaCL := Transitive partial classical logic.

Bibliography

Aerts, D.: 1982, Description of many separated physical entities without the paradoxes encountered in quantum mechanics, Vol. 12, pp. 1131–1170.

Aerts, D.: 1984, Construction of a structure which enables to describe the joint system of a classical and a quantum system, *Reports on Mathematical Physics* **20**, 117–129.

Aerts, D. and Van Steirteghem, B.: 2000, Quantum axiomatics and a theorem of M.P. Solèr, *International Journal of Theoretical Physics* **39**, 497–502.

Aspect, A. and Grangier, P.: 1985, Tests of Bell's inequalities with pairs of low energy correlated photons: an experimental realization of Einstein-Podolsky-Rosen-type correlations, *in* P. Lahti and P. Mittelstaedt (eds), *Symposium on the Foundations of Modern Physics*, World Scientific, Singapore, pp. 51–71.

Aspect, A., Grangier, P. and Roger, G.: 1981, Experimental tests of realistic local theories via Bell's theorem, *Physical Review Letters* **47**, 460–467.

Battilotti, G.: 1998, Embedding classical logic into basic orthologic with a primitive modality, *Logic Journal of the IGPL,* **6**, 383–402.

Battilotti, G. and Faggian, C.: 2002, Quantum logic and the cube of logics, *in* D. Gabbay and F. Guenthner (eds), *Handbook of Philosophical Logic*, Vol. 6, Kluwer Academic Publishers, Dordrecht, pp. 213–226.

Battilotti, G. and Sambin, G.: 1999, Basic logic and the cube of its extensions, *in* A. Cantini, E. Casari and P. Minari (eds), *Logic and Foundations of Mathematics*, Kluwer Academic Publishers, Dordrecht, pp. 165–186.

Bell, J. L. and Slomson, A. B.: 1969, *Models and Ultraproducts: An Introduction*, North-Holland, Amsterdam.

Bell, J. S.: 1966, On the problem of hidden variables in quantum mechanics, *Reviews of Modern Physics* **38**(38), 447–452.

Belluce, L. P.: 1986, Semisimple algebras of infinite valued logic and bold fuzzy set theory, *Canadian Journal of Mathematics* **38**, 1356–1379.

Beltrametti, E. and Bugajski, S.: 1995, A classical extension of quantum mechanics, *Journal of Physics A: Mathematical and General* **28**, 247–261.

Beltrametti, E. and Bugajski, S.: 1997, Effect algebras and statistical physical theories, *Journal of Mathematical Physics* **38**, 3020–3030.

Beltrametti, E. and Cassinelli, G.: 1981, *The Logic of Quantum Mechanics*, Vol. 15 of *Encyclopedia of Mathematics and its Applications*, Addison-Wesley, Reading.

Bennett, M. K.: 1995, *Affine and Projective Geometry*, Wiley-Interscience, New York.

Bennett, M. K. and Foulis, D. J.: 1997, Interval and scale effect algebras, *Advances in Mathematics* **19**, 200–215.

Birkhoff, G.: 1967, *Lattice Theory*, Vol. 25 of *Colloquium Publications*, 3^{rd}(new) edn, American Mathematical Society, Providence.

Birkhoff, G. and von Neumann, J.: 1936, The logic of quantum mechanics, *Annals of Mathematics* **37**, 823–843.

Bruns, G., Greechie, R. J., Harding, J. and Roddy, M.: 1990, Completions of orthomodular lattices, *Order* **7**, 789–807.

Bub, J.: 1999, *Interpreting the Quantum World*, Cambridge University Press, Cambridge.

Bugajski, S.: 1993, Delinearization of quantum logic, *International Journal of Theoretical Physics* **32**, 389–398.

Busch, P.: 1985, Elements of unsharp reality in the EPR experiment, *in* P. Lahti and P. Mittelstaedt (eds), *Symposium on the Foundations of Modern Physics*, World Scientific, Singapore, pp. 343–357.

Busch, P., Grabowski, M. and Lahti, P.: 1995, *Operational Quantum Mechanics*, number m31 in *Lectures Notes in Physics*, Springer, Berlin.

Busch, P., Lahti, P. and Mittelstaedt, P.: 1991, *The Quantum Theory of Measurement*, number m2 in *Lectures Notes in Physics*, Springer, Berlin.

Cattaneo, G.: 1993, Fuzzy quantum logic II: the logics of unsharp quantum mechanics, *International Journal of Theoretical Physics* **32**, 1709–1734.

Cattaneo, G.: 1997, A unified framework for the algebra of unsharp quantum mechanics, *International Journal of Theoretical Physics* **36**, 3085–3117.

Cattaneo, G., Dalla Chiara, M. L. and Giuntini, R.: 1999, How many notions of 'sharp'?, *International Journal of Theoretical Physics* **38**, 3153–3161.

Cattaneo, G., Garola, C. and Nisticò, G.: 1989, Preparation-effect versus question-proposition structures, *Physics Essays* **2**, 197–216.

Cattaneo, G. and Giuntini, R.: 1995, Some results on BZ structures from hilbertian unsharp quantum physics, *Foundations of Physics* **25**, 1147–1182.

Cattaneo, G. and Gudder, S. P.: 1999, Algebraic structures arising in axiomatic unsharp quantum physics, *Foundations of Physics* **29**, 1607–1637.

Cattaneo, G. and Laudisa, F.: 1994, Axiomatic unsharp quantum theory (from Mackey to Ludwig), *Foundations of Physics* **24**, 631–683.

Cattaneo, G. and Nisticò, G.: 1986, Brouwer-Zadeh posets and three-valued Łukasiewicz posets, *Fuzzy Sets and Systems* **33**, 165–190.

Chang, C. C.: 1958, Algebraic analysis of many valued logics, *Transactions of the American Mathematical Society* **88**, 74–80.

Chang, C. C.: 1959, A new proof of the completeness of Łukasiewicz axioms, *Transactions of the American Mathematical Society* **93**, 467–490.

Cignoli, R., D'Ottaviano, I. M. L. and Mundici, D.: 2000, *Algebraic Foundations of Many-valued Reasoning*, Vol. 7 of *Trends in Logic*, Kluwer Academic Publishers, Dordrecht.

Cutland, N. J. and Gibbins, P. F.: 1982, A regular sequent calculus for quantum logic in which $\wedge$ and $\vee$ are dual, *Logique et Analyse - Nouvelle Serie -* **25**(45), 221–248.

Czelakowski, J.: 1975, Logics based on partial Boolean σ-algebras (i), *Studia Logica* **34**, 371–395.

da Costa, N. C. A., French, S. and Krause, D.: 1992, The Schrödinger problem, *in* M. Bibtol and O. Darrigol (eds), *Erwin Schrödinger: Philosophy and the Birth of Quantum Mechanics*, Editions Frontières, pp. 445–460.

Dalla Chiara, M. L.: 1981, Some metalogical pathologies of quantum logic, *in* E. Beltrametti and B. van Fraassen (eds), *Current Issues in Quantum Logic*, Vol. 8 of *Ettore Majorana International Science Series*, Plenum, New York, pp. 147–159.

Dalla Chiara, M. L. and Giuntini, R.: 1989, Paraconsistent quantum logics, *Foundations of Physics* **19**, 891–904.

Dalla Chiara, M. L. and Giuntini, R.: 1994, Unsharp quantum logics, *Foundations of Physics* **24**, 1161–1177.

Dalla Chiara, M. L. and Giuntini, R.: 1995a, The logics of orthoalgebras, *Studia Logica* **55**, 3–22.

Dalla Chiara, M. L. and Giuntini, R.: 1995b, Physical interpretations of the Łukasiewicz quantum logical connectives, *in* C. Garola and A. Rossi (eds), *The Foundations of Quantum Mechanics - Historical Analysis and Open Questions*, Kluwer, Dordrecht, pp. 179–185.

Dalla Chiara, M. L. and Giuntini, R.: 1999, Łukasiewicz theory of truth, from the quantum logical point of view, *in* J. Woleński and E. Köhler (eds), *Alfred Tarski and the Vienna Circle*, Kluwer, Dordrecht, pp. 127–134.

Dalla Chiara, M. L. and Giuntini, R.: 2002, Quantum logics, *in* D. Gabbay and F. Guenthner (eds), *Handbook of Philosophical Logic*, Vol. 6, Kluwer Academic Publishers, Dordrecht, pp. 129–228.

Dalla Chiara, M. L., Giuntini, R. and Leporini, R.: 2003, Quantum computational logics. a survey, *in* V. Hendricks and J. Malinowski (eds), *50 Years of Studia Logica*, Trends in Logic, Kluwer Academic Publishers, Dordrecht, pp. 213–255.

Dalla Chiara, M. L. and Toraldo di Francia, G.: 1993, Individuals, kinds and names in physics, *in* G. Corsi, M. L. Dalla Chiara and G. Ghirardi (eds), *Bridging the Gap: Philosophy, Mathematics, and Physics*, Kluwer Academic Publisher, Dordrecht, pp. 261–283.

Dalla Chiara, M. L. and Toraldo di Francia, G.: 1995, Identity questions from quantum theory, *in* K. G. et al. (ed.), *Physics, Philosophy and the Scientific Community*, Kluwer Academic Publisher, Dordrecht, pp. 39–46.

Davies, E. B.: 1976, *Quantum Theory of Open Systems*, Academic, New York.

Deutsch, D., Ekert, A. and Lupacchini, R.: 2000, Machines, logic and quantum physics, *Bulletin of Symbolic Logic* **3**, 265–283.

Dichtl, M.: 1984, Astroids and pastings, *Algebra Universalis* **18**, 380–385.

Dishkant, H.: 1972, Semantics of the minimal logic of quantum mechanics, *Studia Logica* **30**, 17–29.

Dvurečenskij, A.: 1993, *Gleason's Theorem and its Applications*, number 60 in *Mathematics and its Applications*, Kluwer, Dordrecht.

Dvurečenskij, A.: 1997, Measures and ⊥-decomposable measures of effects of a Hilbert space, *Atti del Seminario Matematico e Fisico dell' Universita di Modena* **45**, 259–288.

Dvurečenskij, A. and Pulmannová, S.: 1994, D-test spaces and difference poset, *Reports on Mathematical Physics* **34**, 151–170.

Dvurečenskij, A. and Pulmannová, S.: 2000, *New Trends in Quantum Structures*, Vol. 516 of *Mathematics and Its Applications*, Kluwer Academic Publishers, Dordrecht.

Einstein, A., Podolsky, B. and Rosen, N.: 1935, Can quantum-mechanical description of reality be considered complete?, *Physical Review* **47**, 777–780.

Ekert, A., Hayden, P. M. and Inamori, H.: 2000, Basic concepts in quantum computation, *in* R. Kaiser, C. Westbrook and F. David (eds), *Coherent Atomic Matter Waves*, Vol. LXXII of *Nato Advanced Study Institute - Les Houches*, Springer Verlag, Berlin, pp. 661–699.

Engesser, K. and Gabbay, D.: 2002, Quantum logic, Hilbert space, revision theory, *Artificial Intelligence* **136**, 61–100.

Faggian, C.: 1998, Classical proofs via basic logic, *in* M. Nielson and W. Thomas (eds), *Computer Science Logic 11th International Workshop, CSL'97*, Vol. 1414 of *Lecture Notes in Computer Science*, Springer Verlag, pp. 203–219.

Faggian, C. and Sambin, G.: 1997, From basic logic to quantum logics with cut-elimination,, *International Journal of Theoretical Physics* **12**, 31–37.

Finch, P. D.: 1970, Quantum logic as an implication algebra, *Bulletin of the Australian Mathematical Society* **2**, 101–106.

Fitting, M.: 1969, *Intuitionistic Logic, Model Theory and Forcing*, North-Holland, Amsterdam.

Foulis, D. J., Greechie, R., Dalla Chiara, M. L. and Giuntini, R.: 1996, Quantum Logic, *in* G. Trigg (ed.), *Encyclopedia of Applied Physics*, Vol. 15, VCH Publishers, pp. 229–255.

Foulis, D. J.: 1999, A half-century of quantum logic, what have we learned?, *in* D. Aerts and J. Pykacz (eds), *Quantum Structures and the Nature of Reality*, Vol. 7, Kluwer Academic Publishers, Dordrecht, pp. 1–36.

Foulis, D. J.: 2000, MV and Heyting effect algebras, *Foundations of Physics* **30**, 1687–1706.

Foulis, D. J. and Bennett, M. K.: 1994, Effect algebras and unsharp quantum logics, *Foundations of Physics* **24**, 1325–1346.

Foulis, D. J. and Greechie, R. J.: 2000, Specification of finite effect algebras, *International Journal Theoretical Physics* **39**, 665–676.

Foulis, D. J. and Munem, M.: 1984, *Calculus With Analytic Geometry*, Worth Publishing, London.

Foulis, D. J. and Randall, C. H.: 1981, Empirical logic and tensor product, *Interpretation and Foundations of Quantum Mechanics*, Vol. 5 of *Grundlagen der exakten Naturwissenschaften*, Bibliographisches Institut, Mannheim, pp. 9–20.

Foulis, D. J. and Randall, C. H.: 1983, Properties and operational propositions in quantum mechanics, *Foundations of Physics* **13**, 843–857.

Garola, C.: 1980, Propositions and orthocomplementation in quantum logic, *International Journal of Theoretical Physics* **19**, 369–378.

Garola, C.: 1985, Embedding of posets into lattices in quantum logic, *International Journal of Theoretical Physics* **24**, 423–433.

Gerelle, E. R., Greechie, R. J. and Miller, F. R.: 1974, Weights on spaces, *in* C. Enz and J. Mehra (eds), *Physical Reality and Mathematical Description*, Reidel, Dordrecht, pp. 169–192.

Ghirardi, G., Rimini, A. and Weber, T.: 1986, Unified dynamics for microscopic systems, *Physical Review* **D 34**, 470–491.

Gibbins, P. F.: 1985, A user-friendly quantum logic, *Logique-et-Analyse.-Nouvelle-Serie* **28**, 353–362.

Gibbins, P. F.: 1987, *Particles and Paradoxes - The Limits of Quantum Logic*, Cambridge University Press, Cambridge.

Girard, J. Y.: 1987, Linear logic, *Theoretical Computer Science* **50**, 1–102.

Giuntini, R.: 1990, Brouwer-Zadeh logic and the operational approach to quantum mechanics, *Foundations of Physics* **20**, 701–714.

Giuntini, R.: 1991a, *Quantum Logic and Hidden Variables*, number 8 in *Grundlagen der exakten Naturwissenschaften*, Bibliographisches Institut, Mannheim.

Giuntini, R.: 1991b, A semantical investigation on Brouwer-Zadeh logic, *Journal of Philosophical Logic* **20**, 411–433.

Giuntini, R.: 1992, Brouwer-Zadeh logic, decidability and bimodal systems, *Studia Logica* **51**, 97–112.

Giuntini, R.: 1993, Three-valued Brouwer-Zadeh logic, *International Journal of Theoretical Physics* **32**, 1875–1887.

Giuntini, R.: 1995a, Quasilinear QMV algebras, *International Journal of Theoretical Physics* **34**, 1397–1407.

Giuntini, R.: 1995b, Unsharp orthoalgebras and quantum MV algebras, *in* C. Garola and A. Rossi (eds), *The Foundations of Quantum Mechanics - Historical Analysis and Open Questions*, Kluwer, Dordrecht, pp. 325–337.

Giuntini, R.: 1996, Quantum MV algebras, *Studia Logica* **56**, 393–417.

Giuntini, R.: 2000, An independent axiomatization of QMV algebras, *in* C. Garola and A. Rossi (eds), *The Foundations of Quantum Mechanics*, World Scientific, Singapore.

Giuntini, R.: n.d., Weakly linear QMV algebras, *Algebra Universalis* .

Giuntini, R. and Greuling, H.: 1989, Toward an unsharp language for unsharp properties, *Foundations of Physics* **19**, 931–945.

Giuntini, R. and Pulmannovà, S.: 2000, 'ideals and congruences in QMV algebras, *Communications in Algebra* **28**, 1567–1592.

Gleason, A. M.: 1957, Measures on the closed subspaces of a Hilbert space, *Journal of Mathematics and Mechanics* **6**, 885–893.

Goldblatt, R.: 1974, Semantics analysis of orthologic, *Journal of Philosophical Logic* **3**, 19–35.

Goldblatt, R.: 1984, Orthomodularity is not elementary, *The Journal of Symbolic Logic* **49**, 401–404.

Goldblatt, R.: 1993, *Mathematics of Modality*, CSLI, Stanford.

Greechie, R. J.: 1968, On the structure of orthomodular lattices satisfying the chain condition, *Journal of Combinatorial Theory* **4**, 210–218.

Greechie, R. J.: 1969, An orthomodular poset with a full set of states not embeddable in Hilbert space, *Caribbean Journal of Mathematics and Science* **1**, 1–10.

Greechie, R. J.: 1971, Orthomodular lattices admitting no states, *Journal of Combinatorial Theory* **10**, 119–131.

Greechie, R. J.: 1974, Some results from the combinatorial approach to quantum logic, *Synthese* **29**, 113–127.

Greechie, R. J.: 1975, On three dimensional quantum proposition systems, *in* L. Castell, M. Drieschner and C. von Weizsäcker (eds), *Quantum Theory and the Structures of Time and Space*, Carl Hanser Verlag, Munchen-Wien, pp. 71–84.

Greechie, R. J.: 1978, Another nonstandard quantum logic (and how I found it), *in* A. Marlow (ed.), *Mathematical Foundations of Quantum Theory*, Academic Press,, London, pp. 71–85.

Greechie, R. J.: 1981, A non-standard quantum logic with a strong set of states, *in* E. Beltrametti and B. van Fraassen (eds), *Current Issues in Quantum Logic*, Vol. 8 of *Ettore Majorana International Science Series*, Plenum, New York, pp. 375–380.

Greechie, R. J. and Foulis, D. J.: 1995, The transition to effect algebras, *International Journal of Theoretical Physics* **34**, 1369–1382.

Gruska, J.: 1999, *Quantum Computing*, McGraw-Hill, London.

Gudder, S. P.: 1979, A survey of axiomatic quantum mechanics, *in* C. A. Hooker (ed.), *The Logico-Algebraic Approach to Quantum Mechanics*, Vol. II, Reidel, Dordrecht, pp. 323–363.

Gudder, S. P.: 1995, Total extensions of effect algebras, *Foundations of Physics Letters* **8**, 243–252.

Gudder, S. P.: 1998, Sharply dominating effect algebras, *Tatra Mountains Mathematical Publications* **15**, 23–30.

Gudder, S. P. and Greechie, R. J.: 1996, Effect algebra counterexamples, *Mathematica Slovaca* **46**, 317–325.

Hájek, P.: 1998, *Metamathematics of Fuzzy Logic*, Vol. 4 of *Trends in Logic*, Kluwer Academic Publishers, Dordrecht.

Halmos, P. R.: 1951, *Introduction to Hilbert Space and the Theory of Spectral Multiplicity*, Chelsea, New York.

Hardegree, G. M.: 1975, Stalnaker conditionals and quantum logic, *Journal of Philsophical Logic* **4**, 399–421.

Hardegree, G. M.: 1976, The conditional in quantum logic, *in* P. Suppes (ed.), *Logic and Probability in Quantum Mechanics*, Reidel, Dordrecht, pp. 55–72.

Hardegree, G. M.: 1981, An axiom system for orthomodular quantum logic, *Studia Logica* **40**, 1–12.

Holland, S. S.: 1995, Orthomodularity in infinite dimensions: a theorem of M. Solèr, *Bulletin of the American Mathematical Society* **32**, 205–232.

Hughes, G. E. and Cresswell, M. J.: 1984, *A Companion to Modal Logic*, Methuen, London.

Hughes, R. I. G.: 1985, Semantic alternatives in partial Boolean quantum logic, *Journal of Philosophical Logic* **14**, 411–446.

Hughes, R. I. G.: 1987, *The Structure and Interpretation of Quantum Mechanics*, Cambridge University Press, Cambridge.

Isham, C. J.: 1995, *Lectures on Quantum Theory: Mathematical and Structural Foundation*, Imperial College Press, London.

Jammer, M.: 1974, *The Philosophy of Quantum Mechanics*, Wiley-Interscience, New York.

Jauch, J. M.: 1968, *Foundations of Quantum Mechanics*, Addison-Wesley, London.

Kalmbach, G.: 1983, *Orthomodular Lattices*, Academic Press, New York.

Keller, H. A.: 1980, Ein nichtklassischer hilbertscher Raum, *Mathematische Zeitschrift* **172**, 41–49.

Kochen, S. and Specker, E. P.: 1965a, The calculus of partial propositional functions, *in* Y. Bar-Hillel (ed.), *Proceedings of the 1964 International Congress for Logic, Methodology and Philosophy of Science*, North-Holland, Amsterdam, pp. 45–57.

Kochen, S. and Specker, E. P.: 1965b, Logical structures arising in quantum theory, *in* J. Addison, L. Henkin and A. Tarski (eds), *The Theory of Models*, North-Holland, Amsterdam, pp. 177–189.

Kochen, S. and Specker, E. P.: 1967, The problem of hidden variables in quantum mechanics, *Journal of Mathematics and Mechanics* **17**, 59–87.

Kôpka, F. and Chovanec, F.: 1994, D-posets, *Mathematica Slovaca* **44**, 21–34.

Kraus, K.: 1983, *States, Effects and Operations*, Vol. 190 of *Lecture Notes in Physics*, Springer, Berlin.

Krause, D.: 1992, On a quasi-set theory, *Notre Dame Journal of Formal Logic* **33**, 402–411.

Ludwig, G.: 1983, *Foundations of Quantum Mechanics*, Vol. 1, Springer, Berlin.

Łukasiewicz, J.: 1970a, Logistic and philosophy, *in* L. Borkowski (ed.), *Selected Work*, North-Holland, Asterdam, pp. 218–235.

Łukasiewicz, J.: 1970b, On determinism, *in* L. Borkowski (ed.), *Selected Work*, North-Holland, Asterdam, pp. 110–128.

Łukasiewicz, J.: 1970c, On three-valued logic, *in* L. Borkowski (ed.), *Selected Work*, North-Holland, Amsterdam.

Mackey, G.: 1957, *The Mathematical Foundations of Quantum Mechanics*, Benjamin, New York.

Mangani, P.: 1973, Su certe algebre connesse con logiche a più valori, *Bollettino Unione Matematica Italiana* **8**, 68–78.

Manin, Y. I.: 1976, Problem of present day mathematics: I (Foundations), *in* F. Browder (ed.), *Proceedings of the Symposia in Pure Mathematics*, Vol. 28, American Mathematical Society, pp. 445–460.

Minari, P.: 1987, On the algebraic and kripkean logical consequence relation for orthomodular quantum logic, *Reports on Mathematical Logic* **21**, 47–54.

Mittelstaedt, P.: 1972, On the interpretation of the lattice of subspaces of Hilbert space as a propositional calculus, *Zeitschrift für Naturforschung* **27a**, 1358–1362.

Mittelstaedt, P.: 1978, *Quantum Logic*, Reidel, Dordrecht.

Mittelstaedt, P.: 1986, *Sprache und Realität in der modernen Physik*, Bibliographisches Institut, Mannheim.

Mittelstaedt, P. (ed.): 1985, *Recent Developments in Quantum Logic*, number 6 in *Grundlagen der exakten Naturwissenschaften*, Bibliographisches Institut, Mannheim.

Morash, R. P.: 1973, Angle bisection and orthoautomorphisms in Hilbert lattices, *Canadian Journal of Mathematics* **25**, 261–272.

Mundici, D.: 1992, The logic of Ulam's game with lies, *in* C. Bicchieri and M. L. Dalla Chiara (eds), *Knowledge, Belief and Strategic Interaction*, Cambridge University Press, Cambridge.

Navara, M.: 1999, Two descriptions of state spaces of orthomodular structures, *International Journal of Theoretical Physics* **38**, 3163–3178.

Neubrunn, T. and Riečan, B.: 1997, *Integral, Measure and Ordering*, Kluwer Academic Publishers, Dordrecht.

Nielsen, M. and Chuang, I.: 2000, *Quantum Computation and Quantum Information*, Cambridge University Press, Cambridge.

Nishimura, H.: 1980, Sequential method in quantum logic, *Journal of Symbolic Logic* **45**, 339–352.

Nishimura, H.: 1994, Proof theory for minimal quantum logic I and II, *International Journal of Theoretical Physics* **33**, 102–113, 1427–1443.

Paoli, F.: 2002, *Substructural Logics: A Primer*, Vol. 13 of *Trends in Logic*, Kluwer Academic Publishers.

Peres, A.: 1995, *Quantum Theory: Concepts and Methods*, Kluwer Academic Publishers, Dordrecht.

Piron, C.: 1976, *Foundations of Quantum Physics*, W. A. Benjamin, Reading.

Pitowsky, I.: 1989, *Quantum Probability - Quantum Logic*, number 321 in *Lectures Notes in Physics*, Springer, Berlin.

Pratt, V.: 1993, Linear logic for generalized quantum mechanics, *Workshop on Physics and Computation (PhysComp'92)*, IEEE, Dallas, pp. 166–180.

Pták, P. and Pulmannová, S.: 1991, *Orthomodular Structures as Quantum Logics*, number 44 in *Fundamental Theories of Physics*, Kluwer, Dordrecht.

Pulmannová, S.: 1995, Representation of D-posets, *International Journal of Theorethical Physics* **34**, 1689–1696.

Putnam, H.: 1969, Is logic empirical?, *in* R. S. Cohen and M. W. Wartofsky (eds), *Boston Studies in the Philosophy of Science*, Vol. 5, Reidel, Dordrecht, pp. 216–241.

Pykacz, J.: 2000, Łukasiewicz operations in fuzzy set theories and many-valued representations of quantum logics, *Foundations of Physics,* **30**, 1503–1524.

Rédei, M.: 1996, Why von Neumann did not like the Hilbert space formalism of quantum mechanics (and what he liked instead), *Studies in the History and Philosophy of Modern Physics* **27**, 493–510.

Rédei, M.: 1998, *Quantum Logic in Algebraic Approach*, Vol. 91 of *Fundamental Theories of Physics*, Kluwer Academic Publishers, Dordrecht.

Redhead, M.: 1987, *Incompleteness, Nonlocality and Realism - A Prolegomenon to the Philosophy of Quantum Mechanics*, Clarendon Press, Oxford.

Reed, M. and Simon, B.: 1972, *Methods of Modern Mathematical Physics*, Vol. I, Academic Press, New York.

Riečanova, Z.: 1999, Subalgebras, intervals, and central elements of generalized effect algebras, *International Journal of Theoretical Physics* **38**, 3209–3220.

Rosenthal, K. I.: 1990, *Quantales and their Applications*, Longman, New York.

Sambin, G.: 1997, Basic logic, a structure in the space of logic. in preparation.

Sambin, G., Battilotti, G. and Faggian, C.: 2000, Basic logic: reflection, symmetry, visibility,, *The Journal of Symbolic Logic* **65**, 979–1013.

Schroeck, F. E.: 1996, *Quantum Mechanics on Phase Space*, Vol. 74 of *Fundamental Theories of Physics*, Kluwer Academic Publishers, Dordrecht.

Solèr, M. P.: 1995, Characterization of Hilbert spaces by orthomodular spaces, *Communications in Algebra* **23**, 219–243.

Stalnaker, R.: 1981, A theory of conditionals, *in* W. Harper, G. Pearce and R. Stalnaker (eds), *Ifs. Conditionals, Belief, Decision, Chance, and Time*, Reidel, Dordrecht, pp. 41–55.

Suppes, P.: 2002, *Representation and Invariance of Scientific Structures*, CSLI, Stanford.

Svozil, K.: 1998, *Quantum Logic*, Springer, Singapore.

Takeuti, G.: 1981, Quantum set theory, *in* E. G. Beltrametti and B. C. van Fraassen (eds), *Current Issues in Quantum Logic*, Vol. 8 of *Ettore Majorana International Science Series*, Plenum, New York, pp. 303–322.

Tamura, S.: 1988, A Gentzen formulation without the cut rule for ortholattices, *Kobe Journal of Mathematics* **5**, 133–150.

Toffoli, T.: 1980, Reversible computing, *in* J. W. de Baker and J. van Leeuwen (eds), *Automata, Languages and Programming*, Vol. 84 of *Lecture Notes in Computer Science*, Springer Verlag, Berlin, pp. 632– 644.

Toraldo di Francia, G.: 1990, *Un universo troppo semplice*, Feltrinelli.

van Fraassen, B.: 1974, The labyrinth of quantum logics, *in* R. Cohen and M. Wartosky (eds), *Logical and Epistemological Studies in Contemporary Physics*, Vol. 13 of *Boston Studies in the Philosophy of Science*, Reidel, Dordrecht, pp. 224–254.

van Fraassen, B.: 1991, *Quantum Mechanics. An Empiricist View*, Clarendon Press, Oxford.

Varadarajan, V. S.: 1985, *Geometry of Quantum Theory*, 2 edn, Springer, Berlin.

von Neumann, J.: 1932, *Mathematische Grundlagen der Quantenmechanik*, Springer, Berlin.

von Neumann, J.: 1996, *Mathematical Foundations of Quantum Theory*, Princeton University Press, Princeton.

Wright, R.: 1990, Generalized urn models, *Foundations of Physics* **20**, 881–903.

Zadeh, L.: 1965, Fuzzy sets and, *Information and Control* **8**, 338–353.

Zierler, N.: 1961, Axioms for non-relativistic quantum mechanics, *Pacific Journal of Mathematics* **11**, 1151–1169.

Index of Symbols

Subject Index

E

F

G

H

Index of Names

TRENDS IN LOGIC

1. G. Schurz: *The Is-Ought Problem.* An Investigation in Philosophical Logic. 1997 ISBN 0-7923-4410-3
2. E. Ejerhed and S. Lindström (eds.): *Logic, Action and Cognition.* Essays in Philosophical Logic. 1997 ISBN 0-7923-4560-6
3. H. Wansing: *Displaying Modal Logic.* 1998 ISBN 0-7923-5205-X
4. P. Hájek: *Metamathematics of Fuzzy Logic.* 1998 ISBN 0-7923-5238-6
5. H.J. Ohlbach and U. Reyle (eds.): *Logic, Language and Reasoning.* Essays in Honour of Dov Gabbay. 1999 ISBN 0-7923-5687-X
6. K. Došen: *Cut Elimination in Categories.* 2000 ISBN 0-7923-5720-5
7. R.L.O. Cignoli, I.M.L. D'Ottaviano and D. Mundici: *Algebraic Foundations of many-valued Reasoning.* 2000 ISBN 0-7923-6009-5
8. E.P. Klement, R. Mesiar and E. Pap: *Triangular Norms.* 2000 ISBN 0-7923-6416-3
9. V.F. Hendricks: *The Convergence of Scientific Knowledge.* A View From the Limit. 2001 ISBN 0-7923-6929-7
10. J. Czelakowski: *Protoalgebraic Logics.* 2001 ISBN 0-7923-6940-8
11. G. Gerla: *Fuzzy Logic.* Mathematical Tools for Approximate Reasoning. 2001 ISBN 0-7923-6941-6
12. M. Fitting: *Types, Tableaus, and Gödel's God.* 2002 ISBN 1-4020-0604-7
13. F. Paoli: *Substructural Logics: A Primer.* 2002 ISBN 1-4020-0605-5
14. S. Ghilardi and M. Zawadowki: *Sheaves, Games, and Model Completions.* A Categorical Approach to Nonclassical Propositional Logics. 2002 ISBN 1-4020-0660-8
15. G. Coletti and R. Scozzafava: *Probabilistic Logic in a Coherent Setting.* 2002 ISBN 1-4020-0917-8; Pb: 1-4020-0970-4
16. P. Kawalec: *Structural Reliabilism.* Inductive Logic as a Theory of Justification. 2002 ISBN 1-4020-1013-3
17. B. Löwe, W. Malzkorn and T. Räsch (eds.): *Foundations of the Formal Sciences II.* Applications of Mathematical Logic in Philosophy and Linguistics, Papers of a conference held in Bonn, November 10-13, 2000. 2003 ISBN 1-4020-1154-7
18. R.J.G.B. de Queiroz (ed.): *Logic for Concurrency and Synchronisation.* 2003 ISBN 1-4020-1270-5
19. A. Marcja and C. Toffalori: *A Guide to Classical and Modern Model Theory.* 2003 ISBN 1-4020-1330-2; Pb 1-4020-1331-0
20. S.E. Rodabaugh and E.P. Klement (eds.): *Topological and Algebraic Structures in Fuzzy Sets.* A Handbook of Recent Developments in the Mathematics of Fuzzy Sets. 2003 ISBN 1-4020-1515-1; Pb 1-4020-1516-X

21. V.F. Hendricks and J. Malinowski: *Trends in Logic*. 50 Years Studia Logica. 2003
ISBN 1-4020-1601-8

22. M. Dalla Chiara, R. Giuntini and R Greechie: *Reasoning in Quantum Theory*. Sharp and Unsharp Quantum Logics. 2004 ISBN 1-4020-1978-5

KLUWER ACADEMIC PUBLISHERS – DORDRECHT / BOSTON / LONDON

FSC
www.fsc.org
MIX
Papier aus verantwortungsvollen Quellen
Paper from responsible sources
FSC® C105338

If you have any concerns about our products,
you can contact us on
ProductSafety@springernature.com

In case Publisher is established outside the EU,
the EU authorized representative is:
Springer Nature Customer Service Center GmbH
Europaplatz 3, 69115 Heidelberg, Germany

Printed by Libri Plureos GmbH
in Hamburg, Germany